GENERAL AND APPLIED MICROBIOLOGY

GENERAL AND APPLIED MICROBIOLOGY

Prof. Harmeet Kour
Deptt. of Botany, Govt. P.G. College for Women,
Gandhinagar, Jammu

Prof. Veena Tandon
Deptt. of Botany, Govt. P.G. College for Women,
Gandhinagar, Jammu

ANMOL PUBLICATIONS PVT. LTD.
NEW DELHI - 110 002 (INDIA)

ANMOL PUBLICATIONS PVT. LTD.
H.O.: 4374/4B, Ansari Road, Daryaganj,
New Delhi-110 002 (India)
Ph.: 23278000, 23261597
B.O.: No. 1015, Ist Main Road, BSK IIIrd Stage
IIIrd Phase, IIIrd Block,
Bangalore - 560 085 (India)
Visit us at: www.anmolpublications.com

General and Applied Microbiology

First Published, 2009

ISBN 978-81-261-4051-0

PRINTED IN INDIA

Printed at Balaji Offset, Delhi.

Contents

Preface

Microbes are the tiny organisms, which are hard to be seen with the naked eye. However, in spite of their minuteness, their role in human life is of considerable importance. They are of different sizes, shapes and with different tasks assigned by nature. They are also called bacteria in general parlance. Some of them are beneficial while some others are harmful for man. Apart from the bacteria related directly to body, there are others which provide delicious nutrition to man. The scope of applied microbiology is expanding so rapidly that it is impossible to present all of it in a single book which a student new to the field can cover in one course.

Capter 1

Microbiology

INTRODUCTION

Microbiology is the study of microorganisms, that is the organisms which are of microscopic dimensions. These organisms are too small to be clearly perceived by the unaided human eye. If an object has a diameter of less than 0.1 mm, the eye can not perceive it at all and very little detail can be perceived in an object with a diameter of I mm. Roughly speaking organisms with a diameter of 1 mm or less are microorganisms and fall into the broad domain of microbiology. Since most microorganisms are only a few thousandths of a millimeter in size. they can only be seen with the aid of microscope.

As a direct consequence of the invisibility of micros to the naked eye and the need for special techniques to study them, microbiology was the last of the three major divisions in biology the other two are botany and zoology to develop. Microbiology is the study of organisms that are too small to be clearly seen by the unaided eye. Roughly speaking, organisms with a diameter of 1 mm or less which can not be seen by the human eye unaided, are called microorganisms. These include protozoa, algae, fungi and bacteria Viruses are ultramicroscopic and have an obligate parasitic relationship, but for practical purposes these still come under the domain of microbiology.

At present, there is general agreement to include five major groups as microorganisms: the subdivisions of virology, bacteriology, mycology, phycology and protozoology

Traditionally, all organisms had been included under the disciplines of either botany or zoology. Bacteria, algae and fungi have in the past been considered to be the part of the Plant Kingdom while protozoa have been included in the Animal Kingdom.

This view cannot be supported from a taxonomic standpoint. Since microbiology encompasses the study of groups of organisms in all three of these divisions of biology it may be argued that it covers a greater biological diver itv than the other two divisions.

Although microorganisms have existed for a long time, their existence was unknown until the invention of the microscope in the 17th century. The year 1674, marks the birth of microbiology when Antony van Leeu-wenhoek, a Dutch cloth merchant, looked at a drop of lake water through a glass lens which he had ground.

What he observed through this simple magnifying lens was an amazing sight since that was perhaps the first time that man ever had a glimpse of the world of the microbes. Subsequently, in a series of letters to the Royal Society of London, Leeuwenhoek describ-ed a variety of microorganisms such as protozoa, algae, yeast and bacteria and the description was so precise that it is now possible to assign them into specific genera without any additicnal descrip-tion. Leeuwenhoek had little formal education but his keen interest in nature made him to examine a variety of materials. Glass grinding and preparation of lenses was his hobby and this led him to the assembly of about 400 simple microscopes. The earliest microscope that he constructed consisted of a spherical lens mounted on two plates.

MICROBES AND CHEMICAL CHANGES

At about this time Liebig the German chemist, had claimed that fermentation and putrefaction were processes caused by the chemical instability of organic molecules. On the other hand, Pasteur had shown that living yeast cells were required for the conversion of sugar into alcohol and had suggested that these fermentations and transformations are brought about by the agency of microorganisms. At this time. the distilleries in France, where alcohol was produced from sugar beet had difficulties in their fermentations and requested Pasteur for help. Pasteur found that the trouble was due to the replacement of the alcoholic fermentation by a lactic fermentation. Instead of finding yeast in the fermenting brew, he found rod-shaped structures.

He concluded that these rod shaped organisms were responsible for the conversion of sugar into lactic acid. Subsequently, he also showed that different fermentations are brought about by specific microorganisms. For example, the alcoholic fermentation was the result of living yeast cells while the lactic acid fermentation was the result of the activity of lactic bacteria.

Interestingly in 1897, after Pasteur, Buchner showed that intact yeast cells were not. required for the conversion of sugar into alcohol and carbon dioxide He prepared a cell free yeast juice by grinding yeast cells with sand and showed that this cell free juice had the ability to carry out the alcoholic fermentation. He concluded that the active agent in the yeast cells responsible for bringing about the fermentation was a protein and he named it as zymase. This finding gave support to Liebig's thesis that living cells were not required to bring about fermentations and marks the beginning of the study of enzyme chemistry.

ANCIENT MICROBIOLOGICAL HISTORY

Ancient man recognized many of the factors involved in disease. Early civilizations on crete, india, pakistan and scotland invented toilets and sewers; lavatories, dating around 2800 bc, have been found on the orkney islands and in homes in pakistan about the same time. One archaeologist has stated that "the high quality of the sanitary arrangements could well be envied in many parts of the world today". In rome, 315 ad, the public lavatories were places where people routinely socialized and conducted business.

Ten to twenty people could be seated around a room, with their wastes being washed away by flowing water; it must have been difficult to *"stand on your dignity"* under such circumstances. The chinese used toilet paper as early as ad 589. In europe moss, hay and straw were used for the same purpose. I can personally attest to the use as late as 1962 of "slick magazines" as toilet paper in certain european camp grounds. The first cities to use water pipes (of clay) were in the indus valley of pakistan around 2700 bc. Metal water pipes were used in egypt (2450 bc) and the palace of knossos on crete around 2000 bc had clay pipes.

Rome built elaborate aqueducts and public fountains throughout its empire to insure a clean supply of water for its citizens. Rome had a "water commissioner" who was responsible for seeing that the water supply was kept adequate and clean; the punishment for pollution of the water supply was death. Lead was commonly used for roman pipes and the subsequent fall of the roman empire has been related by some to the effects of lead on the roman brain.

Most ancient peoples recognized that some diseases were communicable and isolated individuals thought to carry "infections". An example of this is the universal shunning of lepers, which occurs even today. When the black death struck europe, entire villages were abandoned as people fled in an effort to escape the highly infectious plague.

Similarly, in the middle ages the rich of europe fled to their country homes when small pox struck in an effort to escape its terrible consequences. The fact that people who recovered from a particular disease were immune to that disease was probably recognized many different times in many places. Often these survivors were expected to nurse the ill. Greek and roman physicians routinely prescribed diet and exercise as a treatment for ills.

Sadly, we know that this knowledge did not help most of our ancestors and that the human life span was, until the last 200 years, more often than not cut short due to infectious disease. Even today approximately 15,000,000 children die per year, mainly from infectious diseases that are preventable with basic sanitation, immunization and simple medical treatments. One might honestly question just how far we have come in our treatment of disease. An excellent synopsis of the history of microbiology ancient people had certainly seen masses of microbes, such as mold and bacterial colonics, on spoilcd food, but it is doubtful if anyone considered that they were viewing living organisms.

Small boys and maybe a few love-sick adults staring into a clear pond, must have seen tiny specks moving rapidly about and some may have considered them living creatures, but to express this to their friends would be equivalent to us telling our friends that we'd seen a flying saucer.

The first person to report seeing microbes under the microscope was an englishman, robert hooke. Working with a crude compound microscope he saw the cellular structure of plants around 1665. He also saw fungi which he drew. However, because his lens were of poor quality he was apparently unable to "see" bacteria.

Anton van leeuwenhoek was a man born before his time. Although not the first to discover the microscope or to use magnifying lens, he was the first to see and describe bacteria. We know that he was a "cloth merchant" living in delft holland. And that he used magnifying lens to view the quality of the weave of the merchandise he purchased. He traveled to england in 1668 to view english cloth and there he saw drawings of magnifications of cloth much greater than any of the current lens available in holland would do.

He returned to holland and took up lens grinding. Being meticulous, he developed his lens grinding to an art and in the process tested them by seeing how much detail he could observe with a given lens. One can guess that he chanced to look at a sample of pond water or other source rich in microbes and was amazed to see distinct, uniquely shaped organisms going, apparently purposefully, about their lives in a tiny microcosm. He made numerous microscopes from silver and gold and viewed everything he could including the scum on his teeth and his semen.

His best lens could magnify ~300-500 fold which allowed him to see microscopic algae and protozoa and larger bacteria. He clearly had excellent eyesight because he accurately drew pictures of microbes that were at the limit of the magnification of his lens. He used only single lens and not the compound lens of the true microscopes we employ today; which makes his observations all the more amazing. He wrote of his observations to the royal society of london in 1676 and included numerous drawings.

He astonished everyone by claiming that many of the tiny things he saw with his lens were alive because he saw them swimming purposefully about. This caused no end of shock and wonderment and numerous people hurried to delft to see if this dutchman was "in his cups" or if he was really onto something new and wonderful.

A few minutes with one of his numerous microscopes was all it took to convert his visitors to enthusiastic believers in the existence of these tiny beasties living all around them. His discovery was the equivalent of our finding life on mars today.

THE CLEAN NUT AND THE REVENGE OF THE BACTERIA

In the 1800s people began to use hospitals. Hospitals also became centers of physician training. In 1841 young doctor ignaz semmelweis was hired to run a maternity ward in a vienna hospital. There were two birthing wards in his preview, one run by midwives and the other by doctors. Semmelweiss noticed that the death rate among mothers in the doctor's ward ran as high as 18% from the blood infection known as child bed fever or puerperal sepsis, whereas in the midwife ward the death rate was much lower.

When he suggested that the doctors might contribute to this he was fired. Subsequently rehired, he saw a friend die from puerperal sepsis after cutting himself during an autopsy of a patient who had died of puerperal sepsis. He reasoned that there was an invisible agent that caused both deaths and that one could transfer it from the autopsy room to the birthing rooms and thus infect the mothers during birthing.

Acting on this assumption, semmelweis instituted sanitary measures which included having the doctors wash their hands in disinfectant and change from lab coats dripping with pus and blood from the autopsy room to clean lab coats before examining patients or assisting in a birth. The death rate of the mothers dropped by 2/3 in his ward. However, the other doctors objected so strongly to his rules that semmelweis was again fired and left vienna.

He took other hospital jobs where he instituted the same standards of cleanliness which resulted in the same decline in deaths and a revolt of his fellow physicians. Semmelweis ended up dying in an insane asylum from a blood infect that resembled puerperal sepsis.

SPONTANEOUS GENERATION

The mystery of life has puzzled and confounded humans since the first human began to contemplate his world. The religions of ancient societies were built around the seasons, the sun and the renewal of life as these were so clearly tied to survival; both of the human species through birth and death and of the individual in the attainment of sustenance. Spontaneous generation or the idea that life routinely arises from non-life was a common sense explanation of the miracle of life.

It had the advantage of simplicity, ease of understanding and didn't require any waste-of-time thinking. As science and the scientific method grew with the slow accumulation of knowledge, observant individuals began to consider the origin of life more deeply. Simple observations convinced many people that all the larger animals and plants produced life from previous life. Despite this, the mass of humans clung to the comfortable idea of spontaneous generation further, religions saw it as a convenient way to demonstrate the hand of god operating continuously in the world.

Some individuals, such as j.b. Van helment even described how one could make mice from grain, a jar and dirty rags by putting them together in a dark place for a few weeks and soon mice would appear in the jar.

Other, more perceptive individuals, like f. Redi tested the common idea that maggots arise via spontaneous generation on rotting meat. He placed a piece of meat in three jars, one he left open, one he corked tightly and the third he covered with a fine mesh gauze. Maggots only appeared in the open container, no matter how long he left the jars. Redi's experiment was important because of its eloquent simplicity. Anyone could repeat it and obtain the same clear results. Nevertheless, many people clung fiercely to the idea of spontaneous generation, while others designed experiments to test it.

In every case the results of the majority of these experiments indicated that spontaneous generation did not occur. The intellectual ferment this controversy stirred up gradually evolved into the scientific method as the various antagonists questioned each other's assumptions and, more importantly, their experimental design.

These arguments forced the designing of better experiments and eventually persuaded all but the most recalcitrant believers to discard spontaneous generation as an explanation for all higher life forms. Then, in the 1800's the refinement of the microscope, through which people could see tiny life forms that they assumed were simple, gave spontaneous generation proponents new life.

Again, flawed common sense led reasonable people astray. As the existence of microscopic life was accepted, the assumption was that such life must be simple compared to higher, more complex life. The reasoning that followed this erroneous assumption was that since the microbes were small they must be simple & it followed that they were formed by spontaneous generation, hence god was still at work creating micro-life. Small is not simple!

The battle over spontaneous generation raged anew both from the pulpit and the lab. A number of scientists performed elementary experiments in which they treated soups and broth's, which left unheated would team with microbes after a few days, with heat to destroy any life present in them and asked the question: "Would new life arise in these sterile soups"? Spallanzani boiled "Soup" in glass containers and melted the glass closed.

The observation that nothing subsequently grew in this

"heated" soup suggested that spontaneous generation didn't work. His detractors, rightly criticized his experiments, proposing that since air is necessary for life and since he had sealed the flask to air, obviously no life could develop. Others boiled soups and microbes grew, thus apparently supporting spontaneous generation. But again the preponderance of data suggested that spontaneous generation did not even apply to the "simple" microbes.

In 1859 one of the fathers of modern microbiology, l. Pasteur decided to settle the question of spontaneous generation once and for all. A genius at devising definitive experiments, pasteur first drew the necks of glass flasks out so that they remained open to the air, but were bent so that air could only enter by a curved path.

He then added broth and boiled it to destroy contaminating microbes. These flasks were then incubated and observed for months. He reasoned that the microbes in the air that could contaminate the sterile broth would be trapped on the sides of the thin glass necks before they reached the sterile broth. If spontaneous generation didn't occur no growth should take place.

This is exactly what happened, the flasks remained sterile indefinitely, until pasteur tipped the sterile broth up into the curved neck where he predicted the airborne organisms would have settled.

After doing this the broths always grew microbes. These experiments ended the spontaneous generation controversy because these experiment was so elegant and simple and the results so clear, that anyone could repeat them.

Fig. Spore Structure

Later, an earlier problem, in which occasional heated-broths did not remain sterile, was explained with the discovery of the heat resistant bacterial some of which could can survive several hours of boiling without being killed.

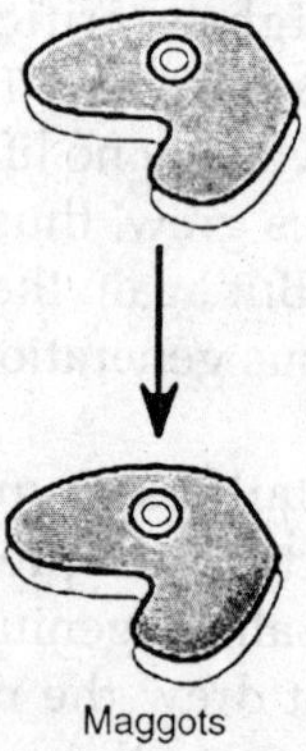

Fig. The Redi Experiment

Pasteur discovered many of the basic principles of microbiology and, along with R. Koch, laid the foundation for the science of microbiology. In 1857 Napoleon III was having trouble with his sailors mutinying because their wine was spoiling after only a few weeks at sea. Naturally Napoleon was distraught because his hopes for world conquest were being scuttled (pardon the pun) over a little spoiled wine, so he begged Pasteur for help.

Pasteur, armed with his trusty microscope, accepted the challenge and soon recognized that by looking at the spoiled wines he could distinguish between the contaminants that caused the spoilage and even predict the taste of the wine solely from his microscopic observations. He then reasoned that if one were to heat the wine to a point where its flavour was unaffected, but the harmful microbes were killed it wouldn't spoil. As we are aware this process, today known as pasteurization, worked exactly the way he predicted and is the foundation of the modern treatment of bottled liquids to prevent their. It is important to realise that pasteurization is NOT the same as sterilization. Pasteurization only kills organisms that may spoil the product, but it allows many microbes to survive, whereas sterilization kills all the living organisms in the treated material.

Pasteur also realized that the yeast that was present in all the produced the alcohol in wine. When he announced this, a number of famous scientists were enraged, because the current theory of wine production was that wine formation was the result of SPONTANEOUS chemical changes that occurred in the grape juice. Pasteur was attacked furiously at scientific meetings, to the point where certain scientists did humorous skits about Pasteur and his tiny little yeast "stills" turning out alcohol. Pasteur had the last laugh however as people all over the world soon realized that if he was right they could control the quality of wine by controlling the yeast that made it. In a short period many others verified his observations and the opposition sank without a sound.

Pasteur discovered many of the basic principles of microbiology and, along with R. Koch, laid the foundation for the science of microbiology. In 1857 Napoleon III was having trouble with his sailors mutinying because their wine was spoiling after only a few weeks at sea. Naturally Napoleon was distraught because his hopes for world conquest were being scuttled (pardon the pun) over a little spoiled wine, so he begged Pasteur for help.

Pasteur, armed with his trusty microscope, accepted the challenge and soon recognized that by looking at the spoiled wines he could distinguish between the contaminants that caused the spoilage and even predict the taste of the wine solely from his microscopic observations. He then reasoned that if one were to heat the wine to a point where its flavour was unaffected, but the harmful microbes were killed it wouldn't spoil. As we are aware this process, today known as pasteurization, worked exactly the way he predicted and is the foundation of the modern treatment of bottled liquids to prevent their. It is important to realise that pasteurization is NOT the same as sterilization. Pasteurization only kills organisms that may spoil the product, but it allows many microbes to survive, whereas sterilization kills all the living organisms in the treated material.

Pasteur also realized that the yeast that was present in all the produced the alcohol in wine. When he announced this, a number of famous scientists were enraged, because the current theory of wine production was that wine formation was the result of SPONTANEOUS chemical changes that occurred in the grape juice. Pasteur was attacked furiously at scientific meetings, to the point where certain scientists did humorous skits about Pasteur and his tiny little yeast "stills" turning out alcohol. Pasteur had the last laugh however as people all over the world soon realized that if he was right they could control the quality of wine by controlling the yeast that made it. In a short period many others verified his observations and the opposition sank without a sound.

MICROBES

A microbe is any living thing that spends its life at a size visible sometimes only with a microscope. It is too tiny to be seen with the naked eye. Microbes are the oldest form of life on Earth. Some types have existed for billions of years. They may live as individuals or cluster together in communities. Microbes live in the water you drink, the food you eat and the air you breathe. Right now, billions of microbes are swimming in your belly and mouth and crawling on your skin! Don't worry, over 95% of microbes are good for you. Microbes include bacteria, viruses, fungi, algae and protozoa. These single-cell organisms are invisible to the eye, but they can be seen with microscopes.

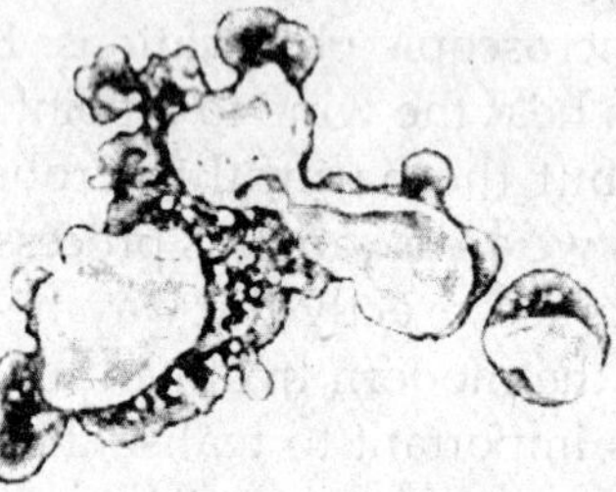

Fig. Microbe

The term microbe is short for microorganism, which means small organism. To help people understand the different types of microbes, they are grouped or classified in various ways. Microbes are very diverse and represent all the great kingdoms of life. In fact, in terms of numbers, most of the diversity of life on Earth is represented by microbes.

CLASSIFICATION OF MICROBES

The term microbe is short for microorganism, which means small organism. To help people understand the different types of microbes, they are grouped or classified in various ways. Microbes are very diverse and represent all the great kingdoms of life. In fact, in terms of numbers, most of the diversity of life on Earth is represented by microbes.

Here is an outline of the major groups of microorganisms:

- Viruses
- Bacteria
- Algae
- Fungi
- Protozoa

Viruses

A virus is too small to be seen without a microscope. A virus is basically a tiny bundle of genetic material carried in a shell called the viral coat. Some viruses have an additional layer around this coat called an envelope. That's basically all there is to viruses. There are thousands of different viruses that come in many shapes. Many are multi-sided or polyhedral. If you've ever looked closely at a cut gem, like the diamond in an engagement ring, you've seen an example of a polyhedral shape. Unlike the diamond in a ring, however, a virus does not taper to a point, but is shaped the same all around. Other viruses are shaped like spiky ovals or bricks with rounded corners. Some are like skinny sticks while others look like pieces of looped string. Some are more complex and shaped like little spaceship landing pods.

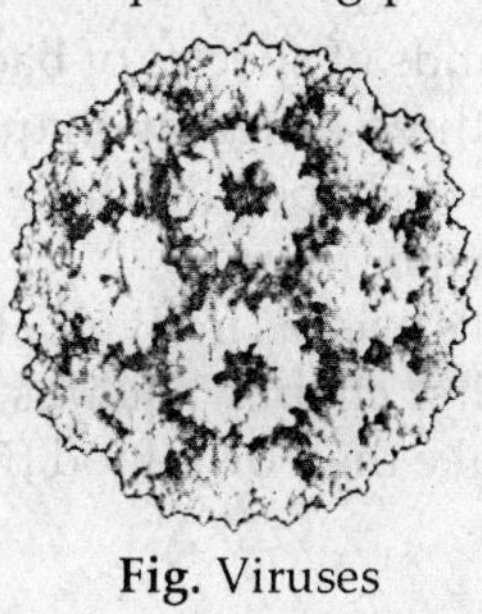

Fig. Viruses

Viruses are found on or in just about every material and environment on Earth from soil to water to air. They're basically found anywhere there are cells to infect. Viruses can infect every living thing.

However, viruses tend to be somewhat picky about what type of cells they infect. Plant viruses are not equipped to infect animal cells, for example, though a certain plant virus could infect a number of related plants.

Sometimes, a virus may infect one animal and do no harm, but cause a great deal of damage when it gets into a different but closely related animal.

Viruses exist to reproduce only. To do that, they have to take over suitable host cells. Upon landing on a suitable host cell, a virus gets its genes inside the cell either by tricking the host cell to pull it inside, or by connecting its viral coat with the host cell wall or membrane and releasing its genes inside, or by injecting their genes into the host cell's DNA.

The viral genes are then copied many times, using the process the host cell would normally use to reproduce its own DNA. The new viral genes then come together and assemble into whole new viruses.

The new viruses are either released from the host cell without destroying the cell or eventually build up to a large enough number that they burst the host cell.

Bacteria

Bacteria consist of only one cell, but they're a very complex group of living things. Some bacteria can live in temperatures above the boiling point and in cold below the freezing point.

There are thousands of species of bacteria, but all of them are basically one of three different shapes. Some are rod- or stick-shaped; others are shaped like little balls. Others still are helical or spiral in shape.

Some bacteria cells exist as individuals while others cluster together to form pairs, chains, squares or other groupings. Some bacteria can make their own food from sunlight, just like plants.

Also like plants, they give off oxygen. Other bacteria absorb food from the material they live on or in. Some of these bacteria can live off iron or sulfur! The bacteria that live in your stomach absorb nutrients from the digested food you've eaten.

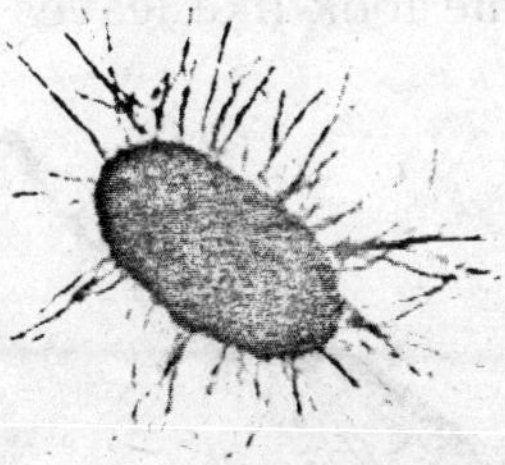

Fig. Bacteria

Some bacteria move about their environment by means of long, whip-like structures called flagella. They rotate their flagella like tiny outboard motors to propel themselves through liquid environments.

They may also reverse the direction in which their flagella rotate so that they tumble about in one place. Other bacteria secrete a slime layer and ooze over surfaces like slugs.

Others stay almost in the same spot. Bacteria live on or in just about every material and environment on Earth from soil to water to air and from your body to the Arctic ice to the Sahara deserts.

Each square centimeter of your skin averages about 100,000 bacteria. A single teaspoon of soil contains more than a billion (1,000,000,000) bacteria.

Algae

Algae are found in fresh and salt water around the world. They can also grow on rocks and trees and in soil when enough water is available.

Most algae are able to make energy from sunlight, like plants do. They produce a large amount of the oxygen we breathe. However, at some stages of their lives, some algae get their nutrients from other living things.

Diatoms are one kind of algae. They have hard shells

made out of glass. When they die, these shells sink to the bottom of their watery environments. We mine deposits of these glass shells that formed hundreds of thousands of years ago to make abrasives, shiny road paint and grit in toothpaste.

Diatoms come in all sorts of shapes—some are round and others are oval. Some look like leaves and others like fat commas.

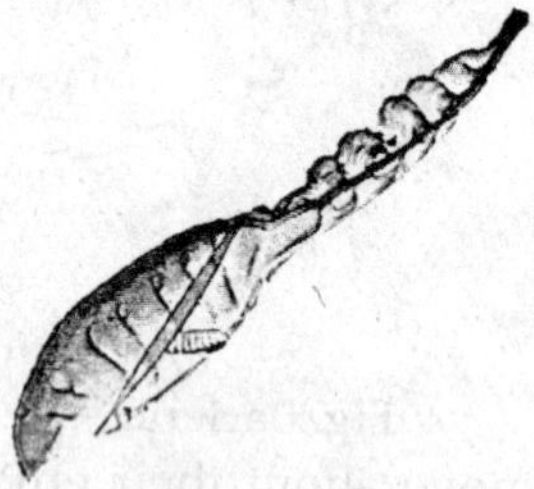

Fig. Algae

Because algae make so much oxygen, these microbes are very helpful.

But sometimes certain kinds of algae can also grow in such large numbers that when they suddenly die off en masse, the breaking down of their cells by bacteria destroys the amount of dissolved oxygen in the water, hurting the animals and plants that live there.

Fungi

Fungi are organisms that scientists once confused with plants. However, scientists have found that, at the cell level, the fungi are more like animals than they are like plants. For one thing, fungi cannot synthesize their own food like plants do, but instead they eat other organisms as do animals.

Fungi come in a variety of shapes and sizes and different types. They can range from single cells to enormous chains of cells that can stretch for miles.

Fungi include single-celled living things that exist individually, such as yeast and multicellular clusters, such as molds or mushrooms.

Yeast cells look round or oval under a microscope. They're too small to see as individuals, but you can see large clusters of them as a white powdery coating on fruits and leaves.

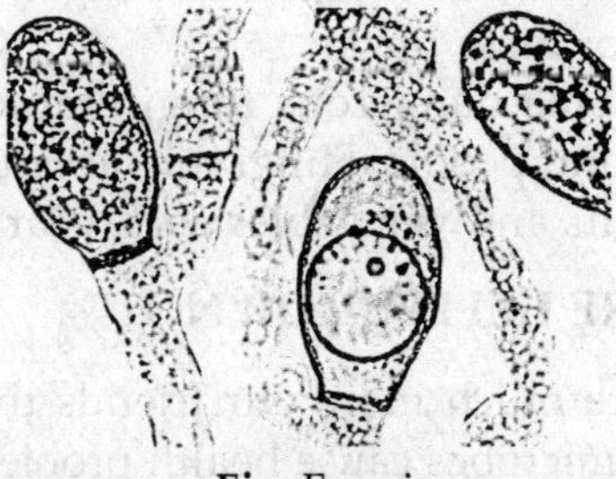

Fig. Fungi

Molds are described as filament-like because they form long filament-like, or thread-like, strands of cells called hyphae. These hyphae are what give mold colonies their fuzzy appearance. They also form the fleshy body, or mushroom, that some species grow. It may seem strange to think of something as big as a mushroom as a microbe. But the cells of the hyphae making up that mushroom are connected in a closer way than the cells of other multicellular living things, like you and me, are.

Fungi usually grow best in environments that are slightly acidic. They can grow on substances with very low moisture. Fungi live in the soil and on your body, in your house and on plants and animals, in freshwater and seawater. A single teaspoon of topsoil contains about 120,000 fungi.

Fungi are basically stationary. But they can spread either by forming reproductive spores that are carried on wind and rain or by growing and extending their hyphae. Hyphae grow as new cells form at the tips, creating even longer chains of cells. Fungi absorb nutrients from living or dead organic matter that they grow on. They absorb simple, easily dissolved nutrients, such as sugars, through their cell walls. They give off special digestive enzymes to break down complex nutrients into simpler forms that they can absorb.

Some fungi are quite useful to us. We've used several kinds to make antibiotics to fight bacterial infections. These antibiotics are based on natural compounds the fungi produce to compete against bacteria for nutrients and space. We use baker's yeast, to make bread rise and to brew beer. Fungi break down dead plants and animals and keep the world tidier.

We're exploring ways to use natural fungal enemies of insect pests to get rid of these bugs.

There are some dangerous fungi that cause diseases in plants, animals and people. Fungi ruin about a quarter to half of harvested fruits and vegetables each year.

MICROBES ARE HELPFUL OR NOT

Microbes are much more our friends than our enemies. Although some microbes cause health problems such as strep throat, chickenpox and the common cold, most microbes make our lives better such as: Bacillus thuringiensis - a common soil bacterium that is a natural pest-killer in gardens and on crops. Arbuscular mycorrhizas - fungus living in the soil that helps crops take up nutrients from the soil. Accharomyces cerevisiae - baker's yeast that makes bread rise.

Escherichia coli - one of many kinds of microbes that live in your digestive system to help you digest your food every day. Streptomyces - bacteria in soil that makes an antibiotic used to treat infections. Pseudomonas putida - one of many microbes that clean wastes from sewage water at water treatment plants. Lactobacillus acidophilus - one of the bacteria that turn milk into yogurt.

There are many other important jobs microbes do. They are used to make medicine. They break down the oil from oil spills. They make about half of the oxygen we breathe. They are the foundation of the food chain that feeds all living things on earth. We've been using microbes for thousands of years to make products we need and enjoy. For example, you can thank fungi for the cheese on your cheeseburger and yeast for your bun. Cheese and bread are two microbe-made foods people have been enjoying since time began. In pollution control, researchers are using bacteria that eat methane gas to clean up hazardous waste dumps and landfills. These methane-eating bacteria make an enzyme that can break down more than 250 pollutants into harmless cells.

By piping methane into the soil, researchers can increase growth of the bacteria that normally live in the polluted soil. More bacteria means faster pollution break up. Also, bacteria is being used as one of the tools to clean up oil spills. These bacteria eat the oil, turning it into carbon dioxide and other harmless by-products. Fungi and bacteria produce antibiotics such as penicillin and tetracycline. These are medicines we use to fight off harmful bacteria that cause sore throats, ear infections, diarrhea and other discomforts. Scientists have changed the genetic material of bacteria and yeasts to turn them into medicine. They inject genes for medicines they want to make into the microbe cells, as if adding new building information to the microbe's cell DNA. The scientists then grow the microbes in huge containers called fermenters where they reproduce into billions, all making new medicines.

Microbes and the Days of Creation

The world of germs and microbes has received much attention in recent years—and for good reason. We frequently hear the term microbe associated with organisms such as Escherichia coli, Salmonella, anthrax bacteria, antibiotic-resistant tuberculosis, MRSA, HIV, malaria, Stachybotrys and other microscopic creatures. But where do microbes fit into the creation account? Were they created along with the rest of the plants and animals in the first week of creation, or were they created later, after the Fall? Are microbes a result of the Curse? These and other questions are some that a group of professional creation microbiologists have been asking and their answers may surprise you. Ongoing research based on the creation paradigm appears to provide some answers to these puzzling questions.

Although I cannot be dogmatic about the details of microbe origin during Creation Week, I believe that a reasonable extrapolation from biological data and Scripture can be made about the nature of microbes in a fully mature creation. Past creation scientists such as Leeuwenhoek, Pasteur and Lister, were blessed by God as He revealed to them critical insight into His creation. So, where do these microbes fit into the very good days of creation? Before answering this question, three terms, microbe, germ and symbiosis, need to be defined. These are relatively new or "modern" terms. First, the Bible does not use these specific terms. These terms were not commonly used until the end of the nineteenth century.

The term microbe was first used in 1878 to describe "extremely minute living beings." Before 1878, scientists including Louis Pasteur, used a variety of terms rather loosely to label the very small organisms that had interested them.

It was not clear whether microbes belonged to the animal or plant kingdom, or to a completely different one. The term microbe was given by Charles E. Sedillot to describe bacteria Later, it would also be used of eukaryotic cells, including algae, fungi, protozoans and slime molds. Some people refer to viruses as microbes, but others do not because viruses are not cells. Viruses have nonliving, as well as living characteristics.

Today, the term germ refers to disease-causing microbes, or pathogens. All germs would have originated after the Fall The Edenic Curse would have profoundly influenced all creation, including viruses, bacteria, fungi and protozoans that would later become pathogens or parasites.

The origin of infectious disease is complex and multifaceted. This topic is further explored in the book. The Genesis of Germs It provides some understanding into the origin of infectious diseases.

From a biblical worldview, infectious diseases and pathogenesis are a secondary state in nature. It is not the way the Creator intended for man and nature. Most microbes are beneficial to man and nature. Only about 5–10% of all bacteria are pathogenic.

Many microbes live in a mutualistic relationship with plants, animals, humans and other microbes. Mutualism is a type of symbiosis. The term symbiosis is used to describe an intimate association between organisms of different species. Some symbiotic relationships between microbes and plants, animals and humans are essential for life on earth.

For this reason most microbiologists maintain that bacteria, fungi, protists and other microbes have been maligned in the news media. Without our intestinal flora, we would not digest food nor acquire vitamins and minerals very efficiently. Without fungi, bacteria, algae and protozoans, life on earth could not last. This is because microbes provide essential "services" (e.g., nitrogen fixation, nitrification and denitrification) in nutrient cycles.

Viruses Fit into Creation

The determination of virus origin is uncertain. It may be that viruses have multiple origins. Some may be degenerate parts from cells after the Curse; still others may have their origin during the days of creation.

Today, we think of viruses (Latin for "poison") only in the context of disease. However, some viruses (or at least virus-like genes) are involved in a positive function in nature. Some groups of viruses, like bacteriophages, play a positive role in controlling bacteria in ecosystems and may play a role in diversity. Another group of viruses play a role in turning off the immune system during pregnancy in mammals and humans. This is a group referred to as endogenous retroviruses (ERVs). ERVs are among a kind of repetitious genetic elements called "retrotransposons". Research has shown that the ERV design prohibits the mother's immune system from damaging the child's body.

These retroviruses cannot fully replicate, only expressed in local immune cells (such as macrophages) of the placenta, thereby preventing them from initiating a full-blown immune response. Thus, the mother's immune system remains competent to respond to other infections but is specifically prevented from mounting an immune response to the developing embryo.

So in creation, the selective ability to turn off the immune system for protection would be a "good" design. Other ERVs also play a positive role in animal and human reproduction. However, since the corruption of creation, the corrupted retrovirus, HIV and various leukemia viruses turn off the entire immune system, leaving the body open to devastating infections. These examples may provide clues to the origin of viruses and how some may have been created during Creation Week by design and how some have been corrupted as a result of the Fall.

Aseptic Technique

Trying to study this mixed population is often difficult and in the tradition of the scientific method; researchers dissect a system and study each piece in isolation. For microorganisms this means separating the organisms and getting them into pure culture. A pure culture is defined as a growth of microorganisms (a culture) that contains one cell type. It is essential in microbiology to be able to obtain and preserve pure cultures. Over 100 years ago, Robert Koch devised methods to achieve this goal and the methods he developed are essentially still used today. The goals of aseptic technique are two-fold. The first objective is to obtain pure cultures and secondly to prevent cross-contamination. Microorganisms in culture must not escape into the environment and microbes in the environment must not get into the cultures we are studying. It is essential that aseptic technique be understood and practiced correctly.

Contaminated cultures are worthless for diagnosis or for doing research on, because it is unclear what microbe is performing any action that is being observed. Aseptic methods commonly used are flame sterilization, tube transfer, streak plates, spread plates and pour plates. Flame sterilization is an easy method to insure sterile transfer of a culture from a source to a growth medium.

Tube transfer is useful for moving inocula from one tube to another. Mechanical dilution by making streak plates is the preferred method for obtaining a pure culture of a microorganism. Finally, spread plates and pour plates are common methods for enumerating microorganisms and are sometimes useful for obtaining isolated colonies.

Flame Sterilization and Tube Transfer

Flame sterilization is a very quick simple method of killing microorganisms on an inoculating loop or needle. The loop or needle is held inside a flame for a few seconds to bring it to redness and then cooled. Once cool, the loop or needle can be used for various culture manipulations. Make sure that the area that contacts the culture is flamed to redness. Also, be patient and let the loop cool down, this usually takes about 15-30 seconds. Learning this technique is essential to everything else you do in microbiology.

Transfer of culture from agar plates to tubes, or from tube to tube, is a common, simple procedure. It is important to perform these transfers in a consistent and rapid manner. The following protocols have been found effective. To transfer a culture from an agar plate to a broth or agar slant:

- Place the Bunsen burner in front of you and assemble all necessary equipment with in arms reach. Position everything so that you will not burn yourself while trying to inoculate your tubes.
- Label the tube of broth or agar to be inoculated with identifying marks. The culture, the date and your initials for example. Place it in a rack in front of you.
- Holding the inoculating loop handle, flame the entire wire to redness.
- When the wire cools (about 15-30 seconds) remove the lid of the plate with your other hand and obtain an inoculum by removing a small portion of the surface growth on the agar plate. In most cases you will be picking an isolated colony. Choose a well isolated one. Do not dig into the agar. Replace the lid of the plate immediately.
- Hold the tube to be inoculated with the free hand. Remove the cotton plug or cap of the tube with the little finger of the hand holding the needle holder. If a cotton-plugged tube is used, the mouth of the tube should be passed briefly through the flame to singe off dust and lint particles. (Dust or lint may fall into the tube and contaminate the medium.)
- Introduce the inoculum into the tube. When inoculating a tube of broth, rub the wire against the glass just above the fluid level and then tip the tube slightly to wash the inoculum into the broth. The wire should not be rattled against the sides of the tube to shake an inoculum into the broth; this is unnecessary and may create a dangerous aerosol. If the transfer is

made to an agar slant, a single mid-line stroke over its surface is made with the wire or loop.

- Replace the cap or plug.
- Flame the inoculating wire again to redness, slowly to avoid spattering. Put the loop holder down after the wire cools.

Making a Medium

Making medium is as simple as cooking and a crude medium can be made in almost any kitchen with a few utensils and a source of heat. Below is described the production of a chicken broth medium that will grow many common microorganisms.

The medium will be sterilized by tyndaliza-tion. Simply boiling a medium once, while it will kill most vegetative cells, does not kill endospores. However, heating encourages the endospores to germinate and a second boiling kills these microbes. The boiling process is repeated a third time to ensure that all spores have germinated and been killed.

- Add 250 ml (about 1 cup) of water into a glass container or some other vessel that can stand boiling water. The container should be something you can cover. Glass bottles that can stand boiling or canning jars work well.
- To this add 15 grams, about 1 tablespoon of instant chicken broth crystals. and stir until dissolved. Cover loosely so that steam can escape, but dust and dirt cannot enter.
- Microwave the broth on high for 3 minutes or until it just begins to boil.
- Let the broth cool and sit for at least 2 hours up to overnight at room temperature. Make sure that the medium vessel is covered to prevent contamination from the air. At this point the vegetative cells are dead and most endospores will germinate.
- Repeat steps 3 and 4 a second time. This will kill the endospores that germinated on day 1.
- Repeat steps 3 and 4 a third time. This will eliminate

the remaining endospores, making the medium sterile.

- Place the medium in a warm place overnight. If your tyndalization was done correctly, your medium should remain clear and free of microbes.
- While you are waiting for your medium, search for a sample you would like to inoculate into it. Almost any sample should work, but natural samples are a good idea for this experiment. Samples coming from food or from your body may contain pathogens. Something you probably want to avoid.
- Check your medium. If it is still clear and sterile, add a pinch of your collected sample to the medium. (The exact amount really does not matter). Again place the medium in a warm place and check it periodically. After a few days, you should start to see the medium become more turbid.
- If you have the equipment, examine a sample of the medium under a microscope. What shapes of microbes are present? Is there more than one species?
- When finished, carefully dispose of the borth. It is basically a spoiled food, but the properties of the spoilage microbe are unknown and should be treated with care.

STREAK PLATES

The streak plate method is a rapid and simple technique of mechanically diluting a relatively large concentration of microorganisms to a small, scattered population of cells. The goal is to obtain isolated colonies on a large part of the agar surface, so that desired species can then be brought into pure culture. Proper streaking of plates is an indispensable tool in microbiology.

In most cases a closed inoculating loop is used for streaking plates. The wire loop should not be badly oxidized or pitted or it will fail to dilute the inoculum and will scratch the surface of the agar. Streak plates can be made from a broth culture, an agar slant or from an agar plate. It is sometimes convenient to suspend a bit of growth from a solid surface in sterile saline and use this as a source of inoculum. Resuspension of colonies or cultures grown on solid surfaces dilutes the culture and makes streak plating easier.

A loopful of inoculum is transferred from the source and put on the agar surface. When using a large inoculum (a turbid culture or growth from a solid surface), a small spot is spread during the initial transfer. If the inoculum is from a lightly turbid suspension, the first phase of the streaking pattern is begun. The three-phase streaking pattern is recommended for beginners because it is most likely to give satisfactory results with suspensions having a wide range of microbial density.

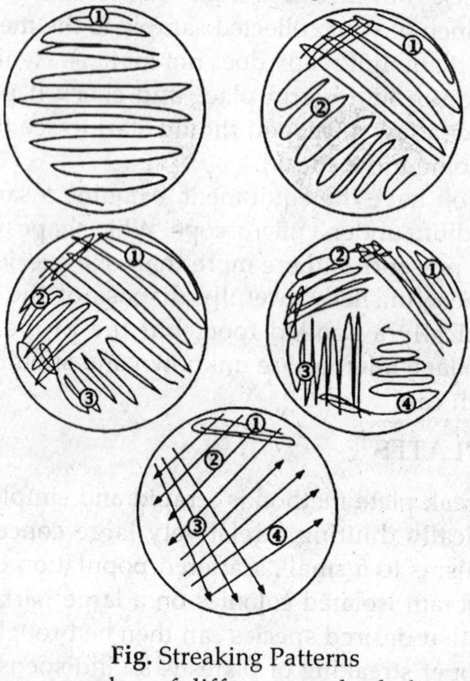

Fig. Streaking Patterns

There are a number of different methods for mechanically diluting microbes on a streak plate. The most common method is spreading microbes across a plate. As the concentration of microbes increases so do the number of phases. Irrespective of the number of phases, loop is flamed between each one. The fifth plate shows an alternative method, where the streaks are not continuous, but are a series of parallel lines. Foobar.

Choosing a streaking pattern is a matter of individual preference and depends upon the number of microorganisms in the sample. Demonstrates the most common patterns, but they are not the only methods. The object of any streaking pattern is the continuous dilution of the inoculum to give many well isolated colonies. For multi-phase streaking it is crucial to flame the loop before starting the next phase. Note the slight overlap into the previous phase to pick up a small inoculum. To streak a plate...

- Flame the loop to sterilize it and let cool.
- Position the plate so that the spot of inoculum is nearest the hand not holding the loop (the opposite hand).
- Lift the plate lid with the opposite hand; just enough to get the loop inside and touch the loop to the inoculum spot. It is often helpful to treat the inoculating loop as if it were a pencil - steadying the loop by resting the heel of the hand against the lab bench.
- Move the loop back and forth across the spot and then gradually continue toward the centre of the plate as you sweep back and forth. Use a very gentle and even pressure.
- When creating each phase, do not worry about keeping each pass across the plate separate from previous ones.
- When about 30% of the plate has been covered by the first streaking phase, remove the loop and flame sterilize it.
- Repeat the above procedure for the second phase, but this time pick up some inoculum by crossing into the first phase 2-3 times and then not passing into it again.
- Repeat as necessary for the third and fourth phases. After streaking the plate, flame sterilize the loop before setting it down.

Spread Plates and Dilution Plating

An absolute requirement for a microbiologist is to be able

to determine the concentration of microorganisms in a given sample. Various particle-counting devices, spectrophotometric methods and microscopic techniques have been used to count cells. However, one drawback to these methods is that they count dead as well as living cells.

The most common method of enumerating viable cells is the plate-count method. Diluting microorganisms and placing them into petri plates (or plates) for incubation is another essential technique for working with microorganisms.

This method suffers from some problems. First, only those organisms which can grow on the medium and at the temperature and atmospheric conditions of incubation, will divide and develop into colonies. Second, each colony may not represent the progeny from one cell, as two or more cells (those in clusters, chains or otherwise close to one another) can give rise to one colony.

For these reasons the counts obtained from the plate-count method are given as the number of colony-forming units (CFU's) per ml (or gram) rather than the number of cells per ml (or gram). Despite these drawbacks, the plate-count method is a powerful means by which concentrations of viable organisms may be estimated.

Also, if it is desirable to count a specific subgroup of microorganisms in a sample, selective media or special incubation conditions can often be used to encourage the growth of only this class of organism.As microbial quantitation involves the use of pipettes (or micropipettes) in preparing dilutions and inoculating plates, the beginning student in microbiology must become familiar with their use.

Due to the possibility of ingesting pathogens and toxic liquids, mouth-pipetting is forbidden in the laboratory! Pipettes are filled and subsequently emptied by the use of propipettes or other pipette bulb. Pay close attention to the demonstration of their use.

Important Safety Consideration: When fitting the pipette and pipette bulb together, use very gentle pressure!! Do not jam these items together! (Force is usually not the answer, a good general rule to live by in this lab and in life for that matter.) The glass pipette will probably break and possibly cause severe injury. Handle the pipette only at the top inch or so. For volumes of 5 ml of less, micropipettes are often the tool of choice.

Instruments are available that are capable of dispensing 5 ml all the way to less than 1 μl (one one-millionth of a liter). Micropipettes have made it possible to miniaturize many experiments and greatly decrease the cost of running them. They are also easy to use and can dispense volumes quickly, increasing the number of experiments that can be performed in a set amount of time. Micropipettes are the tool of choice for small volumes.

MICROBIAL CULTIVATION

When microorganisms are cultivated in the laboratory, a growth environment called a medium is used. The medium may be purely chemical (a chemically defined medium), or it may contain organic materials, or it may consist of living organisms such as fertilized eggs.Microorganisms growing in or on such a medium form a culture. A culture is considered a pure culture if only one type of organism is present and a mixed culture if populations of different organisms are present. When first used, the culture medium should be sterile, meaning that no form of life is present before inoculation with the microorganism.

General microbial media. For the cultivation of bacteria, a commonly used medium is nutrient broth, a liquid containing proteins, salts and growth enhancers that will support many bacteria. To solidify the medium, an agent such as agar is added. Agar is a polysaccharide that adds no nutrients to a medium, but merely solidifies it. The medium that results is nutrient agar.

Many media for microorganisms are complex, reflecting the growth requirements of the microorganisms. For instance, most fungi require extra carbohydrate and an acidic environment for optimal growth. The medium employed for these organisms is potato dextrose agar, also known as Sabouraud dextrose agar. For protozoa, liquid media are generally required and for rickettsiae and viruses, living tissue cells must be provided for best cultivation.

For anaerobic microorganisms, the atmosphere must be oxygen free. To eliminate the oxygen, the culture media can be placed within containers where carbon dioxide and hydrogen gas are generated and oxygen is removed from the atmosphere. Commercially available products achieve these conditions. Anaerobic chambers can also be used within closed compartments and technicians can manipulate culture media within these chambers. To encourage carbon dioxide formation, a candle can be burned to use up oxygen and replace it with carbon dioxide.

Special microbial media. Certain microorganisms are cultivated in selective media. These media retard the growth of unwanted organisms while encouraging the growth of the organisms desired. For example, mannitol salt agar is selective for staphylococci because most other bacteria cannot grow in its high-salt environment. Another selective medium is brilliant green agar, a medium that inhibits Gram-positive bacteria while permitting Gram-negative organisms such as *Salmonella* species to grow.

Still other culture media are differential media. These media provide environments in which different bacteria can be distinguished from one another. For instance, violet red bile agar is used to distinguish coliform bacteria such as *Escherichia coli* from noncoliform organisms. The coliform bacteria appear as bright pink colonies in this media, while noncoliforms appear a light pink or clear. Certain media are both selective and differential.

For instance, MacConkey agar differentiates lactose-fermenting bacteria from nonlactose-fermenting bacteria while inhibiting the growth of Gram-positive bacteria. Since lactose-fermenting bacteria are often involved in water pollution, they can be distinguished by adding samples of water to MacConkey agar and waiting for growth to appear.

In some cases, it is necessary to formulate an enriched medium. Such a medium provides specific nutrients that encourage selected species of microorganisms to flourish in a mixed sample. When attempting to isolate *Salmonella* species from fecal samples, for instance, it is helpful to place a sample of the material in an enriched medium to encourage *Salmonella* species to multiply before the isolation techniques begin.

In order to work with microorganisms in the laboratory, it is desirable to obtain them in pure cultures. Pure cultures of bacteria can be obtained by spreading bacteria out and permitting the individual cells to form masses of growth called colonies. One can then pick a sample from the colony and be assured that it contains only one kind of bacteria. Cultivating these bacteria on a separate medium will yield a pure culture. To preserve microbial cultures, they may be placed in the refrigerator to slow down the metabolism taking place. Two other methods are deep-freezing and freeze-drying.

For deep-freezing, the microorganisms are placed in a liquid and frozen quickly at temperatures below –50°C. Freeze-drying (lyophilization) is performed in an apparatus that uses a vacuum to draw water off after the microbial suspension has been frozen. The culture resembles a powder and the microorganisms can be preserved for long periods in this condition.

STERILIZATION

Wet Heat

Wet heat is the most dependable procedure for the destruction of all forms of microbial life. Steam sterilization generally denotes heating in an autoclave employing saturated steam under a pressure of approximately 15 psi to achieve a chamber temperature of at least 121°C (250°F). The critical factors in insuring the reliability of this sterilization method is:

- Proper temperature and time; and
- The complete replacement of the air with steam (i.e. no entrapment of air).

Some autoclaves utilize a steam activated exhaust valve that remains open during the replacement of air by live steam until the steam triggers the valve to close. Others utilize a pre-cycle vacuum to remove air prior to steam introduction.

Physical controls such as pressure gauges and thermometers are widely used but are considered secondary methods of insuring sterilization. The use of appropriate biological indicators at locations throughout the autoclave is considered the best indicator of sterilization. The biological indicator most widely used for wet heat sterilization is Bacillus stearothermophilus spores.

Dry Heat

Dry heat is less efficient than wet heat sterilization and requires longer times and/or higher temperatures. The specific times and temperatures must be determined for each type of material being sterilized. Generous safety factors are usually added to allow for the variables that can influence the efficiency of this method of sterilization. The moisture of the sterilization environment as well as the moisture history of organisms prior to heat exposure appear to affect the efficiency of dry heat sterilization.

Higher temperatures and shorter times may be used for heat resistant materials. The heat transfer properties and the spatial relation or arrangement of articles in the load are critical in insuring effective sterilization.

The advantage of wet heat is a better heat transfer to and into the cell resulting in overall shorter exposure time and lower temperature. Steam sterilization uses pressurized steam at 121-132° C (250-270° F) for 30 or 40 minutes. This type of heat kills all microbial cells including spores, which are normally heat resistant. In order to accomplish the same effect with dry heat in an oven, the temperature needs to be increased to 160-170° C (320-338° F) for periods of 2 to 4 hours.

General Procedures

All materials, equipment, apparatus contaminated with or containing potentially hazardous organisms should be steam sterilized before being washed and stored, or discarded. Autoclaving is the preferred method. Each individual working with biohazardous materials is responsible for sterilization of materials before disposal.

Biohazardous materials should not be placed in autoclaves overnight in anticipation of autoclaving the next day. To minimize hazard to firemen or disaster crews, all biohazardous materials must be placed in an appropriately marked refrigerator or incubator, sterilized, or otherwise confined at the close of each work day.

All autoclaves must be certified for operating efficiency by the periodic use of biological indicator controls and records maintained for three years. Special precautions should be taken to prevent accidental removal of material from an autoclave before it has been sterilized or the simultaneous opening of both doors on a double door autoclave.

Dry hypochlorites, or any other strong oxidizing material, must not be autoclaved with organic materials such as a paper, cloth, or oil:

Oxidizer + Organic material + Heat = Possible Explosion.

All floors, laboratory benches and other surfaces in buildings where biohazardous materials are handled should be disinfected as often as deemed necessary by the supervisor. The surroundings should be disinfected after completion of operations involving plating, pipetting, centrifuging and similar procedures with biohazardous materials.

It is the responsibility of the supervisor to determine that the disinfectant and the time and method of exposure is effective against the biological agent(s) used in the facility.

Floor drains should be flooded with water or disinfectant at least once each week in order to fill traps and thus prevent the back flow of sewer gases. Floor cleaning procedures which minimize the generation of aerosols should be used. Wet mopping or wet vacuum pick up is recommended.

Water used to mop floors should contain a disinfectant or disinfectant-detergent. (Dry mopping or dusting should be avoided). Where wet procedures are not practical, dry vacuum cleaning with a HEPA filter on the exhaust, sweeping compound used with push brooms, or dry dust mop heads treated to suppress aerosolization may be used. Stock solutions of suitable disinfectants will be maintained in each laboratory for disinfection purposes. NOTE: working dilutions of certain disinfectants have short shelf lives and must be prepared on a regular basis.

Sterilization Procedures

General criteria for sterilization of typical materials are presented below. Supervisors are encouraged to review the type of materials being handled and to establish standard conditions for sterilization. Treatment conditions to achieve sterility will vary in relation to the volume of material treated, volume of the autoclave, the contamination level, the moisture content and other factors.

Steam Autoclave

- *Laundry:* 250°F (121°C) for a minimum of 30 min.
- *Trash:* 250°F (121°C) for at least 45 minutes per bag. Size of the autoclave and of the bags greatly effect sterilization time. Large bags in a small autoclave may require 90 minutes or more.
- *Glassware:* 250°F (121°C) for a minimum of 25 min.
- *Liquids:* 250°F (121°C) for 25 minutes for each gallon.
- *Animals & bedding:* Steam autoclaving not recommended (sterilization time required would be at least 8 hours). Incineration in an approved facility is the recommended treatment of these wastes.

CULTURE COLLECTION [H]

Culture collections of plant-associated microbes provide a needed resource of genotypic and phenotypic variation necessary for effective research in plant pathology. Collections provide a genetic link between the past and present and can provide insight on changes that have occurred since previous epidemics. Recent advances in molecular techniques have shown that pathogen diversity is much more extensive than previously suspected. Reference cultures having phenotypic and genotypic markers can facilitate identification of emerging pathogens, provide data essential for forensic investigation, and be useful in identifying control strategies. Without access to the genetic diversity of microbes, it is impossible to predict durability of plant resistance.

The productivity per given area of photosynthetic microbes such as algae is over ten times greater than that of terrestrial plants. They grow up to 15 times faster than any plant on land. And the biodiversity of photosynthetic microbes is enormous – estimated at more than 100,000 species– and yet most of it remains biochemically and metabolically unexplored. The algae from this collection, as well as additional strains of algae, have been screened through a high-throughput system in order to determine which strain has the best performance characteristics for a given range of environmental conditions. The strains have been tested for different growth conditions such as salinity, pH, and temperature, and for optimum production of product such as algae oil or astaxanthan, a valuable carotenoid pigment.

The Harbor Branch Marine Microbial Culture Collection [sh]

As a part of the drug discovery program, the microbiology group has developed a major culture collection of marine microbes derived in large part from deep-water sponges. The collection consists of over 17,000 microorganisms, with 11,000 from depths greater than 45 meters. Recent funding through the NSF Biotic Surveys and Inventories program allowed us to characterize a subset of the collection beyond the preliminary stage of microscopic, morphological, and Gram-stain identifications and showed extent of the biodiversity that is generated using our microbial isolation techniques.

The HBMMCC also holds over 1,600 fungal isolates. The majority (1,250) of these isolates were derived from marine invertebrates with the remainder isolated from environmental samples such as wood and water samples. The development of this culture collection demonstrates the rich diversity of cultivable microorganisms that is present in the marine environment. Cultures are added to the collection at a rate of 250-500 per year. The Harbor Branch Marine Microbial Database contains information on the subset of the collection that has been identified using molecular techniques. Isolates are grown in liquid culture (fermentation) and extracted to provide material for the drug discovery program. Isolates are also tested for their activities in biotechnological applications.

NEED FOR MICROBIAL TYPE CULTURE COLLECTION CENTRE

Biotechnology in industry and academia has attained mainstay now more than ever before. For the ever-increasing demands of food production and to meet the requirements of growing population it is biotechnology (and its different branches, e.g. agricultural, food and industrial biotechnology) which is giving a hand.

As the name suggests, it needs living organisms (plants, animals and microbes) or their parts, for an end product. Due to human disturbances or natural calamities we have the danger of losing many of the organisms (whether it be plants, animals or microbes). The red data list of the Botanical Survey of India shows a list of plants which are on the verge of becoming extinct. Such plants or those plants which are medicinally important but rare to get can be preserved in the labs through tissue culture and gene banks.

The gene banks will serve not only as backbone for preservation but also help in retrieving if they are completely lost from the system. Each plant species can serve as a reservoir or as a substrate for a host of other organisms including the microbes. Hence, any loss of higher plants might also reflect in the extinction of microorganisms depending on them, particularly if they are host specific1, thus causing irreparable damage to biodiversity. While plants can be preserved through nurseries or tissue culture and gene banks, in the case of microbes, invariably it is raising and maintaining pure cultures under laboratory conditions, followed by preservation. The invention of novel antibiotic compounds or enzymes needs a thorough screening. While a lot of money has to be spent for screening of different microbes, the establishment and maintenance of culture collection centres is a prerequisite.

MICROSCOPES

Since microorganisms are invisible to the unaided eye, the

essential tool in microbiology is the microscope. One of the first to use a microscope to observe microorganisms was Robert Hooke, the English biologist who observed algae and fungi in the 1660s. In the 1670s, Anton van Leeuwenhoek, a Dutch merchant, constructed a number of simple microscopes and observed details of numerous forms of protozoa, fungi and bacteria.

During the 1700s, microscopes were used to further elaborate on the microbial world and by the late 1800s, the sophisticated light microscopes had been developed. The electron microscope was developed in the 1940s, thus making the viruses and the smallest bacteria (for example, rickettsiae and chlamydiae) visible. Microscopes permit extremely small objects to be seen, objects measured in the metric system in micrometers and nanometers. A micrometer (ìm) is equivalent to a millionth of a meter, while a nanometer (nm) is a billionth of a meter. Bacteria, fungi, protozoa and unicellular algae are normally measured in micrometers, while viruses are commonly measured in nanometers. A typical bacterium such as *Escherichia coli* measures about two micrometers in length and about one micrometer in width.

Optical Microscopy

Optical or light microscopy involves passing visible light transmitted through or reflected from the sample through a single or multiple lenses to allow a magnified view of the sample. The resulting image can be detected directly by the eye, imaged on a photographic plate or captured digitally. The single lens with its attachments, or the system of lenses and imaging equipment, along with the appropriate lighting equipment, sample stage and support, makes up the basic light microscope.

Limitations of Optical Microscopy

Limitations of standard optical microscopy (bright field microscopy) lie in three areas;

- The technique can only image dark or strongly refracting objects effectively.

- Diffraction limits resolution to approximately 0.2 micrometre.
- Out of focus light from points outside the focal plane reduces image clarity.

Live cells in particular generally lack sufficient contrast to be studied successfully, internal structures of the cell are colourless and transparent.

The most common way to increase contrast is to stain the different structures with selective dyes, but this involves killing and fixing the sample. Staining may also introduce artifacts, apparent structural details that are caused by the processing of the specimen and are thus not a legitimate feature of the specimen.

These limitations have, to some extent, all been overcome by specific microscopy techniques which can non-invasively increase the contrast of the image. In general, these techniques make use of differences in the refractive index of cell structures. It is comparable to looking through a glass window: you (bright field microscopy) don't see the glass but merely the dirt on the glass.

There is however a difference as glass is a more dense material and this creates a difference in phase of the light passing through.

The human eye is not sensitive to this difference in phase but clever optical solutions have been thought out to change this difference in phase into a difference in amplitude (light intensity).

OPTICAL MICROSCOPY TECHNIQUES

Bright Field Optical Microscopy

Bright field microscopy is the simplest of all the light microscopy techniques.

Sample illumination is via transmitted white light, i.e. illuminated from below and observed from above. Limitations include low contrast of most biological samples and low apparent resolution due to the blur of out of focus material. The simplicity of the technique and the minimal sample preparation required are significant advantages.

Oblique Illumination and what it Means

The use of oblique (from the side) illumination gives the image a 3-dimensional appearance and can highlight otherwise invisible features. A more recent technique based on this method is Hoffmann's modulation contrast, a system found on inverted microscopes for use in cell culture. Oblique illumination suffers from the same limitations as bright field microscopy (low contrast of many biological samples; low apparent resolution due to out of focus objects), but may highlight otherwise invisible structures.

Dark field Optical Microscopy

Dark field microscopy is a technique for improving the contrast of unstained, transparent specimens. Darkfield illumination uses a carefully aligned light source to minimise the quantity of directly-transmitted (un-scattered) light entering the image plane, collecting only the light scattered by the sample. Darkfield can dramatically improve image contrast—especially of transparent objects—while requiring little equipment setup or sample preparation. However, the technique does suffer from low light intensity in final image of many biological samples and continues to be affected by low apparent resolution.

Rheinberg illumination is a special variant of dark field illumination in which transparent, colored filters are inserted just before the condenser so that light rays at high aperture are differently colored than those at low aperture (i.e. the background to the specimen may be blue while the object appears self-luminous yellow). Other colour combinations are possible but their effectiveness is quite variable.

Phase Contrast Optical Microscopy

More sophisticated techniques will show differences in optical density in proportion. Phase contrast is a widely used technique that shows differences in refractive index as difference in contrast. It was developed by the Dutch physicist Frits Zernike in the 1930s The nucleus in a cell for example will show up darkly against the surrounding cytoplasm.

Contrast is excellent; however it is not for use with thick objects. Frequently, a halo is formed even around small objects, which obscures detail.

The system consists of a circular annulus in the condenser which produces a cone of light. This cone is superimposed on a similar sized ring within the phase-objective.

Every objective has a different size ring, so for every objective another condenser setting has to be chosen. The ring in the objective has special optical properties: it first of all reduces the direct light in intensity, but more importantly, it creates an artificial phase difference of about a quarter wavelength.

As the physical properties of this direct light have changed, interference with the diffracted light occurs, resulting in the phase contrast image.

Differential Interference Contrast Microscopy

Superior and much more expensive is the use of interference contrast.

Differences in optical density will show up as differences in relief. A nucleus within a cell will actually show up as a globule in the most often used differential interference contrast system according to Georges Nomarski.

However, it has to be kept in mind that this is an optical effect and the relief does not necessarily resemble the true shape! Contrast is very good and the condenser aperture can be used fully open, thereby reducing the depth of field and maximizing resolution.

The system consists of a special prism in the condenser that splits light in an ordinary and an extraordinary beam. The spatial difference between the two beams is minimal (less than the maximum resolution of the objective). After passage through the specimen, the beams are reunited by a similar prism in the objective.

In a homogeneous specimen, there is no difference between the two beams and no contrast is being generated. However, near a refractive boundary (say a nucleus within the cytoplasm), the difference between the ordinary and the extraordinary beam will generate a relief in the image. Differential interference contrast requires a polarized light source to function; two polarizing filters have to be fitted in the light path, one below the condenser (the polarizer) and the other above the objective (the analyzer).

Note: In cases where the optical design of a microscope produces an appreciable lateral separation of the two beams we have the case of classical interference microscopy, which does not result in relief images, but can nevertheless be used for the quantitative determination of mass-thicknesses of microscopic objects.

Immunofluorescence

Immunofluorescence is the labeling of antibodies or antigens with fluorescent dyes. This technique is often used to visualize the subcellular distribution of biomolecules of interest.

Immunofluorescent-labelled tissue sections or cultures are studied using a fluorescence microscope or by confocal microscopy.

Uses

Most commonly, immunofluorescence employs two sets of antibodies: a primary antibody is used against the antigen of interest; a subsequent, secondary, dye-coupled antibody is introduced that recognizes the primary antibody.

In this fashion the researcher may create several primary antibodies that recognize various antigens, but, because they all share a common constant region, may be recognized by a single dye-coupled antibody.

Typically this is done by using antibodies made in different species.

For example, a researcher might create antibodies in a goat that recognize several antigens, and then employ dye-coupled rabbit antibodies that recognize the goat antibody constant region. This allows re-use of the difficult-to-make dye-coupled antibodies in multiple experiments.

In some cases, it is advantageous to use primary

antibodies directly labelled with a fluorophore. This direct labelling decreases the number of steps in the staining procedure and, more importantly, often avoids cross-reactivity and high background problems.

Fluorescent labelling can be performed in less than one hour with readily available labeling kits. As with most fluorescence techniques, a significant problem with immunofluorescence is photobleaching.

Loss of activity caused by photobleaching can be controlled by reducing the intensity or time-span of light exposure, by increasing the concentration of fluorophores, or by employing more robust fluorophores that are less prone to bleaching.

Many uses of immunofluorescence have been outmoded by the development of recombinant proteins containing fluorescent protein domains, e.g. green fluorescent protein (GFP).

Use of such "tagged" proteins allows much better localization and less disruption of protein function.

Immunofluorescence Method

The purpose of immunofluorescence is to detect the location and relative abundance of any protein for which you have an antibody.

Once you have antibodies to your favorite protein, you can use them to indicate where the protein is located. In this example, we will use antibodies for the calcium ATPase, or pump, that is located in the endoplasmic reticulum (ER) of very cell.

The antibody used here only recognized the chicken calcium ATPase but immunofluorescence can be used on any protein.

The key to this entire process is the ability to visualize the antibody when looking through a microscope. Since antibodies are smaller than calcium ATPases.

Therefore, you have to use a fluorescent dye that is covalently attached to the antibody. When a light illuminates the fluorescent dye, it absorbs the light an emits a different color light which is visible to the investigator and can be photographed.

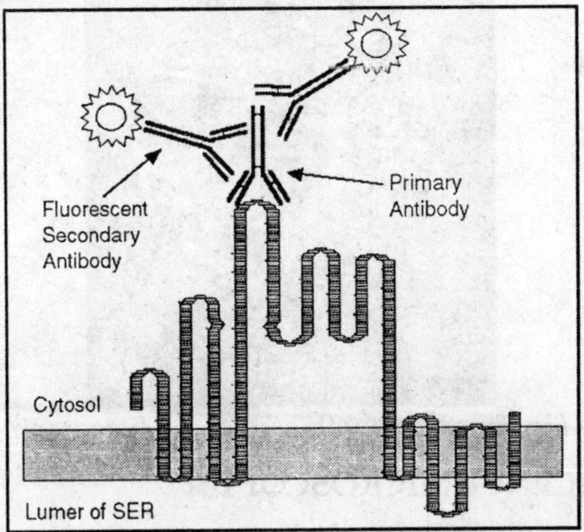

Fig. Immunofluorescence Experiments of Two Antibodies are Employed

The first one, called the primary antibody, is typically generated in a mouse and binds to your favorite protein, which in this case is the chicken calcium ATPase. The secondary antibody was purchased from a company that sells antibodies that bind to mouse antibodies and have a fluorescent dye covalently attached to it.

As illustrated here, the secondary antibodies can bind to multiple sites on the primary antibody and thus produce a brighter signal since more dyes are brought to a single location. The first step is to choose your cells of interest. In this case, we will look at a chicken fibroblast, or skin cell. It was grown in tissue culture and so it appears as an isolated cell witn no visible neighbors.

The cell was fixed with formaldehyde to retain the shape and location of all cellular proteins.

The cell was treated with a mild detergent to disolve small holes in the membranes so the antibodies could have access to the cytoplasm.

Because the calcium ATPase is located in the ER, the antibodies must have access to the cytoplasm or they could not bind to the target protein.

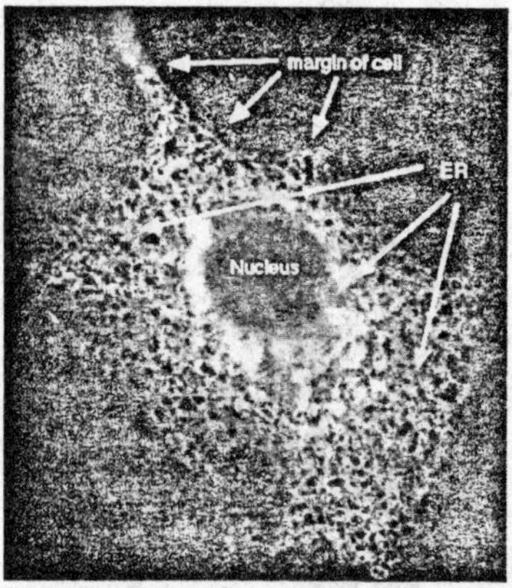

Fig. This Immunofluorescence Micrograph

FLUORESCENCE MICROSCOPY

When certain compounds are illuminated with high energy light, they then emit light of a different, lower frequency. This effect is known as fluorescence. Often specimens show their own characteristic autofluorescence image, based on their chemical makeup.

This method is of critical importance in the modern life sciences, as it can be extremely sensitive, allowing the detection of single molecules. Many different fluorescent dyes can be used to stain different structures or chemical compounds. One particularly powerful method is the combination of antibodies coupled to a fluorochrome as in immunostaining. Examples of commonly used fluorochromes are fluorescein or rhodamine. The antibodies can be made tailored specifically for a chemical compound.

For example, one strategy often in use is the artificial production of proteins, based on the genetic code (DNA). These proteins can then be used to immunize rabbits, which then form antibodies which bind to the protein. The antibodies are then coupled chemically to a fluorochrome and then used to trace the proteins in the cells under study.

Highly-efficient fluorescent proteins such as the green fluorescent protein (GFP) have been developed using the molecular biology technique of gene fusion, a process which links the expression of the fluorescent compound to that of the target protein. Piston DW, Patterson GH, Lippincott-Schwartz J, Claxton NS, Davidson MW Nikon MicroscopyU: Introduction to Fluorescent Proteins. Nikon MicroscopyU. Retrieved on 2007-08-22.

This combined fluorescent protein is generally non-toxic to the organism and rarely interferes with the function of the protein under study. Genetically modified cells or organisms directly express the fluorescently-tagged proteins, which enables the study of the function of the original protein in vivo.

Since fluorescence emission differs in wavelength (colour) from the excitation light, a fluorescent image ideally only shows the structure of interest that was labelled with the fluorescent dye. This high specificity led to the widespread use of fluorescence light microscopy in biomedical research. Different fluorescent dyes can be used to stain different biological structures, which can then be detected simultaneously, while still being specific due to the individual colour of the dye.

To block the excitation light from reaching the observer or the detector, filter sets of high quality are needed. These typically consist of an excitation filter selecting the range of excitation wavelengths, a dichroic mirror and an emission filter blocking the excitation light.

Most fluorescence microscopes are operated in the Epi-illumination mode to further decrease the amount of excitation light entering the detector.

Confocal Laser Scanning Microscopy

Generates the image by a completely different way than the normal visual bright field microscope. It gives slightly higher resolution, but most importantly it provides optical sectioning without disturbing out-of-focus light degrading the image. Therefore it provides sharper images of 3D objects. This is often used in conjunction with fluorescence microscopy.

Deconvolution Microscopy

Fluorescence microscopy is extremely powerful due to its

ability to show specifically labelled structures within a complex environment but also because of its inherent ability to provide three dimensional information of biological structures. Unfortunately this information is blurred by the fact, that upon illumination all fluorescently labeled structures emit light no matter if they are in focus or not.

This means, that an image of a certain structure is always blurred by the contribution of light from structures which are out of focus. This phenomenon becomes apparent as a loss of contrast especially when using objectives with a high resolving power, typically oil immersion objectives with a high numerical aperture.

Fortunately though, this phenomenon is not caused by random processes such as light scattering but can be relatively well defined by the optical properties of the image formation in the microscope imaging system. If one considers a small fluorescent light source (essentially a bright spot), light coming from this spot spreads out the further out of focus one is. Under ideal conditions this produces a sort of "hourglass" shape of this point source in the third (axial) dimension. This shape is called the point spread function (PSF) of the microscope imaging system.

Since any fluorescence image is made up of a large number of such small fluorescent light sources the image is said to be "convolved by the point spread function".

Knowing this point spread function means, that it is possible to reverse this process to a certain extent by computer based methods commonly known as deconvolution microscopy. There are various algorithms available for 2D or 3D Deconvolution. They can be roughly classified in non restorative and restorative methods. While the non restorative methods can improve contrast by removing out of focus light from focal planes, only the restorative methods can actually reassign light to it proper place of origin. This can be an advantage over other types of 3D microscopy such as confocal microscopy, because light is not thrown away but reused. For 3D deconvolution one typically provides a series of images derived from different focal planes (called a Z-stack) plus the knowledge of the PSF which can be either derived experimentally or theoretically from knowing all contributing parameters of the microscope.

Sub-diffraction optical microscopy techniques

It is well known that there is a spatial limit to which light can focus: approximately half of the wavelength of the light you are using. But this is not a true barrier, because this diffraction limit is only true in the far-field and localization precision can be increased with many photons and careful analysis (although two objects still cannot be resolved); and like the sound barrier, the diffraction barrier is breakable. This section explores some approaches to imaging objects smaller than ~250 nm. Most of the following information was gathered (with permission) from a chemistry blog's review of sub-diffraction microscopy techniques.

TYPES OF MICROSCOPES

Various types of microscopes are available for use in the microbiology laboratory. The microscopes have varied applications and modifications that contribute to their usefulness.The light microscope. The common light microscope used in the laboratory is called a compound microscope because it contains two types of lenses that function to magnify an object.

The lens closest to the eye is called the ocular, while the lens closest to the object is called the objective. Most microscopes have on their base an apparatus called a condenser, which condenses light rays to a strong beam. A diaphragm located on the condenser controls the amount of light coming through it. Both coarse and fine adjustments are found on the light microscope.

To magnify an object, light is projected through an opening in the stage, where it hits the object and then enters the objective. An image is created and this image becomes an object for the ocular lens, which remagnifies the image. Thus, the total magnification possible with the microscope is the magnification achieved by the objective multiplied by the magnification achieved by the ocular lens. A compound light microscope often contains four objective lenses: the scanning lens (4X), the low-power lens (10X), the high-power lens (40 X) and the oil-immersion lens (100 X).

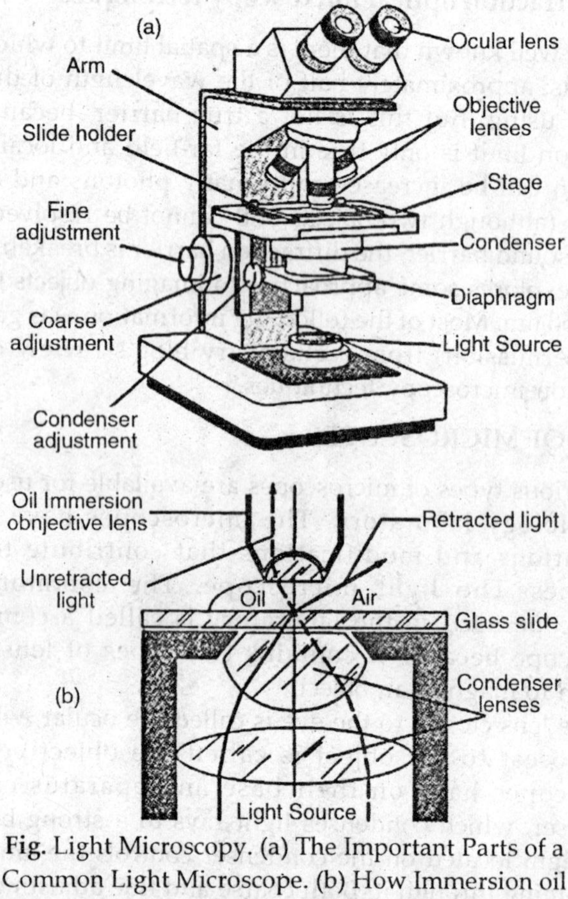

Fig. Light Microscopy. (a) The Important Parts of a Common Light Microscope. (b) How Immersion oil Gathers more Light for use in the Microscope

With an ocular lens that magnifies 10 times, the total magnifications possible will be 40 X with the scanning lens, 100 X with the low-power lens, 400 X with the high-power lens and 1000 X with the oil-immersion lens. Most microscopes are parfocal. This term means that the microscope remains in focus when one switches from one objective to the next objective.

The ability to see clearly two items as separate objects under the microscope is called the resolution of the microscope. The resolution is determined in part by the wavelength of the light used for observing. Visible light has a wavelength of about 550 nm, while ultraviolet light has a wavelength of about 400 nm or less. The resolution of a microscope increases as the wavelength decreases, so ultraviolet light allows one to detect objects not seen with visible light. The resolving power of a lens refers to the size of the smallest object that can be seen with that lens. The resolving power is based on the wavelength of the light used and the numerical aperture of the lens. The numerical aperture (NA) refers to the widest cone of light that can enter the lens; the NA is engraved on the side of the objective lens.

If the user is to see objects clearly, sufficient light must enter the objective lens. With modern microscopes, entry to the objective is not a problem for scanning, low-power and high-power lenses. However, the oil-immersion lens is exceedingly narrow and most light misses it. Therefore, the object is seen poorly and without resolution. To increase the resolution with the oil-immersion lens, a drop of immersion oil is placed between the lens and the glass slide. Immersion oil has the same light-bending ability (index of refraction) as the glass slide, so it keeps light in a straight line as it passes through the glass slide to the oil and on to the glass of the objective, the oil-immersion lens. With the increased amount of light entering the objective, the resolution of the object increases and one can observe objects as small as bacteria. Resolution is important in other types of microscopy as well.

Other Light Microscopes

In addition to the familiar compound microscope, microbiologists use other types of microscopes for specific purposes. These microscopes permit viewing of objects not otherwise seen with the light microscope. An alternative microscope is the dark-field microscope, which is used to observe live spirochetes, such as those that cause syphilis. This microscope contains a special condenser that scatters light and causes it to reflect off the specimen at an angle. A light object is seen on a dark background.

A second alternative microscope is the phase-contrast microscope. This microscope also contains special condensers that throw light "out of phase" and cause it to pass through the object at different speeds. Live, unstained organisms are seen clearly with this microscope and internal cell parts such as mitochondria, lysosomes and the Golgi body can be seen with this instrument.

The fluorescent microscope uses ultraviolet light as its light source. When ultraviolet light hits an object, it excites the electrons of the object and they give off light in various shades of colour.

Since ultraviolet light is used, the resolution of the object increases. A laboratory technique called the fluorescent-antibody technique employs fluorescent dyes and antibodies to help identify unknown bacteria.

Electron Microscopy

The energy source used in the electron microscope is a beam of electrons. Since the beam has an exceptionally short wavelength, it strikes most objects in its path and increases the resolution of the microscope significantly. Viruses and some large molecules can be seen with this instrument.

The electrons travel in a vacuum to avoid contact with deflecting air molecules and magnets focus the beam on the object to be viewed. An image is created on a monitor and viewed by the technologist. The more traditional form of electron microscope is the transmission electron microscope (TEM).

To use this instrument, one places ultrathin slices of microorganisms or viruses on a wire grid and then stains them with gold or palladium before viewing. The densely coated parts of the specimen deflect the electron beam and both dark and light areas show up on the image. The scanning electron microscope (SEM) is the more contemporary form electron microscope. Although this microscope gives lower magnifications than the TEM, the SEM permits three-dimensional views of microorganisms and other objects. Whole objects are used and gold or palladium staining is employed.

Staining Techniques

Because microbial cytoplasm is usually transparent, it is necessary to stain microorganisms before they can be viewed with the light microscope. In some cases, staining is unnecessary, for example when microorganisms are very large or when motility is to be studied and a drop of the microorganisms can be placed directly on the slide and observed.

A preparation such as this is called a wet mount. A wet mount can also be prepared by placing a drop of culture on a cover-slip (a glass cover for a slide) and then inverting it over a hollowed-out slide. This procedure is called the hanging drop. In preparation for staining, a small sample of microor ganisms is placed on a slide and permitted to air dry. The smear is heat fixed by quickly passing it over a flame. Heat fixing kills the organisms, makes them adhere to the slide and permits them to accept the stain.

Simple Stain Techniques

Staining can be performed with basic dyes such as crystal violet or methylene blue, positively charged dyes that are attracted to the negatively charged materials of the microbial cytoplasm.

Such a procedure is the simple stain procedure. An alternative is to use a dye such as nigrosin or Congo red, acidic, negatively charged dyes. They are repelled by the negatively charged cytoplasm and gather around the cells, leaving the cells clear and unstained. This technique is called the negative stain technique.

Differential Stain Techniques

The differential stain technique distinguishes two kinds of organisms. An example is the Gram stain technique. This differential technique separates bacteria into two groups, Gram-positive bacteria and Gram-negative bacteria. Crystal violet is first applied, followed by the mordant iodine, which fixes the stain. Then the slide is washed with alcohol and the Gram-positive bacteria retain the crystal-violet iodine stain; however, the Gram-negative bacteria lose the stain. The Gram-negative bacteria subsequently stain with the safranin dye, the counterstain, used next. These bacteria appear red under the oil-immersion lens, while Gram-positive bacteria appear blue or purple, reflecting the crystal violet retained during the washing step.

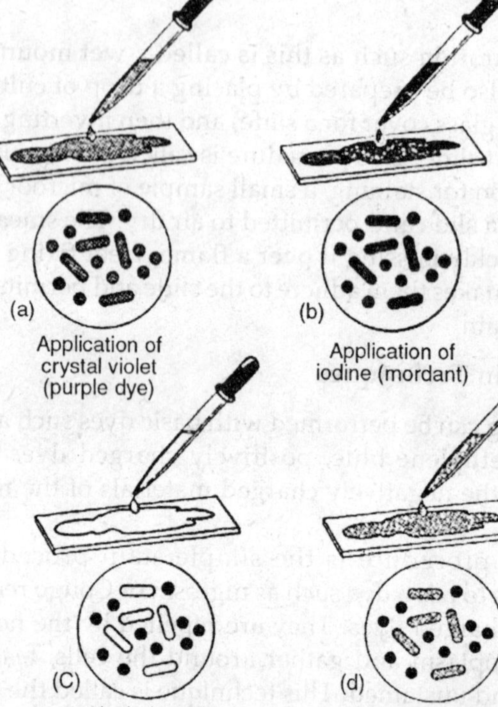

Fig. The Gram Stain Procedure used for Differentiating Bacteria into two Groups.

Another differential stain technique is the acid-fast technique. This technique differentiates species of *Mycobacterium* from other bacteria. Heat or a lipid solvent is used to carry the first stain, carbolfuchsin, into the cells.

Then the cells are washed with a dilute acid-alcohol solution. *Mycobacterium* species resist the effect of the acid-alcohol and retain the carbolfuchsin stain (bright red). Other bacteria lose the stain and take on the subsequent methylene blue stain (blue). Thus, the acid-fast bacteria appear bright red, while the nonacid-fast bacteria appear blue when observed under oil-immersion microscopy.

Other stain techniques seek to identify various bacterial structures of importance. For instance, a special stain technique highlights the flagella of bacteria by coating the flagella with dyes or metals to increase their width. Flagella so stained can then be observed. A special stain technique is used to examine bacterial spores. Malachite green is used with heat to force the stain into the cells and give them colour. A counterstain, safranin, is then used to give colour to the nonsporeforming bacteria. At the end of the procedure, spores stain green and other cells stain red.

LIGHT MICROSCOPY

The light microscope, so called because it employs visible light to detect small objects, is probably the most well-known and well-used research tool in biology. Yet, many students and teachers are unaware of the full range of features that are available in light microscopes. Since the cost of an instrument increases with its quality and versatility, the best instruments are, unfortunately, unavailable to most academic programs. However, even the most inexpensive "student" microscopes can provide spectacular views of nature and can enable students to perform some reasonably sophisticated experiments.

A beginner tends to think that the challenge of viewing small objects lies in getting enough magnification. In fact, when it comes to looking at living things the biggest challenges are, in order,

- Obtaining sufficient contrast
- Finding the focal plane
- Obtaining good resolution
- Recognizing the subject when one sees it

The smallest objects that are considered to be living are the bacteria. The smallest bacteria can be observed and cell shape recognized at a mere 100x magnification. They are invisible in bright field microscopes, though. These pages will describe types of optics that are used to obtain contrast, suggestions for finding specimens and focusing on them and advice on using measurement devices with a light microscope.

Types of Light Microscopes

The bright field microscope is best known to students and is most likely to be found in a classroom. Better equipped classrooms and labs may have dark field and/or phase contrast optics. Differential interference contrast, Nomarski, Hoffman modulation contrast and variations produce considerable depth of resolution and a three dimensional effect.

Fluorescence and confocal microscopes are specialized instruments, used for research, clinical and industrial applications. Other than the compound microscope, a simpler instrument for low magnification use may also be found in the laboratory. The stereo microscope, or dissecting microscope usually has a binocular eyepiece tube, a long working distance and a range of magnifications typically from 5x to 35 or 40x. Some instruments supply lenses for higher magnifications, but there is no improvement in resolution. Such "false magnification" is rarely worth the expense.

Bright Field Microscopy

With a conventional bright field microscope, light from an incandescent source is aimed toward a lens beneath the stage called the condenser, through the specimen, through an objective lens and to the eye through a second magnifying lens, the ocular or eyepiece. We see objects in the light path because natural pigmentation or stains absorb light differentially, or because they are thick enough to absorb a significant amount of light despite being colorless. A *Paramecium* should show up fairly well in a bright field microscope, although it will not be easy to see cilia or most organelles. Living bacteria won't show up at all unless the viewer hits the focal plane by luck and distorts the image by using maximum contrast.

A good quality microscope has a built-in illuminator, adjustable condenser with aperture diaphragm (contrast) control, mechanical stage and binocular eyepiece tube. The condenser is used to focus light on the specimen through an opening in the stage. After passing through the specimen, the light is displayed to the eye with an apparent field that is much larger than the area illuminated. The magnification of the image is simply the objective lens magnification (usually stamped on the lens body) times the ocular magnification.

Students are usually aware of the use of the coarse and fine focus knobs, used to sharpen the image of the specimen. They are frequently unaware of adjustments to the condenser that can affect resolution and contrast. Some condensers are fixed in position, others are focusable, so that the quality of light can be adjusted. Usually the best position for a focusable condenser is as close to the stage as possible. The bright field condenser usually contains an aperture diaphragm, a device that controls the diameter of the light beam coming up through the condenser, so that when the diaphragm is stopped down (nearly closed) the light comes straight up through the centre of the condenser lens and contrast is high. When the diaphragm is wide open the image is brighter and contrast is low.

A disadvantage of having to rely solely on an aperture diaphragm for contrast is that beyond an optimum point the more contrast you produce the more you distort the image. With a small, unstained, unpigmented specimen, you are usually past optimum contrast when you begin to see the image.

Mount the Specimen on the Stage

The cover slip must be up if there is one. High magnification objective lenses can't focus through a thick glass slide; they must be brought close to the specimen, which is why coverslips are so thin. The stage may be equipped with simple clips (less expensive microscopes), or with some type of slide holder. The slide may require manual positioning, or there may be a mechanical stage (preferred) that allows precise positioning without touching the slide.

Optimize the Lighting

A light source should have a wide dynamic range, to provide high intensity illumination at high magnifications and lower intensities so that the user can view comfortably at low magnifications.

Better microscopes have a built-in illuminator and the best microscopes have controls over light intensity and shape of the light beam. If your microscope requires an external light source, make sure that the light is aimed toward the middle of the condenser. Adjust illumination so that the field is bright without hurting the eyes.

Adjust the Condenser

To adjust and align the microscope, start by reading the manual. If no manual is available, try using these guidelines. If the condenser is focusable, position it with the lens as close to the opening in the stage as you can get it. If the condenser has selectable options, set it to bright field. Start with the aperture diaphragm stopped down (high contrast). You should see the light that comes up through the specimen change brightness as you move the aperture diaphragm lever.

FOCUS, LOCATE AND CENTRE THE SPECIMEN

Start with the lowest magnification objective lens, to home in on the specimen and/or the part of the specimen you wish to examine. It is rather easy to find and focus on sections of tissues, especially if they are fixed and stained, as with most prepared slides. However it can be very difficult to locate living, minute specimens such as bacteria or unpigmented protists. A suspension of yeast cells makes a good practice specimen for finding difficult objects.

- Use dark field mode (if available) to find unstained specimens. If not, start with high contrast (aperture diaphragm closed down).
- Start with the specimen out of focus so that the stage and objective must be brought closer together. The first surface to come into focus as you bring stage and objective together is the top of the cover slip. With smears, a cover slip is frequently not used.
- If you are having trouble, focus on the edge of the cover slip or an air bubble, or something that you can readily recognize. The top edge of the cover slip comes into focus first, then the bottom, which should be in the same plane as your specimen.
- Once you have found the specimen, adjust contrast and intensity of illumination and move the slide around until you have a good area for viewing.

Adjust Eyepiece Separation, Focus

With a single ocular, there is nothing to do with the eyepiece except to keep it clean. With a binocular microscope (preferred) you need to adjust the eyepiece separation just like you do a pair of binoculars. Binocular vision is much more sensitive to light and detail than monocular vision, so if you have a binocular microscope, take advantage of it.

One or both of the eyepieces may be a telescoping eyepiece, that is, you can focus it. Since very few people have eyes that are perfectly matched, most of us need to focus one eyepiece to match the other image. Look with the appropriate eye into the fixed eyepiece and focus with the microscope focus knob. Next, look into the adjustable eyepiece (with the other eye of course) and adjust the eyepiece, not the microscope.

Select an Objective lens for Viewing

The lowest power lens is usually 3.5 or 4x and is used primarily for initially finding specimens. We sometimes call it the scanning lens for that reason. The most frequently used objective lens is the 10x lens, which gives a final magnification of 100x with a 10x ocular lens. For very small protists and for details in prepared slides such as cell organelles or mitotic. Typical high magnification lenses are 40x and 97x or 100x. The latter two magnifications are used exclusively with oil in order to improve resolution.

Move up in magnification by steps. Each time you go to a higher power objective, re-focus and re-centre the specimen. Higher magnification lenses must be physically closer to the specimen itself, which poses the risk of jamming the objective into the specimen. Be very cautious when focusing. By the way, good quality sets of lenses are parfocal, that is, when you switch magnifications the specimen remains in focus or close to focused.

Bigger is not always better. All specimens have three dimensions and unless a specimen is extremely thin you will be unable to focus with a high magnification objective. The higher the magnification, the harder it is to "chase" a moving specimen.

The apparent field of an eyepiece is constant regardless of magnification used. So it follows that when you raise magnification the area of illuminated specimen you see is smaller. Since you are looking at a smaller area, less light reaches the eye and the image darkens. With a low power objective you may have to cut down on illumination intensity. With a high power you need all the light you can get, especially with less expensive microscopes.

Bright field microscopy is best suited to viewing stained or naturally pigmented specimens such as stained prepared slides of tissue sections or living photosynthetic organisms. It is useless for living specimens of bacteria and inferior for non-photosynthetic protists or metazoans, or unstained cell suspensions or tissue sections. Here is a not-so-complete list of specimens that might be observed using bright-field microscopy and appropriate magnifications (preferred final magnifications are emphasized).

- *Prepared slides, stained:* Bacteria (1000x), thick tissue sections (100x, 400x), thin sections with condensed chromosomes or specially stained organelles (1000x), large protists or metazoans (100x).

- *Smears, stained:* Blood (400x, 1000x), negative stained bacteria (400x, 1000x).
- *Living preparations (wet mounts, unstained):* Pond water (40x, 100x, 400x), living protists or metazoans (40x, 100x, 400x occasionally), algae and other microscopic plant material (40x, 100x, 400x). Smaller specimens will be difficult to observe without distortion, especially if they have no pigmentation.

Care of the Microscope

- Everything on a good quality microscope is unbelievably expensive, so be careful.
- Hold a microscope firmly by the stand, only. Never grab it by the eyepiece holder, for example.
- Hold the plug (not the cable) when unplugging the illuminator.
- Since bulbs are expensive and have a limited life, turn the illuminator off when you are done.
- Always make sure the stage and lenses are clean before putting away the microscope.
- Never use a paper towel, a kimwipe, your shirt, or any material other than good quality lens tissue or a cotton swab (must be 100% natural cotton) to clean an optical surface. Be gentle! You may use an appropriate lens cleaner or distilled water to help remove dried material. Organic solvents may separate or damage the lens elements or coatings.
- Cover the instrument with a dust jacket when not in use.
- Focus smoothly; don't try to speed through the focusing process or force anything. For example if you encounter increased resistance when focusing then you've probably reached a limit and you are going in the wrong direction.

Chapter 2

Classification of Bacteria

BACTERIAL GROWTH AND DEATH

Bacteria are widely present in soil, water, dust, air, human contamination sources and animal reservoirs. Many are accidental intruders into human processes. The basic differences between prokaryotic cells and eukaryotic cells. A complete understanding of antimicrobials and their mechanism of action is critical for mastery of medical microbiology.

Structure and Growth: Prokaryotes vs Eukaryotes		
Prokaryotes		**Eukaryotes**
Nucleoid	>>>>>>	Nucleus
Operon Gene Structure		Inron & Exon Gene Structure
Free Polysomes	>>>>>>	Ribosomes assoc. with Rough E.R.
Polycistronic mRNA		Monocistronic mRNA
No E.R.	>>>>>>	Rough & Smooth
Free Enymes	>>>>>>	Lysosomes
Proton Motive Force (pmf)	>>>>>>	Mitochondria
n-formyl-Met protein modification	>>>>>>	None

The purpose for categorizing bacteria is to establish phylogeny; genetic and evolutionary relationships and to devise chemotherapy strategies based on better comprehension. Bacterial morphology, biochemistry and physiology (autotrophs vs heterotrophs; aerobic vs anaerobic; culture/survival conditions) and sequence homology (moleculary biology) is the basis for bacterial diversity. Diagnostics is a competitive market built on exploiting these differences among the bacteria. Diagnostics uses rapid integration of new technologies for state of the art, fast, simple medical care to benefit patients. Major concerns include minimizing cost while maximizing reliability and reproducibility.

All bacteria have a lipid bilayer cytoplasmic membrane with embedded proteins that control movement of nutrients and waste products across the cell boundary. Gram-negative cells have a second, outer membrane (OM) that is very different from the cytoplasmic membrane, with porin proteins that limit access of large molecular weight substances to the cell interior. Also present in the outer part of the OM bilayer is lipopolysaccharide (LPS). The OM encloses another unique Gram-negative feature, the periplasmic space where a thin peptidoglycan layer (~3nm thick) and enzymes including the penicillin-binding proteins are located.

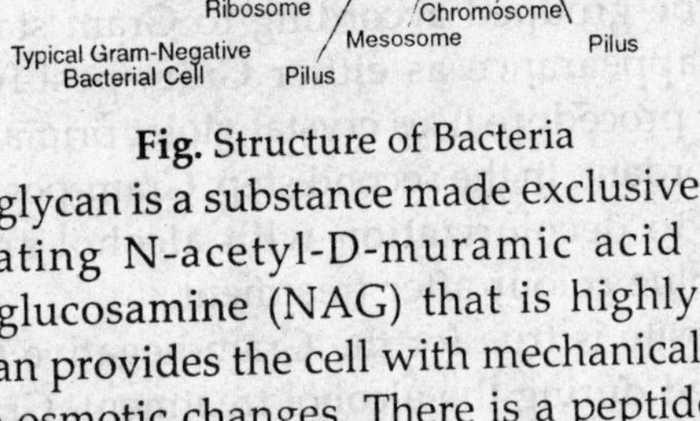

Fig. Structure of Bacteria

Peptidoglycan is a substance made exclusively by bacteria using repeating N-acetyl-D-muramic acid (NAM) and N-acetyl-D-glucosamine (NAG) that is highly crosslinked. Peptidoglycan provides the cell with mechanical strength and resistance to osmotic changes. There is a peptidoglycan layer in Gram-positive bacteria also, but it is much more substantial (10-80nm thick) and more highly cross-linked. Gram-positive bacteria also have teichoic and lipoteichoic acids that set them apart from Gram-negative cells. Bacteria from both groups may have a loose capsule outside the cell wall and the capsule composition is characteristic of the species. Capsules form a protective barrier that makes it difficult for the immune system to r ecognize and opsonize the bacterial cell.

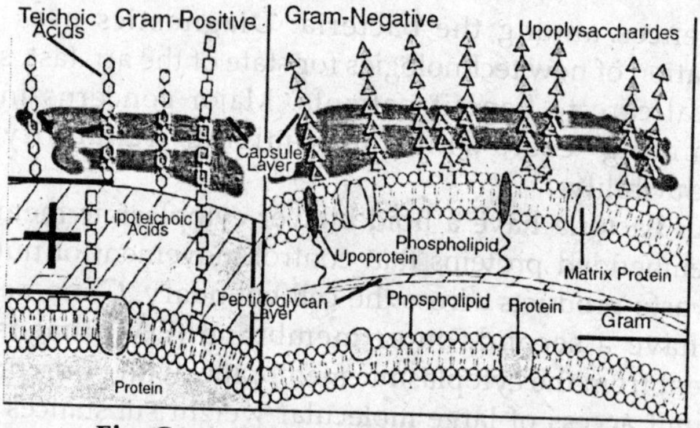

Fig. Gram Positive and Gram Negative

BACTERIAL STAINING

The first step for any bacterial stain is heat fixing the bacterium to the microscope slide. Next, a stain is used to provide contrast between the microbe and the background. Simple stains like the Methylene Blue Stain are taken up by most bacteria and allow visualization without differentiating among the various types of bacteria.

Other stains like the Gram stain and Acid-fast stain allow different staining patterns with the various types of bacteria. Bacteria can be grouped according to Gram stain and light microscopic appearance as either Gram-positive or Gram-negative. The procedure uses crystal violet primary stain with iodine as a mordant. In the second step, Gram-positive bacteria are resistant to decolorization with alcohol and remain a purple/dark blue colour after treatment.

The opposite is true for the Gram-negative bacteria that are decolorized during the alcohol treatment. Gram-negative bacteria show instead the pink colour of the safranin counterstain used in the final staining step. The difference in staining pattern is due to a major distinction in the bacterial cell wall. Some bacteria cannot be stained using the Gram-stain method. The spirochetes (Borrelia, Leptospira and Treponema) do not stain well, but have the Gram-negative cell wall structure. These bacteria can be visualized using silver staining or dark-field microscopy. Chlamydia and Rickettsia species have unconventional cell walls and must be stained using special methods and Mycoplasma species don't have a cell wall at all. Mycobacterium species are essentially Gram-positive, but have a thick, waxy coating and can only be detected using the Acid-Fast Stain (carbol fuchsin).

Bacteria that move about have motility provided by flagellae (singular, flagella) that extend from the cell surface and reach beyond the capsule. Flagellar ultrastructure shows differences in membrane anchoring between the bacterial groups. Bacterial pili (singular pilus, also known as a fimbria) are small, hair-like structures that mediate adherence to target host structures. Pili are made of adhesin proteins and some bacteria possess the F' or sex pilus that allows exchange of DNA with "receptive" cells in a process known as conjugation.

BACTERIAL SHAPES

Cocci (= round; prototype Staphylococcus). Bacilli (= rods; prototype Escherichia coli). Other shapes are intermediate between round and rod, like coccobacillary (Gram-negative short rods - Haemophilus; Bordetella; Legionella) and commashaped rods (Gram-negative, Vibrio, Campylobacter and Helicobacter species).

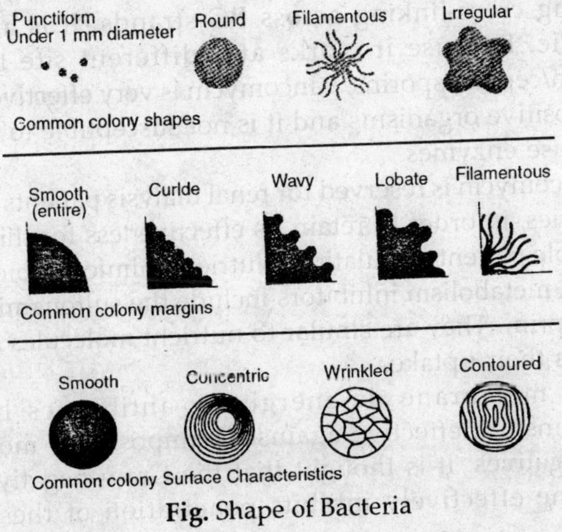

Fig. Shape of Bacteria

ANTIMICROBIAL SITES OF ACTION

Cell wall synthesis inhibitors are a major class of antibiotics, including penicillins and penicillin derivative antibiotics (amoxicillin, oxacillin, etc), the cephalosporins and vancomycin (a peptide antibiotic). The penicillin and cephalosporin drugs localize to the periplasmic space in Gram-negative bacteria, where the bacterial enzymes that synthesize peptidoglycan are found (transpeptidases that crosslink PG strands). This is also where the ^{2}-lactamase enzymes that allow antibiotic resistance are located. Drugs having the ^{2}-lactam ring structure are susceptible to hydrolysis at the highly constrained ring. Cephalosporins have an additional five-membered ring and represent more complex forms of ^{2}-lactam class antibiotics.

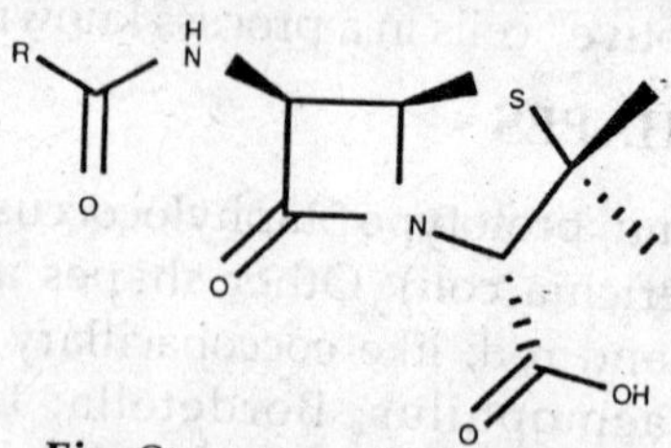

Fig. Structure of Penicilline

Vancomycin targets cell wall cross linking as well, binding to the ends of the PG peptide pentamer and sterically hindering cross-linking across PG strands. It is a peptide antibiotic. Because it works at a different site than the penicillin/cephalosporins, vancomycin is very effective against Gram-positive organisms and it is not susceptible to bacterial ^{2}-lactamase enzymes.

Vancomycin is reserved for renal dialysis patients in many formularies, in order to retain its effectiveness for this highly susceptible patient population. Nutrient mimic antibiotics like folic acid metabolism inhibitors include the sulfonamides and trimethoprim. They are similar to nutrient molecules and this facilitates their uptake.

Cell membrane de-energizing antibiotics like the polymyxins are effective against Grampositives more than Gram-negatives. It is thought that the Gram-negative outer membrane effectively inhibits penetration of these large antibiotics into the interior where the cytoplasmic membrane could be attacked. Nucleic acid synthesis. DNA synthesis is another very effective sight of antibacterial inhibition. DNA gyrase is the target (fluoroquinolones and nalidixic acid are the drugs). For RNA synthesis, DNA-directed RNA polymerase is the target bacterial molecule and Rifampin (rifampicin) is the drug.

These agents are considered to be bactericidal because they inhibit fundamental cellular processes. Protein synthesis is the antibiotic site of action for the remainder of the antibacterial compounds. There are 50S ribosomal subunit inhibitors (like chloramphenicol and erythromycin) and 30S ribosomal subunit inhibitors (streptomycin, gentamicin) and tRNA binding inhibitiors (tetracycline; also binds to the 30S ribosomal subunit).

BACTERIAL RESISTANCE FACTORS

Structure	Effects on Antibiotics
GRAM-NEGATIVES	
LPS of Gram negative	Retards/prevents penetration of bulky, high MW antibiotics (novobiocin, erythromycin).
Lipid bilayer of Gram-negative outer membrane	Penetration of water-soluble drugs is severely hindered.
Hydrophilic pores of Gram-negatives	Allow penetration of water- solubles up to 650 Daltons (sulfonamides)
Nutrient receptor transport proteins of Gram-negative outer membrane	Agents related to nutrient transport (sideromycins) utilize these natural receptors
GRAM-POSITIVES	
Teichoic and teichuronic acids	Strong anionic character of these polymers may affect rate of penetration
ALL BACTERIA	
Nutrient transport proteins of cytoplasmic membrane	Facilitate rapid penetration of agents related to nutrients by structure (D-cycloserine, phosphomycin; tetracyclines)
Mutations in PBPs Absence of high-affinity PBPs	Changed susceptibility to penicillin and derivative β-lactam antibiotics
β-lactamase enzymes	Target site is the β-lactam ring in penicillin and similar antibiotics (β-lactam ring)

BACTERIAL GROWTH

The definition of bacterial growth is increase in biomass. For growth in liquid media, a growth cycle determination can be made. According to the graph (right), lag phase is the horizontal beginning period when the enzymes are being synthesized that will accomplish the bacterial utilization of available nutrients.

The log phase is the period in which a sharp upward slope of the growth curve is noted. The kinetics are exponential, with a synchronous growth pattern quickly established in liquid culture.

In the stationary phase, there is no net increase in biomass, but rather a steady state where dead cells are quickly replaced with newly synthesized cells. The entire cycle is subject to chemostasis and death phase occurs when nutrients are depleted and toxic metabolic by-products build up to inhibit new growth.

Growth of bacteria on solid media or agar allows for enumeration (determination of viable cell number), selection of isolated colonies and determination of colony morphology (many bacteria have characteristic colony appearances). Pure cultures are obtained when a single kind of bacteria is present. The use of selective media allows the growth of bacteria to produce a characteristic change in the growth media that enables an identification to be made.

BACTERIAL DEATH

Death is defined as the loss of viability and loss of ability to produce progeny. Sterilization is used when the inactivation of all microbes is an absolute requirement, such as the sterilization of surgical instruments. Sanitizing a surface results in the reduction of bacterial content and antisepsis is the sanitizing of a body surface.

Sterilization is the removal of all microorganisms while disinfection is the removal of vegetative cells only. Viability is the capability to grow and develop as an independent unit and lethality is irreparable damage to the genome, envelope, or the proteins required for protein synthesis.

Each of these terms is used to describe a bacterial control process, but they are not precisely interchangeable terms. Students of microbiology should be able to use the terms correctly.

THE CATEGORIES OF INFECTION RISK AND METHODS OF CONTAINMENT

HIGH RISK **Surgical Instruments** **Needles, syringes** **Human tissues**	▪ Introduction into sterile body areas ▪ Close contact with mucous membrane or damaged skin ▪ Disposal of infectious waste materials	**STERILIZATION** Dry Heat Moist Heat Autoclaving Irradiation Chemicals (ethylene oxide) Filtration (physical removal)
MEDIUM RISK **Respiration Devices** **Endoscopes** **Thermometers**	▪ Contact with mucous membranes ▪ Prior to use with immunocompromised ▪ Contaminated with pathogenic material	**DISINFECTION** Alcohol Chlorine solution Phenolics Hydrogen peroxide
LOW RISK **Wheel chairs** **Patient Gurneys, Beds** **Walls, Floors, Sinks**	▪ Mainly in contact with healthy skin ▪ Little or no actual patient contact	**SANITIZING (CLEANING)** Surfactants Quatemary ammonium compounds

Physical agents of microbial death include moist heat, autoclaving, dry heat pasteurization and freezing or lyophilization. Moist heat is efficient at protein denaturation. Autoclaving is moist heat intensified by pressure. By definition, autoclaving is 121oC (250oF) for 15-20 minutes at 15 psi pressure.

Dry heat must be applied for 1-2 hours at 160oC to effectively oxidize microbes directly. Both these methods result in complete microbial killing, including spores. Pasteurization is the heating of a liquid to 62oC for 30 minutes and it kills known milk pathogens without completely sterilizing the milk (Lactobacillus species are not killed). Freezing causes microbial death by concentration of solutes. Bacterial cells can be preserved by lyophilization (freeze-drying) or by adding an emulsifying agent. Mechanical agents of microbial death include ultrasonic treatment, filtration (<400 nm filter pore size) or pressure. Chemical agents of microbial death include acid or alkali solutions (pH extreme), salts, heavy metals (silver nitrate), halogens (chlorine), alkylating agents (ethylene oxide; formaldehyde – these cause protein and nucleic acid damage), surfactants (detergents that compromise cytoplasmic membrane) and phenol and alcohols that cause protein denaturation. Chemical antimicrobial modes of action include physical denaturation and inactivation of proteins; de-energizing of the cell membrane and disruption of cellular membrane permeability; and the dissolving of cellular membranes (phenolics).

Sporulation: Bacteria that are resistant to environmental extremes frequently produce spores to allow survival during adverse conditions. The process of condensation of bacterial DNA within a series of membranes that accumulate calcium, dipicolinic acid and protein layers is known as sporulation. Clostidium species are known for the production of spores that allow survival of the organism under severe nutrient deprivation and dehydration conditions. Clostidium botulinum spores survive in the soil and then on harvested produce to germinate later and cause toxin-induced disease. Spores are even resistant to the actions of common antimicrobial agents and treatments.

Clostidium difficle causes disease transmission in the spore-saturated environment of hospital rooms where patients suffer from antibiotic-associated diarrhea. Species in the genus Bacillus are also spore-formers. The danger of Bacillus anthracis to the general public stems from its spore stage form and its potential for weaponization, storage and intentional release. Bacillus cereus spores germinate in warm food (especially associated with rice) and the toxins produced by the newly growing bacteria cause one form of food poisoning.

GROWTH AND REPRODUCTION OF BACTERIA

Unlike multicellular organisms, increases in the size of bacteria (cell growth) and their reproduction by cell division are tightly linked in unicellular organisms. Bacteria grow to a fixed size and then reproduce through binary fission, a form of asexual reproduction. Under optimal conditions, bacteria can grow and divide extremely rapidly and bacterial populations can double as quickly as every 9.8 minutes. In cell division, two identical clone daughter cells are produced. Some bacteria, while still reproducing asexually, form more complex reproductive structures that facilitate the dispersal of the newly-formed daughter cells. Examples include fruiting body formation by Myxobacteria and arial hyphae formation by Streptomyces, or budding. Budding involves a cell forming a protrusion that breaks away and produces a daughter cell.

In the laboratory, bacteria are usually grown using solid or liquid media. Solid growth media such as agar plates are used to isolate pure cultures of a bacterial strain. However, liquid growth media are used when measurement of growth or large volumes of cells are required. Growth in stirred liquid media occurs as an even cell suspension, making the cultures easy to divide and transfer, although isolating single bacteria from liquid media is difficult. The use of selective media (media with specific nutrients added or deficient, or with antibiotics added) can help identify specific organisms.

Most laboratory techniques for growing bacteria use high levels of nutrients to produce large amounts of cells cheaply and quickly. However, in natural environments nutrients are limited, meaning that bacteria cannot continue to reproduce indefinitely. This nutrient limitation has led the evolution of different growth strategies. Some organisms can grow extremely rapidly when nutrients become available, such as the formation of algal (and cyanobacterial) blooms that often occur in lakes during the summer.

Other organisms have adaptations to harsh environments, such as the production of multiple antibiotics by Streptomyces that inhibit the growth of competing microorganisms. In nature, many organisms live in communities (e.g. biofilms) which may allow for increased supply of nutrients and protection from environmental stresses. These relationships can be essential for growth of a particular organism or group of organisms (syntrophy). Bacterial growth follows three phases. When a population of bacteria first enter a high-nutrient environment that allows growth, the cells need to adapt to their new environment. The first phase of growth is the lag phase, a period of slow growth when the cells are adapting to fast growth. The lag phase has high biosynthesis rates, as enzymes and nutrient transporters are produced. The second phase of growth is the logarithmic phase (log phase), also known as the exponential phase. The log phase is marked by rapid exponential growth.

The rate at which cells grow during this phase is known as the growth rate (k) and the time it takes the cells to double is known as the generation time (g). During log phase, nutrients are metabolised at maximum speed until one of the nutrients is depleted and starts limiting growth.

The final phase of growth is the stationary phase and is caused by depleted nutrients. The cells reduce their metabolic activity and consume non-essential cellular proteins. The stationary phase is a transition from rapid growth to a stress response state and there is increased expression of genes involved in DNA repair, antioxidant metabolism and nutrient transport.

PLANT VIRUS

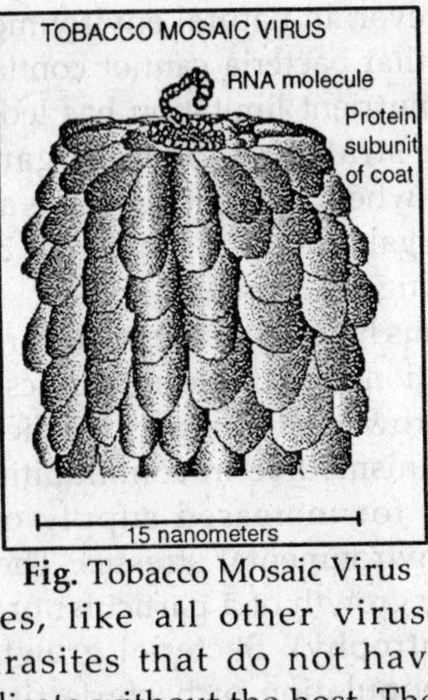

Fig. Tobacco Mosaic Virus

Plant viruses, like all other viruses, are obligate intracellular parasites that do not have the molecular machinery to replicate without the host. The plant viruses are defined as viruses pathogenic to higher plants.

While this article does not intend to list all plant viruses, it discusses some important viruses as well as their uses in plant molecular biology. Although plant viruses are not nearly as well understood as the animal counterparts, one plant virus has become iconic. The first virus to be discovered was *Tobacco mosaic virus* (TMV).

HISTORY OF PLANT VIRUS

The discovery of plant viruses causing disease is often accredited to Martinus Beijerinck who determined, in 1898, that plant sap obtained from tobacco leaves with the "mosaic disease" remained infectious when passed through a porcelain filter. This was in contrast to bacteria microorganisms, which were retained by the filter. Beijerinck referred to the infectious filtrate as a "contagium vivum fluidum", thus the coinage of the modern term "virus". After the initial discovery of the 'viral concept' there was need to classify any other known viral diseases based on the mode of transmission even though microscopic observation proved fruitless. In 1939 Holmes published a classification list of 129 plant viruses. This was expanded and in 1999 there were 977 officially recognized, and some provisional, plant virus species.

The purification (crystallization) of the TMV was first performed by Wendell Stanley, who published his findings in 1935, although he did not determine that the RNA was the infectious material. However, he received the Nobel Prize in Chemistry in 1946. In the 1950s a discovery by two labs simultaneously proved that the purified RNA of the TMV was infectious which reinforced the argument. The RNA was carrying genetic information to code for the production of new infectious particles. More recently virus research has been focused on understanding the genetics and molecular biology of plant virus genomes, with a particular interest in determining how the virus can replicate, move, and infect plants. Understanding the virus genetics and protein functions has been used to explore the potential for commercial use by biotechnology companies. In particular, viral-derived sequences have been used to provide an understanding of novel forms of resistance. The recent boom in technology allowing humans to manipulate plant viruses may provide new strategies for production of value-added proteins in plants.

STRUCTURE OF PLANT VIRUS

Viruses are very small and can only be observed with an electron microscope. The structure of a virus is given by its coat of proteins, which surround the viral genome. Assembly of viral particles takes place spontaneously. Over 50% of known plant viruses are rod shaped (flexuous or rigid).

The length of the particle is normally dependent on the genome but it is usually between 300–500 nm with a diameter of 15–20 nm. Protein subunits can be placed around the circumference of a circle to form a disc. In the presence of the viral genome, the discs are stacked, then a tube is created with room for the nucleic acid genome in the middle.

The second most common structure amongst plant viruses are isometric particles. They are 40–50 nm in diameter. In cases when there is only a single coat protein, the basic structure consists of 60 *T* subunits, where *T* is an integer. Some viruses may have 2 coat proteins are the formation of the particle is analogous to a football.

There are three genera of *Geminiviridae* that possess geminate particles which are like two isometric particles stuck together. A very small number of plant viruses have, in addition to their coat proteins, a lipid envelope. This is derived from the plant cell membrane as the virus particle buds off from the cell.

TRANSMISSION OF PLANT VIRUSES

Through Sap

It implies direct transfer of sap by contact of and wounded plant with a healthy one. Such process occurs during agricultural practices by tools, hands, or by animal feeding on the plant. Generally TMV, potato viruses and cucumber mosaic viruses are transmitted via sap.

Insects

Plant viruses need to be transmitted by a vector, most often insects such as leafhoppers. One class of viruses, the Rhabdoviridae, have been proposed to actually be insect viruses that have evolved to replicate in plants. The chosen insect vector of a plant virus will often be the determining factor in that virus' host range: it can only infect plants that the insect vector feeds upon. This was shown in part when the old world white fly made it to the USA, where it transferred many plant viruses onto new hosts.

Depending on the way they are transmitted, plant viruses are classified as non-persistent, semi-persistent and persistent. In non-persistent transmission, viruses become attached to the distal tip of the stylet of the insect and on the next plant it feeds on, it inoculates it with the virus. Semi-persistent viral transmission involves the virus entering the foregut of the insect.

Those viruses that manage to pass through the gut into the haemolymph and then to the salivary glands are known as persistent. There are two sub-classes of persistent viruses: propergative and circulative. Propergative viruses are able to replicate in both the plant and the insect (and may have originally been insect viruses), whereas circulative can not.

Many plant viruses encode within their genome polypeptides with domains essential for transmission by insects.

In non-persistent and semi-persistent viruses, these domains are in the coat protein and another protein known as the helper component. A bridging hypothesis has been proposed to explain how these proteins aid in insect-mediated viral transmission. The helper component will bind to the specific domain of the coat protein, and then the insect mouthparts — creating a bridge.

In persistent propagative viruses, such as tomato spotted wilt virus (TSWV), there is often a lipid coat surrounding the proteins that is not seen in the other classes of plant viruses. In the case of TSWV, 2 viral proteins are expressed in this lipid envelope. It has been proposed that the viruses bind via these proteins and are then taken into the insect cell by receptor-mediated endocytosis.

Nematodes

Soil-borne nematodes also have been shown to transmit viruses. They acquire and transmit them by feeding on infected roots. Viruses can be transmitted by non-persistently and persistently, but there is no evidence of viruses being able to replicate in nematodes.

Plasmodiophorids

A number of viral genera are transmitted, both persistently and non-persistently, by soil borne zoosporic protozoa. These protozoa are not phytopathogenic themselves, but parasitic. Transmission of the virus takes place when they become associated with the plant roots. An example is *Polymyxa graminis,* which has been shown to transmit a plant viral diseases in ceral crops.

Seed and Pollen Borne Viruses

Plant virus transmission from generation to generation occurs in about 20% of plant viruses. When viruses are transmitted by seeds, the seed is infected in the generative cells and the virus is maintained in the germ cells and sometimes, but less often, in the seed coat. When the growth and development of plants is delayed because of situations like unfavourable weather, there is an increase in the amount of virus infections in seeds. There does not seem to be a correlation between the location of the seed on the plant and its chances of being infected.

Little is known about the mechanisms involved in the transmission of plant viruses via seeds, although it is known that it is environmentally influenced and that seed transmission occurs because of a direct invasion of the embryo via the ovule or by an indirect route with an attack on the embryo mediated by infected gametes. These processes can occur concurrently or separately depending on the host plant. It is unknown how the virus is able to directly invade and cross the embryo and boundary between the parental and progeny generations in the ovule. Many plants species can be infected through seeds including but not limited to the families Leguminoseae, Solanacease, Compositae, Rosaceae, Curcurbitaceae, Gramineae.

TRANSLATION OF PLANT VIRAL PROTEINS

As mentioned above, 90% of plant viruses have genomes that consist of single stranded RNA, meaning that they are in the same sense orientation as messenger RNA. Viruses use the plant ribosomes to produce the 4-10 proteins encoded by their genome. However, since all of the proteins are encoded on a single strand (that is, they are polycistronic) this will mean that the ribosome will either only produce one protein, as it will terminate translation at the first stop codon or that a polyprotein will be produced. Plant viruses have had to evolve special techniques to allow the production of viral proteins by plant cells.

5' Cap

In order for translation to occur eukaryotic mRNAs require a 5' Cap structure. This means that viruses must also have one. This normally consists of 7MeGpppN where N is normally adenine or guanine. The viruses encode a protein, normally a replicase, with a methyltransferase activity to allow this.

Some viruses are cap-snatchers. During this process, a ^{7m}G-capped host mRNA is recruited by the viral transcriptase complex and subsequently cleaved by a virally encoded endonuclease. The resulting capped leader RNA is used to prime transcription on the viral genome.

However some plant viruses do not use cap, yet translate efficiently due to cap-independent translation enhancers present in 5' and 3' untranslated regions of viral mRNA

Production of Sub-genomic RNAs

Some viruses use the production of sub-genomic RNAs to ensure the translation of all proteins within their genomes.

In this process the first protein encoded on the genome, and this the first to be translated, is a replicase. This protein will act of the rest of the genome producing negative strand sub-genomic RNAs then act upon these to form positive strand sub-genomic RNAs that are essentially mRNAs ready for translation.

Segmented Genomes

Some viral families, such as the *Bromoviridae* instead opt to have multi-partite genomes, genomes split between multiple viral particles. For infection to occur, the plant must be infected with all particles across the genome. For instance *Brome mosaic virus* has a genome split between 3 viral particles, and all 3 particles with the different RNAs are required for infection to take place.

Polyprotein Processing

This stratergy is adopted by viral genra such as the Potyviridae and Tymovirus. The ribosome translates a single protein from the viral genome. Within the polyprotein is an ezyme with proteinase function that is able cleave the polyprotein into the various single proteins or just cleave away the replicase, which can then produce sub-genomic RNAs.

VIRUS REPRODUCTION ANDVIRUS DISEASE

ERA in virus diseases of man was opened in 1902 by Reed and Carroll.' They and their associates recorded a number of remarkably precise observations on the etiology of yellow fever that have since been abundantly confirmed. Some of their findings bear directly on ideas which are developed in this communication. Thus, in their now classic studies with human volunteers, which proved the correctness of the theory that yellow fever was transmitted by the bite of an Aedes mosquito, they made the following observations.

Mosquitos became infective when they had taken a blood meal during the first three days of the disease. Mosquitos did not become infective when they bit early in the incubation period or late in the disease. Mosquitos were not infective soon after they had bitten and became so only after a period of 12 days or more. They then remained infective for life. These findings, so puzzling at the time, are now fully understood. Each can be explained on the basis of modem knowledge of virus reproduction and the results thereof.

Yellow fever is still one of the most classic of the virus diseases of man and it may be rewarding to review its pathogenesis briefly. When an infected mosquito bites, a very small amount of yellow fever virus is introduced into the blood. This small amount of virus then promptly disappears. The virus does not reappear in the peripheral blood until just before the disease develops.

Therefore, normal mosquitos biting during the four-day incubation period do not take in virus. However, during the incubation period, the virus is undergoing reproduction in cells of the liver and other organs. When the number of cells damaged by the reproductive process becomes large enough, symptoms and signs of yellow fever rapidly appear. At about this time large numbers of newly formed virus particles are released from the infected tissues into the blood. Now if a mosquito bites, it will take in virus. As the disease progresses, still higher concentrations of virus develop in the liver, lymph nodes and kidneys.

This is the result of continued virus reproduction. In parallel, the concentration of virus in the blood increases rapidly and may reach a level of 107 particles per ml. If a mosquito bites at this time, it will take in a considerable amount of virus. After about the fourth day of disease, neutralizing anitibodies begin to appear and as they increase in coiicentration, the number of infective virus particles in the blood rapidly decreases and finally such particles disappear. Thus, late in the disease a biting mosquito fails to take in any infective virus and so cannot transmit the agent.

Analogous events occur in the mosquito itself. After ingestion of infective blood, the virus is first carried to the stomach of the insect and then is distributed to other organs. Only after the virus has reproduced in numerous cells in the insect does it eventually reach the salivary glands. This process takes about 12 days and during the period the mosquito, although actively supporting virus reproduction, is not mechanically able to transmit the agent. The mosquito does not develop neutralizing antibodies and, being unable to rid itself of the virus, remains infective for life.

Despite extensive virus reproduction, tissue damage apparently is not produced and therefore disease does not develop in the mosquito. Although the details of pathogenesis have been documented more thoroughly for yellow fever than for any other virus disease of man, there is now sufficient information about a number of such infections to support a general conceptual scheme: only a very small number of infective virus particles are needed to initiate virus reproduction in a person who has had no previous contact with the agent.

On theoretical grounds, a single particle should, if it is adsorbed by a susceptible cell, initiate reproduction. This has, in fact, been demonstrated with vaccinia4 and influenza viruses in experimental animals. Each new virus particle from a cell can initiate reproduction in another susceptible cell. Thus, if a cell yields 100 new particles, 100 other cells of similar type can be infected and will produce a similar yield. This appears to be the basis for the logarithmic increase in virtus concentration which is the regular result of reproduction. The duration of the incubation period seems to be dependent upon two independent variables: the number of cells initially infected and the rate of virus reproduction.

The larger the number of cells initially infected, the shorter is the incubation period. The rate of virus reproduction appears to be directly related to the duration of the incubation period; the slower the rate, the longer the period. Experiments with influenza'i and other viruses in animals have demonstrated that when both variables are known the incubation period can be predicted with considerable precision. Virus disease develops only if virus reproduction occurs, but reproduction can occur without leading to disease. As a result of reproduction, the concentration of virus particles increases in infected tissue at a logarithmic rate and reaches a high level before either signs or symptoms of infection appear. This has been found to occur in every instance so far studied.

The severity of the disease may be largely determined by the concentration of virus particles in the infected tissues. Experiments have demonstrated a direct relation between virus concentration and extent of lesions with a number of viruses in animals.' From this it can be deduced that the extent of the lesions is dependent on the number of cells damaged by the reproductive process or its products. Recovery occurs when the rate of the virus reproduction decreases.

The most probable means is through reduction in the number of susceptible cells. Because recovery sometimes coincides with the appearance of antibodies against the virus, it has been held that the development of specific antibodies is responsible for recovery. There is strong evidence against this view, e.g., recovery from recurrent herpes simplex occurs without a change in antibody level, recovery from psittacosis is not associated with the appearance of significant amounts of neutralizing antibody and the injection of immune serum does not definitely lead to recovery in any virus disease.

Persistent immunity, which is of common but not uniform occurrence, is best explained by the persistence of the virus. Evidence for persistence has been obtained in a number of instances. The theory of virus persistence requires that continued reproduction occur even though this be at a very slow rate. However, it is not necessary to assume continued presence of virus in all susceptible cells to explain persistent immunity. The antigenic stimulus provided by a small amount of slowly reproducing virus would maintain antibody production and neutralizing antibodies are hiahly effective in preventing a second infection in the large majority of instances.

In the case of transient immunity, as with influenza and certain other common respiratory infections, the virus may in fact be eliminated, for these are typically surface cell infections. Although virus disease does not develop in the absence of virus reproduction, it is not a regular or even the most common sequel of the reproductive process.

Virus disease is attributable to damage of cells and the

development of damage depends on the nature of the virus, its rate of reproduction and the concentration reached as well as upon the nature of the cell itself. Alteration of any one of these variables may yield a system in which reproduction occurs but disease does not develop. The easiest variable to alter is, of course, the type of cell infected.

Thus, reproduction of mumps virus in the human parotid gland or meninges leads to obvious disease. Yet the same virus on comparable reproduction in the chicken embryo causes no damage." Similarly, the reproduction of influenza viruses in the upper respiratory tract of man or the ferret often induces disease, but does not in the mouse until so-called adaptation has occurred. The effect of the rate and extent of reproduction is evident from studies with virus variants.

So-called avirulent or attenuated viruses, which do not induce obvious disease, reproduce at a slower rate and yield lower concentrations of virus than virulent strains of the same agent. Direct evidence that virus reproduction is an essential prerequisite for the development of virus disease is secured when the reproductive process is inhibited by suitable compounds. As an example, certain bacterial polysaccharides inhibit the reproduction of some small viruses in experimental animals and thereby prevent the development of disease.'

It seems clear now that the crucial problem in virus disease is, in fact, virus reproduction. If the mechanism of virus reproduction can be worked out and the process controlled, means for the management of virus diseases may be at hand. The important questions are: How do viruses multiply inside living cells and how does this process sometimes damage the cells? What are the biochemical events necessary for synthesis of new virus particles and how do they alter cell metabolism?

REPRODUCTION OF BACTERIAL VIRUSES

There is now a comprehensive scheme to explain the reproduction of these agents. It may be of value to review this in brief and then to determine how it fits in with present information on the reproduction of viruses that infect man. Bacteriophage particles resemble tiny sperm and have both a head and a tail-like structure.' The shell of the head and the tail are composed largely of protein. Inside the head there is nucleic acid and this is entirely of the desoxyribose type. A single phage particle can infect a bacterium.

Only the tail penetrates the bacterial wall and nucleic acid from the head is introduced. The phage particle itself does not enter the bacterium. In fact, the head-shell and part of the tail can be removed at this stage and yet phage reproduction will still occur in the bacterium.' This phenomenon appears to be unique in biology. The phage particle, in the process of initiating reproduction, disintegrates and its chief contribution to the bacterium is nucleic acid.

Shortly after this there are surprising developments. The bacterium stops synthesizing bacterial components and begins to produce phage nucleic acid and phage protein. The two are made separately, not as intact phage particles. After a time tailless heads appear in the bacterial cytoplasm.'

Later the final assembly of the phage particle is achieved by the addition of the tail and the introduction of nucleic acid. The new particle is now infective. The bacterial cell which supported this process is so damaged that it disintegrates and releases more than 100 new phage particles. Each new particle can initiate this complex reproductive cycle as soon as it finds another suitable uninfected bacterium.

Thus, with phage, reproduction is dependent on a series of discrete biosynthetic processes which produce the required precursor materials separately and.finally put them together as new phage particles. In supporting this remarkable process, the bacterium is driven to commit suicide.

If this concept is valid, it should be possible to control phage reproduction by inhibiting the production of any one of the required precursor materials or by preventing the final assembly of mature phage particles. Such control has, in fact, been accomplished. Chloramphenicol and analogues of tryptophane inhibit protein synthesis in bacteria. As would be expected, they prevent phage reproduction when applied during the stage of phage protein formation.

Certain pyrimidine analogues may inhibit nucleic acid

synthesis in bacteria and some inhibit phage reproduction when applied early in the process. Proflavine does not prevent the synthesis of either phage protein or nucleic acid but does prevent final assembly of the mature particle.

Only tailless heads are produced when this compound is applied late in the reproductive cycle and these are not infective. Apparently, however, the damage to the bacterium occurs while phage protein and nucleic acid are being formed and proflavine, although preventing assembly of the particle, does not protect the bacterial cell from destruction. It disintegrates at the expected time but does not yield infective phage particles."'

Comparison With Animal Viruses

One may now ask, in what respects does the reproduction of animal viruses correspond with the reproduction of bacterial viruses? It appears that there may be some differences, but their importance is not yet clear. Viruses which infect man resemble tiny spheres. No tail-like structures have been positively identified. They are composed largely of protein and nucleic acid, like phage particles, but some contain the ribose type, others the desoxyribose type and certain may contain both. Some have central dense areas resembling tiny nuclei." One virus particle can often initiate reproduction in a susceptible cell, but precisely how this is accomplished is not yet known. If the intact particle actually enters the cell, some change occurs, for it cannot be recovered later. Whether or not the particle actually disintegrates, as does phage, is not yet clear.

There then occurs an eclipse period during which there is no direct evidence as to intracellular developments. Some indirect evidence supports the idea that required precursor materials are formed during this period, as with phage, but their demonstration has not yet been achieved. Later, formed elements resembling virus particles begin to appear in the cytoplasm and then they increase in number. Recent evidence suggests that some viruses may develop limiting membranes before release from the cell.'

Unlike phage, the reproduction of many animal viruses does not lead to destruction of the cell. Presumably, the biochemical alterations produced in the cell are often not as severe as those caused by phage reproduction. The number of new particles produced varies from a few to more than 1,000 per cell, depending on the virus. As with phage, each new particle after release can initiate another reproductive cycle when it finds another susceptible cell. So far there are no data that make the phage scheme clearly inapplicable to the reproduction of animal viruses. If the process is like that of phage, dependent on the production of precursor materials during the eclipse period, it would be expected that reproduction should be inhibited by substances analogous to those that inhibit phage reproduction.

There is evidence in support of this idea. Thus, certain amino acid analogues, such as methoxinine or ethionine, may inhibit protein synthesis in cells. When applied early in the eclipse period, they can inhibit influenza' and poliomyelitis virus reproduction.' Moreover, some pyrimidine or purine analogues, such as diaminopurine, may inhibit nucleic acid synthesis in cells. Early in the eclipse period they can inhibit the reproduction of vaccinia and encephalitis viruses.3 Acridine compounds closely related to proflavine, which prevents assembly of mature phage particles, can inhibit the reproduction of influenza and mumps virus."

Finally, certain glycosides of benzimidazole inhibit the reproduction of influenza viruses only when they are applied during the first half of the eclipse period." This is well before any new infective particles have appeared. In addition, unlike most other inhibitory compounds, some glycosides of benzimidazole do not cause demonstrable cell damage even though they are highly active as inhibitors of influenza and mumps virus reproduction.:"

Inhibition of Virus Reproduction

Thus, it appears that even as judged from the effects of chemical inhibitors of reproduction, the available evidence does not require the constrtuction of a scheme for animal virus reproduction different from that for phage reproduction. Because virus disease is dependent on virus reproduction and since reproduction can be inhibited by chemical compounds, it might be held that effective chemotherapy awaits only the finding of less toxic and more potent compounds.

This idea seems sufficiently logical and so well supported by experimental evidence that an obvious difficulty may be overlooked.

This difficulty was implicit even as early as 1902 in the work of Reed and Carroll' on yellow fever. It will be recalled that mosquitos became infective when they bit patients on the very first day of yellow fever. This was because a large amount of virus was already present in the blood. The virus had reached the blood only after extensive reproduction in the liver and other organs. This process occurred during the incubation period. Comparable findings have been obtainied consistently when human volunteers were inoculated with various viruses. Much virus reproduction occurs before the signs and symptoms of disease develop and on the last day of the incubation period the concentration of virus in the infected tissue is usually high, often as high as during the full-blown disease.

Experiments in animals with a variety of viruses have served merely to affirm and considerably to extend this general principle. Therefore, it appears that although inhibition of virus reproduction is a potentially feasible means for controlling virus diseases, it may need to be accomplished before the disease is fullblown if it is to be effective. To be most effective, it may prove to be necessary to inhibit virus reproduction during the incubation period.

BACTERIAL REPRODUCTION AND GENETICS

Most bacteria reproduce by a relatively simple asexual process called binary fission: each cell increases in size and divides into two cells. During this process there is an orderly increase in cellular structures and components, replication and segregation of the bacterial DNA and formation of a septum or cross wall which divides the cell into two. The process is evidently coordinated by activities associated with the cell membrane. The DNA molecule is believed to be attached to a point on the membrane where it is replicated.

The two DNA molecules remain attached at points side-by-side on the membrane while new membrane material is synthesized between the two points. This draws the DNA molecules in opposite directions while new cell wall and membrane are laid down as a septum between the two chromosomal compartments. When septum formation is complete the cell splits into two progeny cells. The time interval required for a bacterial cell to divide or for a population of cells to double is called the generation time. Generation times for bacterial species growing in nature may be as short as 15 minutes or as long as several days.

Genetic Exchange in Bacteria

Although procaryotes do not undergo sexual reproduction, they are not without the ability to exchange genes and undergo genetic recombination. Bacteria are known to exchange genes in nature by three fundamental processes: conjugation, transduction and transformation. Conjugation involves cell-to-cell contact as DNA crosses a sex pilus from donor to recipient. During transduction, a virus transfers the genes between mating bacteria. In transformation, DNA is acquired directly from the environment, having been released from another cell. Genetic recombination can follow the transfer of DNA from one cell to another leading to the emergence of a new genotype (recombinant).

It is common for DNA to be transferred as plasmids between mating bacteria. Since bacteria usually develop their genes for drug resistance on plasmids (called resistance transfer factors, or RTFs), they are able to spread drug resistance to other strains and species during genetic exchange processes. The genetic engineering of bacterial cells in the research or biotechnology laboratory is often based on the use of plasmids as vectors. The genetic systems of the Archaea are poorly characterized at this point, although the entire genome of Methanosarcina has been sequenced which opens up the possibilities for genetic analysis of the group.

EVOLUTION OF BACTERIA AND ARCHAEA

For most procaryotes, mutation is is a major source of variability that allows the species to adapt to new conditions. The mutation rate for most procaryotic genes is in the neighborhood of 10^{-8}. This means that if a bacterial population doubles from 10^8 cells to 2×10^8 cells, there is likely to be a mutant present for any given gene.

Since procaryotes grow to reach population densities far in excess of 10^9 cells, such a mutant could develop from a single generation during 15 minutes of growth. The evolution of procaryotes, driven by such Darwinian principles of evolution (mutation and selection) is called vertical evolution.

However, as a result of the processes of genetic exchange described above, the bacteria and archaea can also undergo a process of horizontal evolution. In this case, genes are transferred laterally from one organism to another, even between members of different Kingdoms, which allows the recipient to experiment with a new genetic trait. Horizontal evolution is becoming realized to be a significant force in driving cellular evolution. The combined effects of fast growth rates, high concentrations of cells, genetic processes of mutation and selection and the ability to exchange genes, account for the extraordinary rates of adaptation and evolution that can be observed in the procaryotes.

ECOLOGY OF BACTERIA AND ARCHAEA

Bacteria and Archaea are present in all environments that support life. They may be free-living, or living in associations with "higher forms" of life (plants and animals) and they are found in environments that support no other form of life. Procaryotes have the usual nutritional requirements for growth of cells, but many of the ways that they utilize and transform their nutrients are unique. This bears directly on their habitat and their ecology.

Nutritional Types of Organisms

In terms of carbon utilization a cell may be heterotrophic or autotrophic. Heterotrophs obtain their carbon and energy for growth from organic compounds in nature. Autotrophs use CO_2 as a sole source of carbon for growth and obtain their energy from light (e.g. photoautotrophs) or from the oxidation of inorganic compounds (e.g. lithoautotrophs).

Most heterotrophic bacteria are saprophytes, meaning that they obtain their nourishment from dead organic matter. In the soil, saprophytic bacteria and fungi are responsible for biodegradation of organic material. Ultimately, organic molecules, no matter how complex, can be degraded to CO_2 (plus H_2 and H_2O). Probably no naturally-occurring organic substance cannot be degraded by the combined activities of the bacteria and fungi.

Hence, most organic matter in nature is converted by heterotrophs to CO_2, only to be converted back into organic material by autotrophs that die and nourish heterotrophs to complete the carbon cycle. Lithotrophic procaryotes have a type of energy-producing metabolism which is unique. Lithotrophs (also called lithoautotrophs or chemoautotrophs) use inorganic compounds as sources of energy, i.e., they oxidize compounds such as H_2 or H_2S or NH_3 to obtain electrons to feed in to an electron transport system and to produce ATP. Lithotrophs are found in soil and aquatic environments wherever their energy source is present.

Most lithotrophs are autotrophs so they can grow in the absence of any organic material. Lithotrophic species are found among the Bacteria and the Archaea. Sulfur-oxidizing lithotrophs convert H_2S to S^o and S^o to SO_4. Nitrifying bacteria convert NH_3 to NO_2 and NO_2 to NO_3; methanogenic archaea strip electrons off of H_2 as a source of energy and add them to CO_2 to form CH_4 (methane). Lithotrophs have an obvious impact on the sulfur, nitrogen and carbon cycles in the biosphere. Photosynthetic bacteria convert light energy into chemical energy for growth. Most phototrophic bacteria are autotrophs so their role in the carbon cycle is analogous to that of plants. The planktonic cyanobacteria are the "grass of the sea" and their form of oxygenic photosynthesis generates a substantial amount of O_2 in the biosphere.

However, among the photosynthetic bacteria are types of

photosynthetic metabolism not seen in eucaryotes, including photohetero-trophy (using light as an energy source while assimilating organic compounds as a source of carbon), anoxygenic photosynthesis and unique mechanisms of CO_2 fixation (autotrophy). Photosynthesis has not been found to occur among the Archaea, but one archaeal species employs a light-driven non photosynthetic means of energy generation based on the use of a chromophore called bacteriorhodopsin.

ADAPTATIONS TO ENVIRONMENTAL CONDITIONS

Most procaryotes, whether they have been cultured and studied in the laboratory, or observed growing in their natural habitats, seem to be highly adapted to their specific environment by means of their macromolecular structure and/or their physiologic (metabolic) capabilities. The nutritional quality of the environment determines whether a particular organism will be present, but so do various physical parameters such as the availability of light and O_2, as well as the pH, temperature and salinity of the environment.

As examples, the range of procaryotic responses to oxygen and temperature. Procaryotes vary widely in their response to O_2 (molecular oxygen). Organisms that require O_2 for growth are called obligate aerobes; those which are inhibited or killed by O_2 and which grow only in its absence, are called obligate anaerobes; organisms which grow either in the presence or absence of O_2 are called facultative anaerobes. Whether or not a particular organism can exist in the presence of O_2 depends upon the distribution of certain enzymes such as superoxide dismutase and catalase that are required to detoxify lethal oxygen radicals that are always generated by living systems in the presence of O_2.

Procaryotes also vary widely in their response to temperature. Those that live at very cold temperatures (0 degrees or lower) are called psychrophiles; those which flourish at room temperature (25 degrees) or at the temperature of warm-blooded animals (37 degrees) are called mesophiles; those that live at high temperatures (greater than 45 degrees) are thermophiles.

The only limit that seems to be placed on growth of certain procaryotes in nature relative to temperature is whether liquid water exists. Hencc, growing procaryotic cells can be found in supercooled environments (ice does not form) as low as -20 degrees and superheated environments (steam does not form) as high as 120 degrees. Archaea have been detected around thermal vents on the ocean floor where the temperature is as high as 320 degrees!

SYMBIOSIS

The biomass of procaryotic cells in the biosphere, their metabolic diversity and their persistence in all habitats that support life, ensures that these microbe will play a crucial role in the cycles of elements and the functioning of the world ecosystem. However, the procaryotes affect the world ecology in another significant way through their inevitable interactions with insects, plants and animals. Some bacteria are required to associate with insects, animals or plants for the latter to survive.

For example, the sex of offspring of certain insects is determined by endosymbiotic bacteria. Ruminant animals (cows, sheep, etc.), whose diet is mainly cellulose (plant material), must have cellulose-digesting bacteria in their intestine to convert the cellulose to a form of carbon that the animal can assimilate. Leguminous plants grow poorly in nitrogen-deprived soils unless they are colonized by nitrogen-fixing bacteria which can supply them with a biologically-useful form of nitrogen.

BACTERIAL PATHOGENICITY

Some bacteria are parasites of plants or animals, meaning that they grow at the expense of their eucaryotic host and may damage, harm, or even kill it in the process. Such bacteria that cause disease in plants or animals are pathogens. Human diseases caused by bacterial pathogens includc tuberculosis, whooping cough, diphtheria, tetanus, gonorrhea, syphilis, pneumonia, cholera and typhoid fever, to name a few. The bacteria that cause these diseases have special structural or biochemical properties that determine their virulence or pathogenicity. These include:

- Ability to colonize and invade their host;
- Ability to resist or withstand the antibacterial defenses of the host; ability to produce various toxic substances that damage the host. Plant diseases, likewise, may be caused by bacterial pathogens. More than 200 species of bacteria are associated with plant diseases, but a very small handful of genera are involved.

RETROVIRUS

Retroviruses belong to the Retroviridae family of viruses. The genetic material of retroviruses consists of ribonucleic acid (RNA), instead of deoxyribonucleic acid (DNA). Viruses of this type also contain reverse transcriptase. Retroviruses are known to lead to certain types of cancers in both humans and animals, as well as a range of viral infections. Human Immunodeficiency Virus (HIV), the virus that causes Acquired Immune Deficiency Syndrome (AIDS), is one example of a retrovirus. Retroviruses are unique in that they reproduce by transcribing themselves into DNA.

Reverse transcriptase, an enzyme within a retrovirus, makes it possible for the retrovirus' RNA to perform as a template of sorts for the transcription process. Once transcription has taken place, the viral DNA gains access to the DNA of a cell, reproducing along with the cell and its offspring. Within the cell's offspring, referred to as daughter cells, the viral DNA creates RNA replicas of itself. Finally, the RNA replicas leave the daughter cells after coating themselves with a protein.

Retroviruses reverse the normal cell process, which uses RNA to synthesize DNA. By reversing this process, retroviruses take up permanent residence in the genetic material of the infected cell. In some cases, retroviruses destroy the cells they change. Such is the case with the retrovirus HIV. Other retroviruses cause cells to become cancerous. This is what occurs with certain types of leukemia.

Retroviruses are prone to mutation. For this reason,

viruses in this family often become resistant to antiviral drugs within a relatively short period of time. This level of mutability is one of the reasons cited for the difficulty scientists face in trying to develop a safe and effective HIV vaccine. Antibiotics are not effective against retroviruses. They are helpful for battling bacterial-based infections, but are useless against viruses.

Instead, anti-viral medications must be developed and used to combat retroviruses. Vaccination can be used to prevent diseases caused by viruses, including retroviruses. Unfortunately, developing effective retrovirus vaccines is complicated work that may take years, or even generations, to accomplish. Despite the fact that retroviruses are indicated in some potentially deadly diseases, they are vulnerable to something as simple as ordinary soap and water. Handwashing can render a retrovirus inactive. Physical barriers can be useful in preventing the spread of retroviruses, as well. Such barriers include condoms, rubber gloves, and facial masks.

DESCRIPTION OF VIRUS

The virus itself stores its nucleic acid, in the form of a +mRNA (including the 5′cap and 3′PolyA inside the virion) genome and serves as a means of delivery of that genome into cells it targets as an obligate parasite, and constitutes the infection. Once in the host's cell, the RNA strands undergo reverse transcription in the cytosol and are integrated into the host's genome, at which point the retroviral DNA is referred to as a provirus. It is difficult to detect the virus until it has infected the host.

Virion Structure

Virions of retroviruses consist of enveloped particles about 100 nm in diameter. The virions also contain two identical single-stranded RNA molecules 7-10 kilobases (kb) in length. Although virions of different retroviruses do not have the same morphology or biology, all the virion components are very similar.

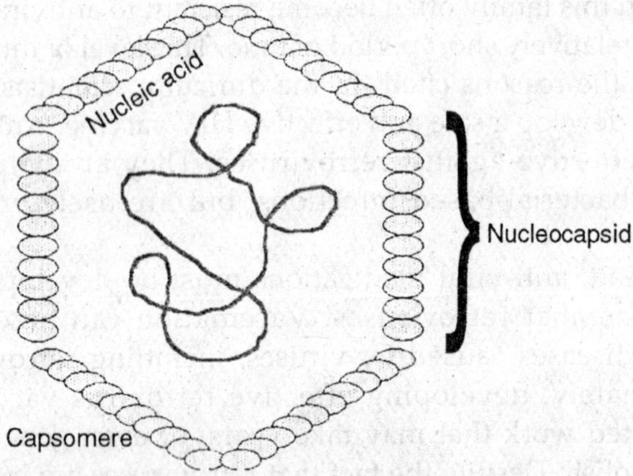

Fig. Structure of Virion

The main virion components are:

- *Envelope:* Composed of a lipid bilayer which is obtained from the host plasma membrane during budding process.
- *RNA:* Consisted of a dimer RNA. It has a cap at 5' end and polyadenyle at 3' end. The RNA genome also has terminal noncoding regions which is important in replication and internal regions that encode virion proteins for gene expression. The 5' end includes four regions which are R, U5, PB and L. R region is a short repeated sequence at each end of the genome during the reverse transcription in order to ensure correct end-to end transfer in growing chain. U5, on the other hand, is a short unique sequence between R and PB. PB consists of 18 bases complementary to 3' end of tRNA primer. L region is an untranslated leader region that gives signal for packaging of genome RNA. The 3' end includes 3 regions which are PP (polypurine), U3 and R. PP is primer for plus-strand DNA synthesis during reverse transcription. U3 is a sequence between PP and R which has signal that provirus can use in transcription. Lastly, R is the terminal repeated sequence at 3' end.

Proteins: Consisted of gag proteins, protease (PR), pol

proteins and env proteins. Gag proteins are major components of the viral capsid which are about 2000-4000 copies per virion. Protease, on one hand, is expressed differently in different viruses. It functions in proteolytic cleavages during virion maturation to make mature gag and pol proteins. Pol proteins are responsible for synthesis of viral DNA and integration into host DNA after infection. Finally, env proteins play role in association and entry of virion into the host cell.

Multiplication

When retroviruses have integrated their own genome into the germ line, their genome is passed on to a following generation. These endogenous retroviruses, contrasted with exogenous ones, now make up 5-8% of the human genome. Most insertions have no known function and are often referred to as "junk DNA". However, many endogenous retroviruses play important roles in host biology, such as control of gene transcription, cell fusion during placental development in the course of the germination of an embryo, and resistance to exogenous retroviral infection.

Endogenous retroviruses have also received special attention in the research of immunology-related pathologies, such as autoimmune diseases like multiple sclerosis, although endogenous retroviruses have not yet been proven to play any causal role in this class of disease. The role of endogenous retroviruses in human gene evolution is explored in a 2005 peer-reviewed article.

While transcription was classically thought to only occur from DNA to RNA, reverse transcriptase transcribes RNA into DNA. The term "retro" in retrovirus refers to this reversal (making DNA from RNA) of the central dogma of molecular biology. Reverse transcriptase activity outside of retroviruses has been found in almost all eukaryotes, enabling the generation and insertion of new copies of retrotransposons into the host genome. These inserts are transcriped by host's enzymes into new RNA molecules which enter the cytosol. Next, some of these RNA molecules are translated into viral proteins. For example, the gag gene is translated into molecules of the capsid protein, the pol gene is transcribed into molecules of reverse transcriptase, and the env gene is translated into molecules of the envelope protein. It is important to note that a retrovirus must "bring" its own reverse transcriptase in its capsid, otherwise it is unable to utilize the infected cell's enzymes to carry out the task, due to the unusual nature of producing DNA from RNA.

Industrial drugs that are designed as protease and reverse transcriptase inhibitors can quickly be proved ineffective because the gene sequences that code for the protease and the reverse transcriptase can undergo many substitutions. These substitutions of nitrogenous bases, which make up the DNA strand, can make either the protease or the reverse transcriptase difficult to attack. The amino acid substitution enables the enzymes to evade the drug regiments because mutations in the gene sequences can cause physical or chemical change which makes them harder to detect by the drug.

When the drugs that are supposed to attack enzymes, such as protease, are designed, the manufacturers target specific sites on the enzyme. One way to attack these targets can be through hydrolysis of molecular bonds, which means that the drug will add molecules of H_2O (water) to specific bonds. By adding molecules of water at a site on the virus, the drug breaks the previous bonds that were linked to each other. If several of these breaks occur, the result can lead to lysis, the death of the virus.

Because reverse transcription lacks the usual proofreading of DNA replication, a retrovirus mutates very often. This enables the virus to grow resistant to antiviral pharmaceuticals quickly, and impedes the development of effective vaccines and inhibitors for the retrovirus.

Genes

Retrovirus genomes commonly contain these three open reading frames that encode for proteins that can be found in the mature virus:

- *Group-specific Antigen* (gag) codes for core and structural proteins of the virus,

- *Polymerase* (pol) codes for reverse transcriptase, protease and integrase,
- *Envelope* (env) codes for the retroviral coat proteins.

Provirus

This DNA can be incorporated into host genome as a provirus that can be passed on to progeny cells. In this way some retroviruses can convert normal cells into cancer cells. Some provirus remains latent in the cell for a long period of time before it is activated by the change in cell environment.

DEVELOPMENT OF RETROVIRUSES

Studies of retroviruses led to the first demonstrated synthesis of DNA from RNA templates, a fundamental mode for transferring genetic material that occurs in both eukaryotes and prokaryotes.

It has been speculated that the RNA to DNA transcription processes used by retroviruses may have first caused DNA to be used as genetic material. In this model, the RNA world hypothesis, cellular organisms adopted the more chemically stable DNA when retroviruses evolved to create DNA from the RNA templates.

CLASSIFICATION

Exogenous

The following genera are included here:

- Genus *Alpharetrovirus*; type species: *Avian leukosis virus*
- Genus *Betaretrovirus*; type species: *Mouse mammary tumour virus*
- Genus *Gammaretrovirus*; type species: *Murine leukemia virus*; others include *Feline leukemia virus*
- Genus *Deltaretrovirus*; type species: *Bovine leukemia virus*; others include the cancer-causing *Human T-lymphotropic virus*
- Genus *Epsilonretrovirus*; type species: *Walleye dermal sarcoma virus*
- Genus *Lentivirus*; type species: *Human immuno-*

deficiency virus 1; others include *Simian, Feline* immunodeficiency viruses
- Genus *Spumavirus*; type species: *Chimpanzee foamy virus*

These were previously divided into three subfamilies (*Oncovirinae, Lentivirinae,* and *Spumavirinae*), but with current knowledge of retroviruses, this is no longer appropriate.

Endogenous

Endogenous retroviruses are not formally included in this classification system, and are broadly classified into three classes, on the basis of relatedness to exogenous genera:

- Class I are most similar to the gammaretroviruses
- Class II are most similar to the betaretroviruses and alpharetroviruses
- Class III are most similar to the spumaviruses

TREATMENT

Antiretroviral drugs are medications for the treatment of infection by retroviruses, primarily HIV. Different classes of antiretroviral drugs act at different stages of the HIV life cycle. Combination of several (typically three or four) antiretroviral drugs is known as highly active anti-retroviral therapy (HAART).

Retroviruses Replication

Retroviruses use RNA instead of DNA as their nucleic acid core. They also contain the enzyme reverse transcriptase, which will detranscribe the RNA sequence into a DNA strand.. Once the retroviral RNA and reverse transcriptase are inside the host cell the enzyme reverses transcription by making a single stranded DNA from the retroviral RNA.

Viral DNA can be integrated into the host DNA. It remains in the genome and is replicated whenever the host DNA replicates.

If viral DNA is transcribed, new viruses are produced by biosynthesis, maturation, and release by budding. Retroviruses include HIV and also cause certain forms of cancer.

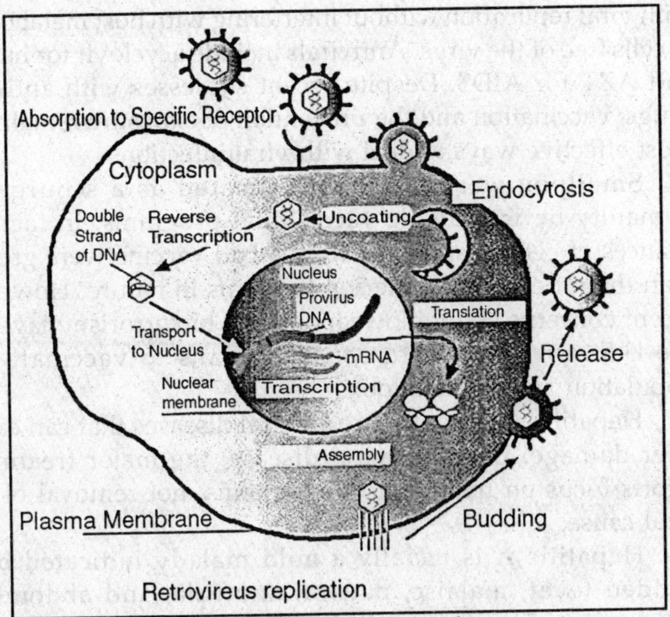

Fig. Replication of a Retrovirus.

Viruses and Diseases

Viruses cause a variety of diseases among all groups of living things. Viral plant diseases can be controlled solely by burning those plants that show symptoms of disease. Viral diseases in humans are controlled by preventing transmission, administering vaccines, and only recently by the administration of antiviral drugs. Virally caused human diseases include the flu, common cold, herpes, measles, chicken pox, small pox, and encephalitis.

Antibiotics are not effective against viruses. Vaccination offers protection for uninfected individuals. Frequent hand washing and condom use may help prevent transmission. Vaccines are substances that stimulate an immune response without causing the illness. Commonly used virus vaccines include polio, measles, and mumps.

Antibiotics do not cure viral infections because viruses use enzymes produced by the host cell, rather than produce not their own. A few antiviral drugs are available that interfere with viral replication without interfering with host metabolism in cells free of the virus. Antivirals include acyclovir for herpes and AZT for AIDS. Despite recent successes with antiviral drugs, vaccination and the prevention of exposure remain the most effective ways to deal with viral infections.

Smallpox was effectively elimated as a scourge on humanity by worldwide vaccination programs. In fact the chances of getting smallpox from a bad vaccine were greater than that of ever encountering the virus in nature. However, recent concerns about bioweapons and bioterrorism have led the U.S. government to consider plans to vaccinate the population against smallpox.

Hepatitis A, B, and C are all viral diseases that can cause liver damage. Like any viral disease, the major treatment efforts focus on treatment of symptoms, not removal of the viral cause.

Hepatitis A is usually a mild malady indicated by a sudden fever, malaise, nausea, anorexia, and abdominal discomfort. The virus causing Hepatitis A is primarilly transmitted by fecal contamination, although contaminated food and water also can promote transmission.

A rare disease in the United States, hepatitis B is endemic in parts of Asia where hundreds of millions of individuals are possibly infected. The Hepatitis B virus (HBV) may be transmitted by blood and blood products as well as sexual contact. The blood supply in developed countries has been screened for the virus that causes this disease for many years, and transmission by blood transfusion is rare. The risk of HBV infection is high among promiscuous homosexual men although it is also transmitted hetereosexually.

Correct use of condoms is thought to reduce or eliminate the risk of transmission. Effective vaccines are available for the prevention of Hepatitis B infection. Some individuals with chronic hepatitis B may develop cirrhosis of the liver. Individuals with chronic hepatitis B are at an increased risk of developing primary liver cancer.

Although this type of cancer is relatively rare in the United States, it is the leading cause of cancer death in the world, primarily because the virus causing it is endemic in eastern Asia.

The virus is transmitted primarily by blood and blood products. Most infected individuals have either received blood transfusions prior to 1990 or have used intravenous drugs. Sexual transmission can occur between monogamous couples (rare) but infection is far more common in those who are promiscuous.

In rare cases, Hepatitis C causes acute disease and even liver failure. About twenty percent of individuals with Hepatitis C who develop cirrhosis of the liver will also develop severe liver disease. Cirrhosis caused by Hepatitis C is presently the leading cause of the need for liver transplants in the United States. Individuals with cirrhosis from Hepatitis C also bear increased chances of developing primary liver cancer. All current treatments for Hepatitis C employ of various preparations of the potent antiviral interferon alpha. However, not all patients who have the disease are good candidates for treatment, so infected individuals are urged to regularly consult their physician.

Vaccination

Vaccination is a term derived from the Latin *vacca*. A vaccine stimulates the antibody production and formation of memory cells without causing of the disease. Vaccines are made from killed pathogens or weakened strains that cause antibody production but not the disease. Recombinant DNA techniques can now be used to develop even safer vaccines.

The immune system can develop long-term immunity to some diseases. Man can use this to develop vaccines, which produce induced immunity. Active immunity develops after an illness or vaccine. Vaccines are weakened (or killed) viruses or bacteria that prompt the development of antibodies. Application of biotechnology allows development of vaccines that are the protein (antigen) which in no way can cause the disease. Passive immunity is the type of immunity when the individual is given antibodies to combat a specific disease. Passive immunity is short-lived.

Emergent Viruses

Viruses are usually quite specific as to their hosts and even to the types of cells they infect in a multicellular host. Recently, some viruses appear to have shifted their host: HIV, hantavirus, and ebola appear to be either viruses shifting to a new (human) host or else viruses whose existence and effects are just now being realized by scientists and the general public.

Viroids and Prions

Viruses would appear to be the simplest form of infectious particle. The discovery of viroids, nucleic acid without a protein capsule, and prions, infectious proteins, subtracts another level of complexity. Both viroids and prions can cause diseases, the most famous of which is mad cow disease (caused by a prion).

METABOLIC DIVERSITY AMONG MICROORGANISMS

MICROBIAL DIVERSITY

Microbial diversity constitutes the most extraordinary reservoir of life in the biosphere that we have only just begun to explore and understand. Diversity is composed of two elements: richness and evenness, so that the highest diversity occurs in communities with many different species present (richness) in relatively equal abundance (evenness)

Microorganisms represent the richest repertoire of molecular and chemical diversity in nature, as they comprise the most diverse forms of life. Over millennia, they have adapted to extremely diverse environments and have developed an extensive range of metabolic pathways. This metabolic wealth has traditionally been exploited by man in processes such as fermentation, production of antibiotics, vitamins, etc. More recently, this largely unexplored reservoir of resources has begun to be harnessed for innovative applications useful to mankind. These include the use of microorganisms for bioproduction of novel as well as difficult-to-synthesize compounds, monitoring pollutant levels and biodegradation of xenobiotic pollutants. The bane of industrial progress has been the generation and release into the environment of huge amounts of toxic compounds which have caused widespread contamination of land and water.

Large amounts of chemicals such as herbicides, insecticides and fertilizers are used in agricultural activities and synthetic chemicals such as plasticizers, dyes, pigments, agrochemicals, solvents, pharmaceuticals, hydraulics, fire retardants, halogenated compounds, etc. are produced by industrial activities. Over the last few decades, significant quantities of these chemicals have been released into the environment creating countless number of contaminated sites. The major contaminants are halogenated and nitrated alicylic, aliphatic, aromatic and polyaromatic chemical compounds. Removal of these xenobiotics involves physical and chemical processes like landfilling, excavation, incineration, etc., which are expensive and sometimes difficult to execute.

Microorganisms are nature's original recyclers, converting toxic organic compounds to harmless products, often carbon dioxide and water. Ever since it was discovered that microbes have the ability to transform and/or degrade xenobiotics, scientists have been exploring the microbial diversity, particularly of contaminated areas in search for organisms that can degrade a wide range of pollutants. In this context, the present review deals with microbial diversity of degradative organisms with specific emphasis on microorganisms degrading hydrocarbons such as aliphatic, aromatic, polycylic aromatics, nitroaromatics and halogenated organic compounds.

Biodegradation of Xenobiotics: Role of Microbial Diversity

Bioremediation, a biodegradation process in which sites contaminated with xenobiotics are cleaned up by means of bacterial bio-geochemical processes, preferably *in situ*, exploits the ability of microorganisms to reduce the concentration and/or toxicity of a large number of pollutants. It is an economical, versatile, environment-friendly and efficient treatment strategy, and a rapidly developing field of environmental restoration. Bioremediation utilizes the microbial ability to degrade and/or detoxify chemical substances such as petroleum products, aliphatic and aromatic hydrocarbons (including polycyclic aromatic hydrocarbons and polychlorinated biphenyls), industrial solvents, pesticides and their metabolites, and metals.

The presence of a large number of diverse bacterial species in nature expands the variety of chemical pollutants that can be degraded and the extent to which pollutant sites can be decontaminated.

A well-known example of bioremediation which highlighted the usefulness of this treatment strategy and accelerated its development is the biological clean up (in addition to physico-chemical methods) of a large accidental oil spill by the tanker *Exxon Valdez* which ran aground on Bligh reef in the Gulf of Alaska in March 1989, spilling approximately 41,000 m_3 of crude oil and contaminating about 2000 km of coastline.

The *Exxon Valdez* and other similar incidents demonstrated the usefulness of bioremediation as a fairly complete solution to oil contamination. Thus, the use of microorganisms for degradation of pollutants is now being increasingly applied as the technology of choice for clean up or restoration of polluted sites as it can be self-sustaining and inexpensive. Microbial diversity offers an immense field of environmentfriendly options for mineralization of contaminants or their transformation into less harmful non-hazardous compounds. There is a general interest in studying the diversity of indigenous microorganisms capable of degrading different pollutants because of their varied effects on the environment.

Efforts have been made to characterize bacterial communities and their responses to pollutants, to isolate potential degraders and to identify the genes involved in degradation processes. The detailed analysis of microbial diversity within an environment can be divided into two broad categories: culture-dependent studies and culture independent studies. It has been established that contaminated environments harbour a wide range of unidentified pollutant-degrading microorganisms that have crucial role in bioremediation that can be assessed only by the culture independent techniques.

Conventional characterization of microbial strains has been subjected to debate, as it is dependent on the ability of the strains to grow under specific environmental conditions. In the past two decades, molecular tools exemplified by 16S rRNA analyses have facilitated the study of natural microbial populations without cultivation, which has made quantitative assessment of microbial diversity now conceivable. However, no single tool allows definitive assessment of the soil microbial community.

Therefore, the use of a polyphasic approach involving a combination of molecular biology techniques, microbiological methods and geochemical techniques or microsensors is necessary to obtain a better understanding of the interaction between the microorganisms and their natural environment. Molecular biology methods are now being employed to study bioremediation since a comprehensive understanding of microbial ecology is required to gain maximum benefits from this process. A number of studies have reported the occurrence of conjugative gene transfer between bacteria in soil through appropriate plasmid-borne catabolic genes into competitive indigenous bacterial populations during bioremediation of contaminants at the industrial sites.

Therefore, we need to understand the role of catabolic genes by molecular cloning and characterization for degradation of a particular organic compound so that it can be applied for bioremediation of contaminated ground water and soil. The relatively new field of molecular microbial ecology which makes use of molecular biology techniques to study microbial ecology is increasing our understanding of microbial diversity as it has revealed the presence of viable but non-culturable bacteria.

Molecular microbial ecology describes the microbial diversity based on DNA sequences without cultivation. DNA is extracted directly from environmental samples so that the cultivation bias is eliminated. The ability to monitor expression of specific catabolic genes may be of great importance in determining the feasibility of bioremediation *in situ*.

Expanding the Catabolic Repertoire of Bacteria: Genetic and Metabolic Engineering

In recent years, a number of compounds previously considered non-degradable are also being degraded by microorganisms, suggesting that under selective pressure of environmental pollution, microbes develop the ability to degrade recalcitrant xenobiotics. However, the fact that many pollutants are still persistent in the environment reflects the inadequacy of the current microbial catabolic capacity to deal with such pollutants.

This is where molecular biologists step in to engineer bacteria for more efficient remediation of contaminated areas, and to degrade hitherto recalcitrant organic pollutants. Advances in genetic engineering have enabled a far more rational approach for the development of microbes for the degradation of xenobiotics, than the classical approach via mutagenesis and screening.

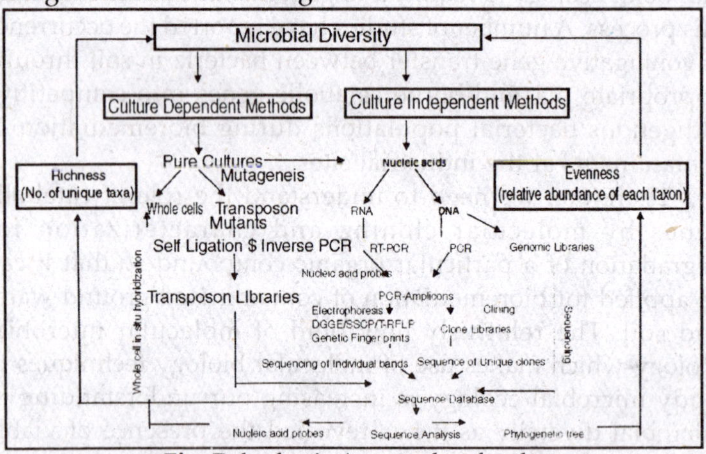

Fig. Polyphasic Approaches for the Analysis of Microbial Diversity.

This has led to the development of a new field of metabolic

engineering that involves the improvement of cellular activities by manipulation of enzymatic, transport and regulatory functions of the cell using recombinant DNA technology.

Metabolic engineering has emerged in the past decade as an interdisciplinary field aiming to improve cellular properties by using modern genetic tools to modify pathways. With rapid developments in new analytical techniques and cloning procedures, it is now possible to introduce directed genetic changes in microbes and subsequently analyse the consequences of the introduced changes at the cellular level. It may therefore involve continuous improvements of the cellular properties through several rounds of genetic engineering.

Advances in the field of genetic engineering, sequencing of whole genomes of several organisms and developments in bioinformatics have speeded up the process of gene cloning and transformation. Furthermore, a number of very powerful analytical techniques have been developed for metabolic pathway analysis and analyses of cellular function, e.g. gas chromatography, gas chromatography–mass spectrometry (GC–MS), nuclear magnetic resonance (NMR), two-dimensional gel electrophoresis, matrix-assisted laser desorption ionization-time of flight (MALDI-TOF), liquid chromatography-mass spectrometry (LC-MS) and DNA chips. Metabolic engineering is therefore concerned with modifying pathways and assessing the physiological outcome of such genetic modifications in an effort to improve the degradative abilities of microorganisms.

BIODEGRADATION OF ORGANIC COMPOUNDS

Nitroaromatic Compounds

The versatile chemistry of the nitro group makes nitroaromatic compounds (NACs) important industrial feedstocks. They are extensively used in many industrial processes as solvents, as precursors for amino aromatic derivatives and in the synthesis of dyes, plasticizers, pharmaceuticals, pesticides (parathion, methyl parathion, dinoseb, dinitrocresole, nitrofen) and explosives {2,4,6-trinitrotoluene (TNT), hexahydro- 1,3,5-trinitro-triazine (RDX) and octahydro-1,3,5,7- tetranitro-1,3,5,7-tetrazine (HMX). However, the very properties that make NACs valuable to industry, viz. stability, persistence and toxicity, render them hazardous when they are released into the environment. On entering mammalian systems, NACs are readily reduced to more reactive and potentially more carcinogenic or mutagenic derivatives due to conversion of nitro groups to more harmful nitroso and hydroxylamino groups.

Owing to their high toxicity, a number of NACs including nitrobenzenes, dinitrotoluenes and mono- and dinitrophenols have been classified as priority pollutants. Presence of NACs in the environment creates intense selective pressure which has led to the evolution of microorganisms capable of degrading a growing number of NACs.

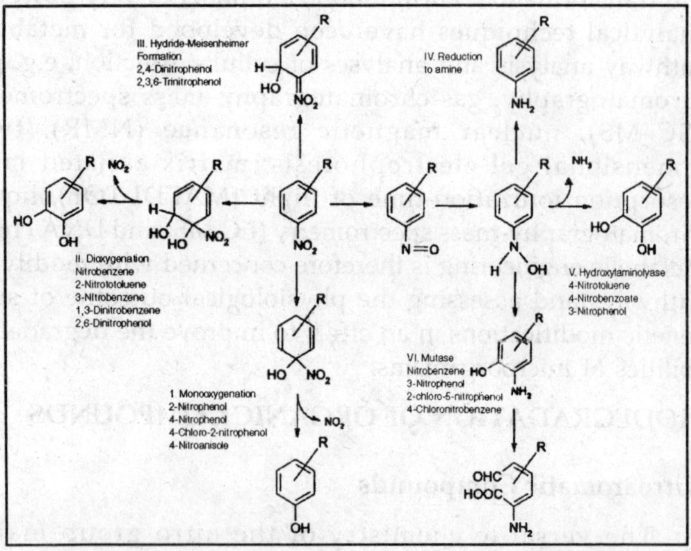

Fig. General Mechanisms for the Degradation of Nitroaromatic Compounds

Successful bioremediation requires not only the knowledge of which microorganisms can degrade a particular compound, but also an understanding of the pathways involved in degradation both at physiological and molecular levels. The degradative pathways which have evolved toward the catabolism of nitro-substituted aromatic rings are catalysed by common enzyme types (mono- and dioxygenases, dehydrogenases) produced by common microbial groups belonging to *Pseudomonas, Nocardia* and *Arthrobacter*.

However, there is a great diversity observed in funneling mono- and di-nitro aromatic molecules through oxygenase-based pathways. The majority of NACs being anthropogenic have been present in the environment for a relatively short time period giving bacteria little time to evolve pathways for their degradation.

Nevertheless, during the past decade or so, bacteria capable of degrading a number of NACs by aerobic pathways, once thought unlikely have been discovered. Our research group has been exploring the microbial diversity of different geographical regions of India in search of bacteria that are capable of mineralizing/degrading NACs.

Soil samples from agricultural fields sprayed with parathion/methyl parathion (nitro-pesticides) were used for isolation of a bacterial strain identified as *Arthrobacter protophormiae* which was capable of utilizing nitroaromatic compounds *o*-nitrobenzoate (ONB), *p*-nitrophenol (PNP) and 4-nitrocatechol (NC) as the sole source of carbon, nitrogen and energy. The pathway for ONB degradation in this organism was elucidated and found to be a novel one and was determined to be encoded on a large 65 kb plasmid.

The pathways for degradation of PNP and NC were also elucidated in this strain and the genes involved in degradation of these compounds were found to reside on the same 65 kb plasmid. Thus, this organism could be useful for bioremediation of sites contaminated with multiple NACs. Another PNP- and NC-degrading organism was isolated from these soil samples and identified as *Burkholderia cepacia*. The PNP-degradation pathway in this organism was elucidated and it was seen that degradation of PNP in both *B. cepacia* and *A. protophormiae* occurs via formation of benzoquinone and hydroquinone. The oxidative degradation of 4-nitrocatechol was found to occur by a novel pathway in this strain involving a reductive dehydroxylation step which is a wellknown reaction in anaerobic metabolism of aromatic compounds, but was reported to be operative aerobically in *B. cepacia* for the first time. Involvement of plasmid in PNP and NC metabolism in this strain was also established.

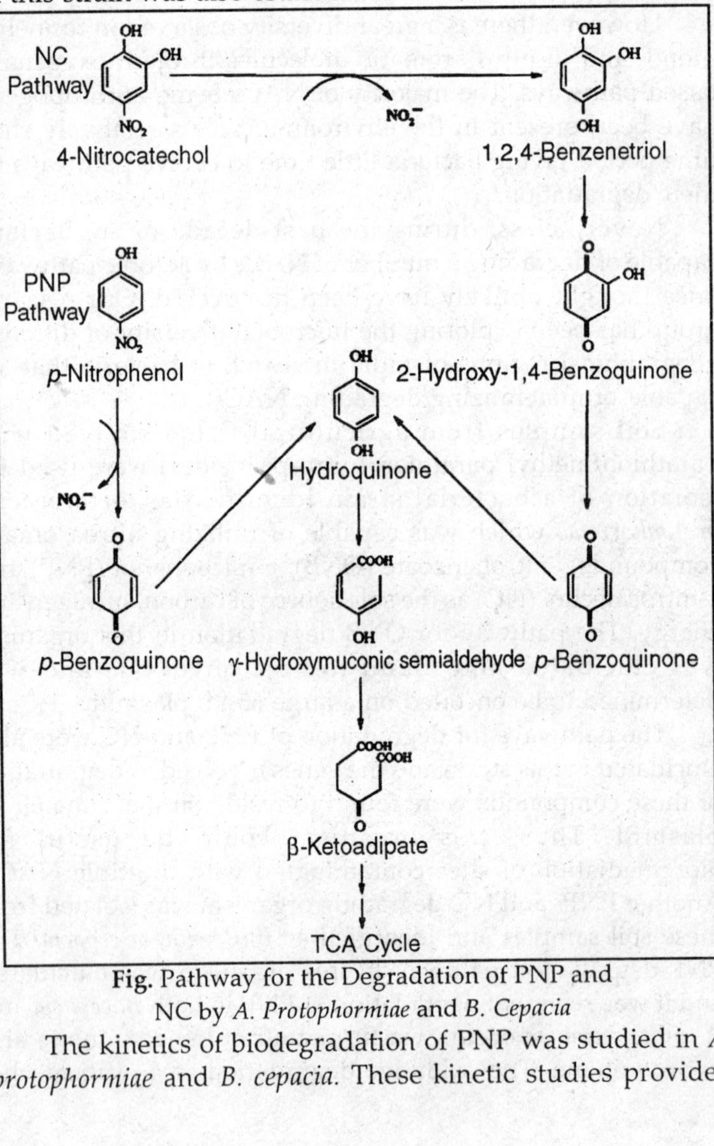

Fig. Pathway for the Degradation of PNP and NC by *A. Protophormiae* and *B. Cepacia*

The kinetics of biodegradation of PNP was studied in *A. protophormiae* and *B. cepacia.* These kinetic studies provided useful information which could be used in the designing of biological treatment plants.

Bioremediation studies have been carried out to study the efficacy of *A. protophormiae* in degrading PNP in soil microcosms and also in small-scale field studies. This organism has shown promising potential for bioremediating PNP under natural environmental conditions (unpublished data). One aspect of bioremediation of a contaminated area by the addition of exogenous microorganisms is the possible effect of added organisms on the native microbial diversity of the area being treated.

Role of Chemotaxis in Biodegradation

The *in situ* biodegradation of a contaminant is a function of the catabolic activity of bacteria and bioavailability of the contaminant to bacteria. Chemotaxis, i.e. the movement of bacteria under the influence of a chemical gradient has been postulated to play an important role in enhancing biodegradation as it increases bioavailability of pollutants to bacteria. Some toxic organic compounds are chemoattractant for different bacterial species, which can lead to improved biodegradation of these compounds.

A *Ralstonia* sp. was isolated by a chemotactic enrichment technique in our laboratory from a pesticide-contaminated agricultural field of Assam. It was chemotactic towards different NACs, i.e. PNP, 3-methyl 4-nitrophenol, 4-nitrocatechol, ONB and *p*-nitrobenzoate (PNB) which were subsequently degraded completely by this organism. It was also chemotactic towards *o*-dinitrobenzoate, *m*-dinitrobenzoate, 2,4-dinitrophenol, 2,5-dinitrophenol, 2,6-dinitrophenol and 3,5-dinitrobenzoate which do not serve as sole source of carbon and energy but are co-metabolized. These NACs are partially transformed in the presence of an alternate carbon source such as succinate. On the other hand, this *Ralstonia* strain did not degrade *o*-nitrophenol, *p*-nitroaniline, 2,3-dinitrotoluene, naphthalene, phenanthrene or salicylic acid, nor was it chemotactic towards these compounds, clearly indicating a relation between chemotaxis and biodegradation.

CRUDE OIL CONSTITUENTS

Crude oil is a complex mixture of hydrocarbons, basically composed of aliphatic, aromatic and asphaltene fractions along with nitrogen, sulfur and oxygen-containing compounds. The constituent hydrocarbon compounds are present in varied proportion resulting in great variability in crude oils from different sources. There are several reports indicating the recalcitrance and potential health hazards of the different constituents of crude oil. These compounds have been reported to be carcinogenic, mutagenic and have immunomodulatory effects on humans, animals and plant life.

The sites contaminated with hydrocarbons are ecologically important locations as one may encounter microbial flora of diverse nature, which may be potential candidates for important industrial processes. There is a plethora of cultivable microbes with the ability to utilize hydrocarbons as sole source of carbon or to transform them to a less toxic form. Earlier reports based on cultivable bacteria suggested that hydrocarboncontaminated soil is predominated by Gram-negative bacteria. The culture independent studies from contaminated soil samples have revealed existence of new bacterial lineages. Studies with molecular tools indicate that specific groups of bacteria commonly occur in oil-contaminated environment.

Aliphatic Hydrocarbon Degradation

The aliphatic fraction is the major constituent of crude oil. The aliphatic fraction of hydrocarbons consists of straight chain, branched chain and cyclic chain carbon moieties. This fraction contains the most readily degraded hydrocarbon compounds and most of the reported hydrocarbon- degrading bacteria have shown relatively good growth on these compounds.

There are reports on degradation of short carbon chain (C8–C16)40 to very long carbon chain (C44) hydrocarbons. The bacterial species of *Acinetobacter, Pseudomonas, Alcaligenes, Burkholderia, Arthrobacter, Flavobacterium, Bacillus,* etc. are a few of the well-known degraders of aliphatic fractions of petroleum hydrocarbons. There are several reports on the isolation of different hydrocarbon-degrading bacteria from petroleum-contaminated sites. Herman *et al.* mentioned the role of various surface-active compounds released by hydrocarbon degrading bacteria, which further enhanced the bioremediation potential by overcoming mass transfer limitation and dissolution rates of these hydrocarbon compounds.

Studies on cycloalkanes suggest that these cyclic hydrocarbons are most resistant to microbial degradation due to their toxic effects on cellular structures. There are reports on utilization of cycloalkanes by *Pseudomonas citronellolis, Brevibacterium erythrogenes* and *Saccharomyces cerevisiae.* Yakimov *et al.* reported a novel hydrocarbon degrading and surfactantproducing bacterium, *Alcanivorax borkumensis.* The sequence analysis of 16S rDNA indicated the presence of different species of *Ralstonia, Commamonas* and *Flavobacterium* along with the common soil bacteria from contaminated groundwater sources.

Aromatic Hydrocarbon Degradation

The aromatic hydrocarbon compounds basically contain benzene ring in their structure and are ubiquitous in the environment. Aromatic hydrocarbons being derivatives of benzene are very stable and this leads to the relative inertness of these molecules to different fate processes. However, many bacterial species have evolved to use these compounds as source of energy. Kerr and Capone have reviewed the microbial degradation of aromatic fraction of petroleum hydrocarbons. It was observed that the light aromatics were most susceptible to microbial attack. Owing to hydrophobicity, higher aromatic hydrocarbons get rapidly associated with organic material rendering them less available to biological uptake. Watanabe *et al.* reported the presence of phylotypes associated with *a* subclass of proteobacteria in sites contaminated with crude oil.

The bacteria affiliated to ?-proteobacterial subgroup were reported to be associated with petroleumcontaminated groundwater. As petroleum hydrocarbons are persistent under anaerobic conditions, contamination of groundwater is a serious environment problem. Stapleton *et al.* demonstrated significant bioremediation potential of degradative bacteria from hydrocarbon-contaminated aquifers which had phylogenetic resemblance with common soil bacteria. This included members from the group *Pseudomonas, Ralstonia, Burkholderia, Sphingomonas, Flavobacterium* and *Bacillus*.

Kniemeyer *et al.* also demonstrated anaerobic degradation of aromatic hydrocarbons by marine sulfate reducing bacteria that use sulfate as the electron acceptor. While studying hydrocarbon degradation by archaeal population, Ficker *et al.* came across a genus *Methanospirillum* that was unrelated to any previously described genus. Another study revealed that the culturable methanogenic archaeal population was much less than the total anaerobic organisms in the biomass.

Polycyclic Aromatic Hydrocarbons

Among the aromatic fraction, polycyclic aromatic hydrocarbons (PAHs) are compounds of intense public concern owing to their persistence in the environment and potential deleterious effects on human health. They represent a unique class of petroleum hydrocarbons because of their pyrogenic nature and the complexity of the assemblages in which they occur.

There has been intense research pertaining to bacterial biodegradation of PAHs and degradation of PAHs composed of three rings has been well documented. Many bacterial, fungal and algal strains have been shown to degrade a wide variety of PAHs.

The most commonly reported bacterial species include *Acinetobacter calcoaceticus, Alcaligens denitrificans, Mycobacterium* sp., *Pseudomonas putida, Pseudomonas fluorescens, Pseudomonas vesicularis, Pseudomonas cepacia, Rhodococcus* sp., *Corynebacterium renale, Moraxella* sp., *Bacillus cereus, Beijerinckia* sp., *Micrococcus* sp., *Pseudomonas paucimobilis* and *Sphingomonas*. Though most of these bacterial species are reported to degrade lowmolecular weight PAHs, there are limited reports on degradation of high-molecular weight PAHs with more than four benzene rings. In general, high-molecular weight PAHs are degraded slowly by indigenous or augmented microorganisms as persistence of PAHs increases with their molecular size.

Early reports are available on the ability of the microorganisms to degrade PAHs dating back to 1975 where Gibson *et al.* showed the involvement of *Beijerinkia* sp. in oxidation of benzo(*a*)pyrene and benzo(*a*)anthracene to dihydrodiols. However, it was not until late 1980s when Mahaffey, Heitkamp and Cerniglia and Mueller reported biodegradation of high-molecular weight (HMW) PAHs.

Mueller demonstrated for the first time the utilization of PAH containing four or more aromatic rings as a sole source of carbon and energy by bacterial isolates. They showed that a seven-member bacterial consortium isolated from creosole-contaminated soil was capable of utilizing fluoranthene and was also capable of co-metabolically biotransforming other HMW PAHs. Boldrin reported fluoranthene mineralization by a consortium of *Mycobacterium, Pseudomonas* and *Alcaligens* sp.

They also postulated that the growth rates were not dependent on the solubility of the substrates, contradicting the general observations that degradation rates depend upon the dissolution rate of hydrocarbon. Although many of the PAH-degrading bacteria described are actinomycetes, a variety of non-actinomycete bacteria have also been reported to metabolize fluoranthene, pyrene, chrysene, and benz[*a*]anthracene.

P. putida, P. aeruginosa, Flavobacterium sp. and an unidentified strain were isolated from a soil-derived mixed culture which was capable of metabolizing fluoranthene and pyrene when supplemented with other forms of organic carbon. The aromatic hydrocarbon-degrading ability of *Sphingomonas yanoikuyae* was recently reviewed and it has been shown to oxidize chrysene. The analysis of 16S DNA sequence types represented organisms closely related to known HMW PAH-degrading bacteria (*Burkholderia, Sphingomonas* and *Mycobacterium*).

Lindstrom in their 20-year experiment on long-term effects

on microbial communities observed that communities exposed to petroleum hydrocarbons exhibited diminished genetic diversity relative to pristine reference communities, but displayed increased physiological tolerance and substrate utilization capabilities.

It was also observed by Vila that a *Mycobacterium* strain, which could use pyrene as a sole source of carbon, could also grow on hexadecane, phenanthrene and fluoranthene and could transform several other PAHs. Hedlund isolated a novel PAHdegrading species of *Neptunomonas naphthovorans*. A *Pseudomonas putida* strain isolated from oil fields of Gujarat could use naphthalene as sole source of carbon and energy and mineralized naphthalene via salicylate.

This strain was also chemotactic towards these compounds. An 83 kb mega plasmid was found to contain all the genes responsible for degradation of naphthalene and salicylate in this strain. The plasmid was transferred into a plasmid-free strain of *P. putida* KT2442 which resulted in the acquisition of chemotaxis and degradation properties by this strain, thus establishing the role of this plasmid in chemotaxis associated with complete mineralization of these two compounds. Another organism, *Arthrobacter sulfureus,* which was also isolated from Gujarat oil fields, could utilize phenanthrene as the sole source of carbon and energy.

The pathway for degradation of this compound was determined in this strain along with that in three other strains, *Acidovorax delafieldi* P4-1, *Brevibacterium* sp. HL4 and *Pseudomonas* sp. DLC-P11. Novel pathways for phenanthrene degradation in *Brevibacterium* HL4 and *Pseudomonas* DLC-P11 were proposed. These organisms could utilize several other PAHs such as acenaphthene, anthracene, fluorine, fluoranthene, naphthalene and pyrene, making them potentially useful in cleaning up petroleumcontaminated soils containing a mixture of different PAHs.

Pseudomonas putida is a good candidate for metabolic engineering and genetic manipulation applications for expression of genes encoding several degradative enzymes. Therefore, a *P. putida* strain was engineered to increase the efficiency of degradation of naphthalene and salicylate. In this strain, utilization of glucose (a simple carbon source) was blocked through metabolic engineering making it primarily dependent on the more complex carbon source, i.e. the organic compound in question (naphthalene/ salicylate) thus degrading it more efficiently.

HALOGENATED ORGANIC COMPOUNDS

Halogenated organic compounds constitute one of the largest and diverse groups of environmental chemicals. Although some of these chemicals are generated by naturally occurring biotic and abiotic processes in the oceans and atmosphere, the widespread use of halogen-based chemistry in industrial-scale chemical processing over the past 100 years has introduced many additional man-made halocarbons into the environment. The occurrence of incomplete pathways and the accumulation of toxic intermediates are important factors in the apparent recalcitrance of environmental pollutants such as chloroaromatics, haloalkanes and chloroethenes.

The recalcitrant nature of these compounds is due to low electron density at the aromatic ring, and the enzyme oxygenase that usually initiates the degradation of other aromatic compounds, cannot attack these halogenated compounds. This reaction does not proceed through an electrophilic attack but via a nucleophilic attack. A variety of microbial enzyme systems have been found to effect cleavage of carbon– halogen bonds, providing the means for these compounds to be utilized as carbon sources or as alternative electron acceptors.

Halogenated Aliphatic Hydrocarbons

These are prevalent groundwater contaminants and are significant components of hazardous wastes and landfill leachates. Many hazardous halogenated aliphatic compounds released from industrial, commercial and agricultural sources are chlorinated or brominated alkanes and alkenes that contain one to three carbon atoms such as halogenated alkanoic acids (HAA), haloalkanes, trichloroethane and ethylene dibromide

(EDB). A number of soil microorganisms which synthesize dehalogenases are capable of utilizing HAA. Hardman examined four *Pseudomonas* and two *Alcaligenes* species capable of growth on 2-monochloropropionic acid and monochloroacetic acid. They found that all isolates contained a single plasmid of molecular size of 53 kb (pUU204) or more. Chlorinated ethenes such as vinyl chloride, trichloroethylene (TCE) and tetrachloroethylene have been frequently detected in drinking water aquifers.

TCE is one of the major industrial solvents used for degreasing and cleaning metals and electronic components. Currently, there is much concern regarding the microbial metabolism of TCE. Bouwer and McCarty demonstrated that under anerobic conditions in the laboratory, TCE and tetrachloroethylene could be degraded during vigorous methanogenesis supported by growth on acetate. Kleopfer demonstrated that reductive dechlorination of TCE to 1,2-dichloroethylene occurred in soil. Recently, an aerobic methane-oxidizing bacterium that degrades TCE in pure cultures was isolated. TCE biodegradation in this bacterium appeared to be a co-metabolic process.

Another aerobic microorganism degraded TCE in the presence of phenol. Freitas dos Santos isolated a mixed culture that exhibited slow growth on 1,2-dibromoethane under aerobic conditions. After repeated subcultivation, during which the growth rate increased slowly, a pure culture was isolated that utilized 1,2-dibromoethane as a sole source of carbon. The organism was classified as *Mycobacterium* sp. strain GP1, making it the first *Mycobacterium* sp. capable of growth on a short chain haloalkane. The dehalogenase genes in this strain encode enzymes that are very similar to dehalogenases found in other microorganisms.

EDB is a brominated hydrocarbon that has been used as a soil fumigant and in anti-knock gasoline. Pignatello has reported the microbial degradation of EDB in aquatic environments. However, no information is available on microorganisms and their genetic traits involved in EDB transformation. A general trend with regard to the degradation of chlorinated aliphatic hydrocarbons is that greater the chlorinated the aliphatic hydrocarbon, the higher the relative rate of reduction with lesser the chlorination, higher the rate of oxidation.

Chlorinated Aromatic Compounds

Chlorinated aromatic compounds are major environmental pollutants because they are often released in substantial quantities, are toxic, resistant to degradation and accumulate in sediment and biota. Chlorinated phenols and their derivatives are widely used in the environment. The toxicity of these compounds tends to increase with the degree of chlorination. The chlorinated phenolics are contaminated with chlorinated dibenzofuran and dioxins which are highly toxic compounds produced as contaminants during the manufacture of pesticides and during incineration.

The pulp and paper mill is a major industrial sector utilizing huge amounts of lignocellulosic materials and water during the manufacturing process, and releases chlorinated lignosulphonic acids, chlorinated resin acids, chlorinated phenols and chlorinated hydrocarbons in the effluent. Tetrachlorodibenzo*p*- dioxin and dibenzofuran are major contaminants of coloured wastewater of pulp and paper mill effluent. Mineralization of chlorinated dioxins or dibenzofurans by single bacterial strains has not been reported so far. Strain *Sphingomonas* sp. RW1 is able not only to degrade the unchlorinated structures but also to transform lower chlorinated congeners.

The transformation of lower chlorinated congeners by strain RW1, however, leads to the accumulation of intermediates, which prevent further transformation of the substrates. A strategy for reducing environmental pollution by chlorinated dioxins and dibenzofurans was developed in the chemostat in which microbial populations in a sediment core contaminated by chlorinated aromatics were enriched and a stable microbial consortium was obtained in the presence of 4-chlorosalicylic acid as sole carbon source.

The stable consortium comprising *Klebsiella pneumoniae,*

Pseudomonas fluorescens, P. mendocina and *P. cichhori* had the potency to degrade chlorinated benzoic acid, salicylic acid, phenoxyacetic acid and dibenzofuran. These strains degraded chlorinated aromatic compounds by *ortho* and *meta* ring cleavage. Pentachlorophenol (PCP) is a general biocide used primarily as preservative of wood, leather, textile and related commercial products. The intermediary metabolites formed during aerobic and anaerobic degradation of PCP are toxic and recalcitrant. The PCP-degrading organisms have been isolated such as *Pseudomonas* sp. strain IST 103 which was capable of utilizing PCP as a carbon source.

The first step in PCP degradation by bacteria is oxidative dehalogenation to tetrachlorohydroquinone, which then dechlorinates to chlorohydroquinone. The enzyme PCP- 4 monooxygenase was responsible for the dechlorination of PCP.

PCP-degrading microbial strains have been isolated but they proved ineffective for *in situ* bioremediation, therefore, two bacterial consortia were developed by continuous enrichment of sediment core of pulp and paper mill and tannery, contaminated by PCP in the chemostat. The microbial consortium from paper mill pulp contained *E. coli, Pseudomonas aeruginosa* and *Acinetobacter* sp. and that from tannery contained *Serratia* sp. (three isolates) and *Pseudomonas fluorescens,* identified on the basis of 16S rDNA sequence analysis.

Another microbial consortium was developed by continuous enrichment of microbial population of sediment core of pulp and paper mill effluent under alkaline conditions in the presence of 4- chlorosalicylic acid. The consortium contained three isolates identified as *Micrococcus diversus, Deinococcus radiophilus* and *Alloiococcus otitis.*

One of the strains, *Micrococcus diversus,* which produces xylanase and lignin peroxidase (no cellulase) is currently under investigation for biopulping and biobleaching. The genes involved in PCP degradation have been reported from genomic DNA.

There is limited information regarding involvement of plasmid in degradation of PCP. Two different-sized plasmids were found to be responsible for carrying genes for degradation of PCP in the strain IST.

The plasmid isolated from the bacterium was subjected to shotgun cloning and transformed into *E. coli* XLBlue1*a*. The recombinant clones having higher potentiality to degrade PCP were selected by utilization of PCP as the sole source of carbon and it released various intermediary metabolites during degradation. The recombinant clones contained an insert of 3 kb that was found to have a homology with *thdf* gene of monooxygenase of thiophene and furan. Southern blot analysis performed by developing gene probes indicated presence of PCP-monooxygenase gene in plasmid of the bacterium.

Recently, bacteria such as *Alcaligenes* sp. and *Pseudomonas* sp., that utilize chlorobenzene 1,2-dichlorobenzene 1,3-dichlorobenzene and 1,4-dichlorobenzene as the sole source of carbon have been isolated or constructed. However, the genetic basis of degradation of chlorobenzenes in these microorganisms has not been described. Pierce reported that *P. cepacia* and several other *Pseudomonas* strains were capable of utilizing mono- and di-chlorinated toluenes as the sole source of carbon and energy. These strains contained plasmids of about 72 MDa which coded for chlorotoluene degradation.

Chlorinated Polycyclic Hydrocarbons

Polycyclic chlorinated hydrocarbons occur as natural constituents and combustion products of fossil fuels and are widespread environmental contaminants. Out of the chlorinated polycyclic hydrocarbons, DDT, polychlorinated biphenyls (PCBs) and *p*-chlorobiphenyls (*p*-CBs) are of general interest because of their widespread occurrence in the environment. It has been reported that *p*-CBs and PCBs are subject to biodegradation. Furukawa and Chakrabarty isolated an *Acinetobacter* sp. capable of utilizing *p*-CB. They showed that the degradation of this compound is encoded on an 82-kb plasmid (pKF1).

Several other *p*-CB-degrading plasmids such as SS50 (53 kb) from *Alcaligenes* sp., have also been isolated. Another *p*-CB degrading plasmid was isolated from *Klebsiella pneumoniae*.

Barton and Crawford have isolated a *Pseudomonas* sp. that is capable of utilizing *p*-CB as the sole carbon and energy source. No genetic information about this species is available. PCBs are water insoluble, non-polar, lipid soluble, inert and highly toxic compounds.

Both aerobic and facultative anaerobic bacteria capable of utilizing PCBs have been isolated from the environment. Takase isolated a *Pseudomonas cruciviae* strain that could grow on more than 10 biphenyl related compounds including *p*- CB. Physiological and genetic studies have shown that the genes for the degradation of PCBs may be plasmid encoded or present on the chromosome.

The examples cited above once again illustrate the diversity of solutions that microorganisms have found to deal with hostile conditions and turn a threat, a toxic compound into a benefit, a growth substrate.

Chapter 3

Reproduction and Nutrition of Algae

ALGAE

Algae are a diverse group of simple, plantlike organisms. Like plants, most algae use the energy of sunlight to make their own food, a process called photosynthesis. However, algae lack the roots, leaves and other structures typical of true plants. Algae are the most important photosynthesizing organisms on Earth. They capture more of the sun's energy and produce more oxygen (a byproduct of photosynthesis) than all plants combined. Algae form the foundation of most aquatic food webs, which support an abundance of animals.

Algae vary greatly in size and grow in many diverse habitats. Microscopic algae, called phytoplankton, float or swim in lakes and oceans. Phytoplankton are so small that 1000 individuals could fit on the head of a pin. The largest forms of algae are seaweeds that stretch 100 m (300 ft) from the ocean bottom to the water's surface. Although most algae grow in fresh water or seawater, they also grow on soil, trees and animals and even under or inside porous rocks, such as sandstone and limestone. Algae tolerate a wide range of temperatures and can be found growing in hot springs, on snow banks, or deep within polar ice.

Algae also form mutually beneficial partnerships with other organisms. For example, algae live with fungi to form lichens—plantlike or branching growths that form on boulders, cliffs and tree trunks. Algae called zooxanthellae live inside the cells of reef-building coral. In both cases, the algae provide oxygen and complex nutrients to their partner and in return they receive protection and simple nutrients. This arrangement enables both partners to survive in conditions that they could not endure alone. The earliest life-forms on this planet are thought to be early ancestors of cyanobacteria, a type of algae formerly called blue-green algae. Fossilized cyanobacteria have been found in rocks more than 3 billion years old.

These early algae formed when there was no oxygen in the atmosphere and scientists theorize that as the algae photosynthesized, they released oxygen as a byproduct, which eventually accumulated in the atmosphere. Algae were probably the first organisms capable of photosynthesis and, until the appearance of plants on earth, the only photosynthesizers for billions of years.

CLASSIFICATION OF ALGAE

While *Cyanobacteria* have been traditionally included among the Algae, recent works usually exclude them due to large differences such as the lack of membrane-bound organelles, the presence of a single circular chromosome, the presence of peptidoglycan in the cell walls, and ribosomes different in size and content from those of the Eukaryotes.

Rather than in chloroplasts, they conduct photosynthesis on specialized infolded cytoplasmic membranes called thylakoid membranes. Therefore, they differ significantly from the Algae despite occupying similar ecological niches.

By modern definitions Algae are Eukaryotes and conduct photosynthesis within membrane-bound organelles called chloroplasts. Chloroplasts contain circular DNA and are similar in structure to Cyanobacteria, presumably representing reduced cyanobacterial endosymbionts.

The exact nature of the chloroplasts is different among the different lines of Algae, reflecting different endosymbiotic events. The table below describes the composition of the three major groups of Algae.

Their lineage relationships. Many of these groups contain some members that are no longer photosynthetic. Some retain plastids, but not chloroplasts, while others have lost plastids entirely.

Supergroup Affiliation	*Members*	*Endosymbiont*	*Summary*
Primoplantae/ Archaeplastida	• Chlorophyta • Rhodophyta • Glaucophyta	Cyanobacteria	These Algae have *primary* chloroplasts, i.e. the chloroplasts are surrounded by *two membranes* and probably developed through a single endosymbiotic event. The chloroplasts of Red Algae have chlorophylls *a* and *d* (often), and phycobilins, while those of Green Algae have chloroplasts with chlorophyll *a* and *b*. Higher plants are pigmented similarly to Green Algae and probably developed from them, and thus Chlorophyta is a sister taxon to the plants; sometimes they are grouped as Viridiplantae.
Excavata and Rhizaria	• Chlorarachniophytes • Euglenids	Green Algae	These groups have green chloroplasts containing chlorophylls *a* and *b* . Their chloroplasts are surrounded by *four and three membranes*, respectively, and were probably retained from ingested Green Algae. Chlorarachniophytes, which belong to the phylum Cercozoa, contain a small nucleomorph, which is a relict of the algae's nucleus. Euglenids, which belong to the phylum Euglenozoa, live primarily in freshwater and have chloroplasts with only three membranes. It has been suggested that the endosymbiotic Green Algae were acquired through myzocytosis rather than phagocytosis.
Chromista and Alveolata	• Heterokonts • Haptophyta • Cryptomonads • Dinoflagellates	Red Algae	These groups have chloroplasts containing chlorophylls *a* and *c*, and phycobilins. The latter chlorophyll type is not known from any prokaryotes or primary chloroplasts, but genetic similarities with the Red Algae suggest a relationship there. In the first three of these groups (Chromista), the chloroplast has four membranes, retaining a nucleomorph in Cryptomonads, and they likely share a common pigmented ancestor, although other evidence casts doubt on whether the Heterokonts, Haptophyta, and Cryptomonads are in fact more closely related to each other than to other groups. The typical dinoflagellate chloroplast has three membranes, but there is considerable diversity in chloroplasts within the group, and it appears there were a number of endosymbiotic events. The Apicomplexa, a group of closely related parasites, also have plastids called apicoplasts. Apicoplasts are not photosynthetic but appear to have a common origin with Dinoflagellate chloroplasts.

RELATIONSHIP TO HIGHER PLANTS

The first plants on earth evolved from shallow freshwater algae much like *Chara* some 400 million years ago. These probably had an isomorphic alternation of generations and were probably heterotrichous. Fossils of isolated land plant spores suggest land plants may have been around as long as 475 million years ago.

Physical Characteristics

With the exception of the cyanobacteria, algae are eukaryotes—that is, the insides of their cells are organized into separate membrane-wrapped organelles, including a nucleus and mitochondria. An important organelle found in eukaryotic algae is the chloroplast, which contains the light-absorbing pigments responsible for capturing the energy in sunlight during photosynthesis. In most algae the primary pigment is chlorophyll, the same green pigment used in plants.

Many algae also contain secondary pigments, including the carotenoids, which are brown or yellow and the phycobilins, which are red or blue. Secondary pigments give algae their colorful hues. The cyanobacteria are prokaryotes—that is, relatively simple unicellular organisms lacking a nucleus and other membrane-bound organelles.

As their modern name implies, the cyanobacteria have many characteristics that resemble bacteria. Like plants, most algae have rigid cell walls composed largely of cellulose. An exception is the diatom, whose cell wall is composed primarily of silica, which provides rigidity and also produces elaborately sculpted patterns of grooves that serve as identifying features.

Many eukaryotic algae have whiplike appendages called flagella attached to their cell walls. By beating flagella in a rotary movement, these algae are able to move through water with considerable speed. A few algae that are devoid of rigid cell walls are able to protrude one part of the body ahead of the other to crawl on solid surfaces in an amoeba-like fashion.

Algae come in a variety of shapes and forms. The simplest form is the single, self-sufficient cell, such as Euglena, dependent only on sunlight and carbon dioxide and minerals from the water. Numerous one-celled algae may clump together to form a colony.

Although these cells are attached to one another, each cell within a colony continues to function independently. Still other algae are multicellular organisms. In the simplest multicellular algae, the cells are joined end to end, forming filaments, both branched and unbranched. More complex structures may be shaped like a small disc, tube, club, or even a tree. The most complex algae have highly specialized cells. Some seaweeds, for instance, have a variety of specialized tissues, including a rootlike holdfast, a stipe, which resembles a plant stalk and a leaf-like blade.

While most algae create their own food through photosynthesis, some are unable to photosynthesize. These algae ingest food from external sources by absorbing simple nutrients through the cell membrane. To absorb more complex nutrients, algae that lack rigid walls are able to engulf food particles and digest them. Some of the algae known as dinoflagellates extend a feeding tube, called a peduncle, to suck in food. Other dinoflagellates use special harpoonlike structures to snare their food. Some algae are parasites, living in or on another organism from which they get their food. Some parasitic red algae live off other red algae and parasitic dinoflagellates live in the intestines of some marine animals, such as copepods and annelids.

Reproduction

Algae reproduce in astoundingly diverse ways. Some reproduce asexually, others use sexual reproduction and many use both. In asexual reproduction an individual reproduces without combining its genetic material with that from another individual. The simplest form of asexual reproduction is binary fission, in which a unicellular organism simply divides into two new individuals. Some multicellular algae, including Sargassum, reproduce asexually through fragmentation, in which fragments of the parent develop into new individuals. In a similar process called budding, special buds detach from multicellular algae and develop into new individuals, commonly found in Sphacelaria.

Many algae produce special cells called spores that are capable of growing into new individuals. If these spores move about using flagella, they are known as zoospores. In sexual reproduction, genetic material from two individuals is combined. The simplest form of sexual reproduction in algae is conjugation, in which two similar organisms fuse, exchange genetic material and then break apart. For example, in Spirogyra, which produces both asexually and sexually, two long, unbranched filaments join via conjugation tubes, through which genetic material is exchanged between cells.

Most multicellular algae undergo a more complex form of sexual reproduction involving the union of special reproductive cells, called gametes, to form a single cell, known as a zygote. Many algae incorporate both sexual and asexual modes of reproduction. This is well demonstrated in the life cycle of the alga Chlamydomonas.

The mature alga is a single haploid cell—that is, it contains only one set of chromosomes. During asexual reproduction the cell undergoes mitosis, a type of cell division that produces genetically identical offspring. Four daughter cells are created that emerge from the enclosing parent cell as spores. The spores develop into mature haploid cells that are genetically identical to the parent cell.

Certain environmental conditions, such as lack of nutrients or moisture, may trigger the haploid daughter cells to undergo sexual reproduction. Instead of forming into spores, the haploid daughter cells form gametes that have two different mating strains. These two strains are structurally similar and are called plus and minus strains.

Opposite mating strains fuse in a process known as isogamy to form a diploid zygote, which contains two sets of chromosomes. After a period of dormancy, the zygote undergoes meiosis, a type of cell division that reduces the genetic content of a cell by half. This cell division produces four genetically unique haploid cells that eventually grow into mature cells. Some multicellular green algae, such as Ulva, follow a distinct pattern of reproduction called alternation of generations, in which it takes two generations—one that reproduces sexually and one that reproduces asexually—to complete the life cycle.

The two mature forms of the algae, alternating between diploid and haploid individuals, are identical in appearance, or isomorphic. The haploid form, called a gametophyte, undergoes mitosis to produce haploid gametes. These gametes unite to form a diploid zygote, which develops into the diploid form called a sporophyte. The sporophyte undergoes meiosis to form haploid spores that, in turn, form gametophytes.

Not all algae that undergo alternation of generations have haploid and diploid forms that look alike. In the life cycle of the seaweed Laminaria, the gametophyte and the sporophyte are distinct in appearance, or heteromorphic. The Laminaria sporophyte appears as long, bladelike structures that grow on rocks just below the water in intertidal zones. The gametophyte is short, with branched filaments.

TYPES OF ALGAE

The most common classification system distributes algae in more than one kingdom. Most algae are classified in the Kingdom Protista, along with other eukaryotic organisms that lack true specialized tissues. The cyanobacteria, however, are classified with the bacteria in the Kingdom Prokaryotae, which consists of prokaryotic organisms. This classification system continues to be intensely debated as new research increases our understanding of the way that these organisms are related. This article uses a classification scheme proposed in 1997 that divides algae, including the cyanobacteria, into 11 different phyla, of which the 5 largest.

Green Algae

Green algae form the phylum Chlorophyta and are named for their green chloroplasts, which are similar in composition to the chloroplasts found in land plants. Green algae range in shape from unicellular plankton that grow in lakes and oceans to colonial filaments of pond scum to leaflike seaweeds that grow along rocky and sandy intertidal areas. Some green algae also live on tree trunks and soil.

Several green algal species are symbiotic, forming lichens with fungi or living with corals. Green algae may also be found inside freshwater sponges, giving the sponges a bright green colour and in permanent snow banks, where a secondary pigment masks the chlorophyll and turns the snow a reddish colour. More than 500 genera and 8000 species of green algae have been identified. Some familiar green algae include the genus Spirogyra, known for its spiral-shaped chloroplasts and the desmids, recognized by their characteristic shape—two symmetrical halves, joined by a small bridge. The green algae known as Stoneworts often grow several feet in length.

Their name comes from calcium crusts that make them feel like stone. Most green algae reproduce both sexually and asexually. Alternation of generations, where the algae alternates between gametophyte and sporophyte generations, is common among the multicellular green algae. Members of the largest phylum of the algae, numbering some 6000 to 7000 species. They are commonly known as green or grass-green algae because of their bright green colour, which is imparted by two chlorophylls, a and b.

Among the oldest of all organisms—the first green algae appeared more than 2 billion years ago in the fossil record—they are believed to be the most immediate relatives of the green land plants. Green algae may occur as single cells (which may be either motile or nonmotile), in colonies (more often nonmotile) and as multicellular filaments (nonmotile). Most have cell walls made up of two layers: an inner cellulose layer and an outer layer of pectin.

The unicellular forms assume a virtually endless variety of shapes; colonial forms may be loose aggregates of single cells or may have these cells arranged in a characteristic pattern. Some filamentous types bear a superficial resemblance to higher plants. The motile unicellular organisms are free-swimming, moving by means of whip-like flagella (usually two in number). Even the nonmotile species may produce motile reproductive cells (zoospores).

Green algae are also found on damp soil, attached to land plants (a few are parasitic) and even in snow and ice. The marine forms are often visible on coastal rocks exposed at low tide. Some terrestrial species combine with fungi in symbiotic associations called lichens. Green algae reproduce vegetatively, by fragmentation and by cell division; asexually, by means of spores and zoospores that develop directly into new plants; and sexually, by the fusion of pairs of sex cells (gametes). Many species exhibit alternation of generations.

In such instances, the gametophyte generation is usually dominant, unlike seed plants, in which the sporophyte phase predominates. Green algae are extremely important as a source of food for other aquatic organisms and also make a major contribution to the world's oxygen supply.

They can, however, have negative effects, as when large populations produce an unpleasant taste and odor in drinking water or clog filtration equipment. In freshwater lakes and ponds polluted by nitrates and phosphates, algal populations sometimes increase suddenly in an "Algal bloom," forming a dense, malodorous scum and drastically decreasing the oxygen supply available to other life forms. Scientific classification: Green algae make up the phylum Chlorophyta. The oldest green algae are classified in the genus Gunflintia.

Red Algae

Red algae form the phylum Rhodophyta with approximately 500 genera and 6000 species. Found in warm coastal waters and in water as deep as 260 m (850 ft), red algae species adapt to varied water depths by having different proportions of pigments. Their red colour is due to a red pigment, phycoerythrin, which is well suited to absorb the blue light that penetrates deeper into water than the other colors of light. Red algae found in deep water may be almost black due to a high concentration of phycoerythrin. At moderate depths red algae appear red, while in shallow water they may appear green because a smaller proportion of phycoerythrin is unable to mask the green of chlorophyll. Most red algae are multicellular and come in a variety of shapes, including filaments, which are shaped like a blade of grass and seaweed shapes. Unlike most other eukaryotic algae, red algae have no flagella. Red algae use diverse strategies to reproduce, including fragmentation and spore production.

One unusual strategy, found in many species including those in the genus Polysiphonia, involves the alternation among three generations. A diploid sporophyte produces diploid spores that germinate into another diploid sporophyte that looks completely different from the first one. Meiosis occurs in the second sporophyte, producing haploid spores that germinate into gametophytes.

Surprisingly, in some species, the gametophytes look nearly identical to the second sporophyte. Almost all red algae live in marine habitats, although some species are found in fresh water or damp soil. Many types of seaweed are red algae, typically found growing along the coast and attaching firmly to the seafloor using a rootlike holdfast. In some species, called coralline algae, the cell walls become hardened with calcium carbonate.

Coralline algae are important members of coral reefs, producing new material and cementing together other organisms. Members of the phylum Rhodophyta, a large group of aquatic algae with approximately 6000 species. The red algae are characterized by reddish phycobilin pigments—phycoerythrin and phycocyanin—that mask the colour of the chlorophylls. Most species grow near tropical and subtropical shores below the low-tide mark. A few are found in fresh water.

Most red algae are small to medium-sized multicellular organisms. The bodies of some are relatively complex, resembling those of kelp. Sexual structures and reproductive cells are highly specialized. Red algae vary greatly in shape; plate-like, coralline, crust-like, leathery and featherlike forms are known. Coralline species accumulate lime as they grow—appearing as flat pink coverings on stones, or fanlike growths resembling true coral—and contribute much of the lime in coral reef deposits. Fossils of red algae have been found in rocks 500 million years old. Red algae are unique among the algae in that no flagellated cells are formed during the life cycle. Some species reproduce by vegetative fragmentation or spore formation, but most undergo a complex life cycle involving alternation of generations.

Sexual plants (gametophytes) produce either male sex organs (antheridia or spermatangia) or female sex organs (carpogonia). The small male sex cells are carried by water currents to the elongated tip (the trichogyne) of a carpogonium, where fertilization occurs. The resulting zygote may divide directly but more commonly gives rise to numerous filaments. They produce spores that develop into an asexual plantlike growth, or sporophyte. The spores then develop into gametophytes. Some red algae are of economic importance.

Golden-Brown Algae, Brown Algae and Diatoms

Golden-brown algae, brown algae and diatoms form the

large and complex phylum Heterokontophyta, with organisms ranging in size from a fraction of a millimeter to more than 100 m (300 ft) long. Heterokontophyta have carotenoid secondary pigments that tend to mask the green of the primary chlorophyll pigment, giving them a golden or golden-brown appearance. Flagellated cells in this phylum have two types of flagella: One is smooth, while the other has two rows of stiff hairs running down opposite sides of the flagellum.

Algae in this phylum typically have an eyespot that can detect light. The golden-brown algae, also known as the yellow-brown algae, include about 200 genera and 1000 species that receive their characteristic coloring from the carotenoid pigment fucoxanthin. These algae are mostly unicellular or colonial, swimming or floating in lakes and oceans as phytoplankton. In shallow ponds that dry up in summer or freeze completely in winter, golden-brown algae survive by forming protective cysts that can withstand the harsh conditions. About 1500 species of almost exclusively marine, brown-colored algae, known as seaweeds, that make up the brown algae phylum in the protist kingdom. They are found mainly in the tidal zones of temperate to polar seas, but some exist in the deep ocean. Brown algae are the largest of the algae; well-known forms include the giant kelp and the free-floating sargassum weed. Their brown colour is derived from the presence of the pigment fucoxanthin, which along with other xanthophyll pigments, masks the green colour of the chlorophyll in the algal cells.

The brown algae are multicellular and have differentiated structures that, in some species, bare a superficial resemblance to the roots, stalks and leaves of true plants. These structures, however, are quite different internally. The cell walls of the algae are made of cellulose similar to that found in red algae; a gelatinous pectic compound called algin covers the outsides of the walls. The plants undergo an alternation of generations; the diploid phase (two sets of genes in a cell) is microscopic and brief and the haploid phase (one set of genes in a cell) is macroscopic and comparatively long-lived.

Brown algae such as kelp are harvested for use as an

emulsion stabilizer, an ingredient of ice cream; as a fertilizer; as a vitamin-containing food source; and for iodine. Scientific classification: Brown algae make up the phylum Phaeophyta in the kingdom Protista. Kelp are classified in the order Laminariales. Sargassum weed is classified in the genus Sargassum When favorable conditions return, the algae emerge from the cysts. Like so many other algae, the unicellular algae tend to reproduce through fission, while the multicellular and colonial forms reproduce either through fragmentation or through spore production.

Diatoms are best known for their glasslike cell wall made of silica. The cell wall has ornate ridge patterns. A diatom consists of two overlapping halves that fit together like a shoebox or a petri dish, with the lid slightly larger and fitting over the base. During asexual cell division, the two glass walls separate and serve as the lids for two new glass bases. The new diatom that grew from the lid is the same size as the parent diatom, while the diatom that grew from the smaller base is slightly smaller than its parent.

Sexual reproduction occurs when the succeeding generations shrink to a critical size. These smallest diatoms form gametes that shed their glass walls. Upon fertilization, the zygotes absorb water to swell and then secrete new, full-sized silica coverings. A very large class with more than 250 genera and 8000 species, diatoms are found floating in freshwater and seawater, growing attached to the seafloor, or growing on soil. The cells are either unicellular or form colonial chains of round cells. When an organism dies, its silica cell wall remains intact. Over time these shells have accumulated to form layers of soft rock in some geologic formations. This diatomaceous earth is mined and quarried for use in filters and bleaching agents, as an abrasive powder for cleaning and polishing metals and for insect pest control (the broken cell walls of silica tear the insect gut). Brown algae include over 260 genera and 1500 species.

Multicellular algae, they may range from tiny filaments to the largest and most complex algae, such as the kelps, with leaflike blades and stems that can be up to 100 m (300 ft) long.

Most brown algae grow in marine waters near the coast, attached to rocks either along the shoreline or underneath the ocean surface. One type, Sargassum, forms huge floating masses in the middle of the Sargasso Sea. The brown or olive colour is due to the pigment fucoxanthin. The life cycles of brown algae vary considerably, but most demonstrate alternation of generations.

DINOFLAGELLATES

Dinoflagellates of the phylum Dinoflagellata are mostly unicellular organisms that may be covered with stiff cellulose plates that resemble armored helmets. Many species have unusual ornamentation, such as horns, spines, or wings. A narrow groove encircles the armor and a second groove runs perpendicular to the first groove.

Flagella beat within these grooves, causing the dinoflagellates to spin like tops as they move through the water. Most of the 130 genera and 2000 species in this phylum are planktonic and live in saltwater, although there are many freshwater planktonic representatives as well. Many dinoflagellate species lack chloroplasts and are dependent on other species for their food. Some are parasites, but most are carnivores, using special harpoonlike structures called trichocysts to capture other organisms to eat.

In contrast, several of the photosynthetic species live inside the tissues of invertebrate animals, such as corals and giant clams. These dinoflagellates share the food they photosynthesize with their host and in return, receive protection and some nutrients. Under favorable environmental conditions, some dinoflagellate species experience population explosions, known as blooms.

If the species involved in the bloom have red pigments, their concentration can be high enough to turn the seawater red, forming red tides. Dinoflagellate blooms can be quite destructive. During the night when photosynthesis halts, such a high concentration of individuals can deplete the oxygen in the water, suffocating fish. Some dinoflagellates release toxins, some of which kill fish, while other toxins are passed up the food chain to humans, where they can cause paralytic shellfish poisoning and ciguatera fish poisoning.

CYANOBACTERIA

Unlike other algae, the cyanobacteria are prokaryotes—single-celled organisms with characteristics that cause biologists to debate whether they are really algae or bacteria. Cyanobacteria are found nearly everywhere, occurring in typical aquatic and terrestrial habitats as well as in such extreme sites as hot springs with temperatures as high as 71° C (160° F) and crevices of desert rocks. Cyanobacteria make up the phylum Cyanophyta and this phylum contains about 150 genera and 2000 species worldwide.

Like other bacteria, cyanobacteria do not have organelles such as nuclei, mitochondria, or chloroplasts. Cyanobacteria are distinguished from bacteria by the presence of internal membranes, called thylakoids, that contain chlorophyll and other structures involved in photosynthesis. While higher plants have two kinds of chlorophyll, called a and b, cyanobacteria contain only chlorophyll a. Cyanobacteria colour varies from blue-green to red or purple and is determined by the proportions of two secondary pigments, c-phycocyanin (blue) and c-phycoerythrin (red), which tend to mask the green chlorophyll present in the thylakoids. Cyanobacteria reproduce asexually by binary fission, spore production, or fragmentation, forming singular cells, colonies, filaments, or gelatinous masses.

Although most lack flagella and are nonmotile, filamentous forms such as Oscillatoria rotate in a screwlike manner and the gelatinous forms glide along their slimy mucus. Cyanobacteria may be both beneficial and harmful to humans: Some act as natural fertilizers in some habitats, especially rice paddies, whereas others produce toxins. Mild cyanobacteria toxins in lakes and oceans cause a rash known as swimmer's itch, while powerful neuromuscular toxins released by other cyanobacteria can kill fish living in the water or the animals that drink the water. In certain conditions, cyanobacteria may form dense blooms, which may produce toxins that make seafood poisonous to humans. Even if the cyanobacteria do not produce toxins, blooms can cause water to have an unpleasant taste and odor.

ALGAE USES

Human ingenuity has found many uses for algae. Algae provide food for people and livestock, serve as thickening agents in ice cream and shampoo and are used as drugs to ward off diseases. More than 150 species of algae are commercially important food sources and over $2 billion of seaweed is consumed anually by humans, mostly in Japan, China and Korea. Red algae Porphyra, (nori), is the most popular food product. After harvesting, nori is dried, pressed into sheets and used in soups, sauces, sushi and condiments.

Algae are nutritious because of their high protein content and high concentrations of minerals, trace elements and vitamins. The high iodine content of many edible algae contribute to low rates of goiter in countries where people frequently eat algae. In coastal areas of North America and Europe, seaweeds are fed to farm animals as a food supplement. Cyanobacteria species that are high in protein, such as Spirulina, are grown commercially in ponds and used mostly as a health food and cattle dietary supplement. Seaweeds also are applied to soils as a fertilizer and soil conditioner, as their high concentrations of potassium and trace elements improve crop production. Some species of cyanobacteria can turn atmospheric nitrogen into ammonia, a form that can then be used by plants as a nutrient. Farmers in tropical countries grow cyanobacteria in their flooded rice paddies to provide more nitrogen to the rice, increasing productivity as much as tenfold.

Seaweeds are a critical source of three chemical extracts used extensively in the food, pharmaceutical, textile and cosmetic industries. Brown algae yield alginic acid, which is used to stabilize emulsions and suspensions; it is found in products such as syrup, ice cream and paint. Different species of red algae provide agar and carrageenan, which are used for the preparation of various gels used in scientific research.

Bacteria, fungi and cell cultures are commonly grown on agar gels. Agar is also used in the food industry to stabilize pie fillings and preserve canned meat and fish. Carrageenan is also used as a thickening and stabilizing agent in products such as puddings, syrups and shampoos.

Algae have been used for centuries, especially in Asian countries, for their purported powers to cure or prevent illnesses as varied as cough, gout, gallstones, goiter, hypertension and diarrhea. Recently, algae have been surveyed for anticancer compounds, with several cyanobacteria appearing to contain promising candidates. Diatoms also have been used in forensic medicine, as their presence in the lungs can indicate a person died due to drowning.

Algae can also serve as indicators of environmental problems in aquatic ecosystems. Because algae grow quickly and are sensitive to changing environmental conditions, they are often among the first organisms to respond to changes. For example, depletion of the diatom community in the Florida Everglades provided strong evidence of phosphorus-related changes in this unique ecosystem. Algal blooms may deplete oxygen concentrations in water and smother fish and plant life, as well as prevent light exposure. Through seaweeds, the earth's sea-blood strengthens our own sea-blood. Seaweeds are an excellent source of trace minerals in our diet.

As our air and water become more acidified, through pollution, minerals are leached and depleted from our land fields and they wash down to the sea, where the wild seaweeds incorporate them. When we eat seaweeds, we take these minerals back into our bodies and they help us maintain an alkaline condition in our bloodstream, which is a healthy condition. An alkaline bloodstream is resistant to fatigue and stress. Seaweeds have admirable qualities: they are flexible, they are tenacious, they are prolific and they are the oldest family of plants on earth.

These plants link us to the primitive vitality of the sea. An average family of seaweed eaters will consume 3+ pounds = 30 wet pounds = one bushel of wet plants. This is a very concentrated food. Anyone who has walked on the slippery rocks of the seashore will be aware of organisms, which are recognizable as belonging to a group known as algae.

A very different form of alga will be familiar to swimming pool owners. The seashore algae are large, slippery and firmly attached to the rocks, while the swimming pool algae are small, often unicellular and either float freely in the water coloring it green, or coat the sides of the pool with a green or brownish film. Algae, as a group, are notoriously difficult to define. Biologically, the name "algae" is given to a group of organisms of mixed affinity. The word itself has no taxonomic significance whatsoever. The algae do, however, share a number of features, which allow them to be treated as a group, although as a group with a very mixed evolutionary history.

Algae are constructed fairly simply. They generally do not have vascular tissue and they do not show the high level of organ differentiation of the familiar, more complex plants. It might be argued however that this distinction doesn't apply to the more advanced brown algae which do have a certain degree of organ differentiation and which even have a type of vascular tissue. Algae have naked reproductive structures. This means that there are no protective layers of cells surrounding reproductive structures. None of the algae have reached even the level of organization in reproductive structures shown by the archegoniate plants, such as the mosses & liverworts.

Even this distinction breaks down in the case of the female reproductive structures of the more advanced red algae. Most algae are photoautotrophic, which means that they can make their own food materials through photosynthesis by using sunlight, water and carbon dioxide. A few algae are not photoautotrophic, but they belong to groups that are usually autotrophs. The word algae is plural so we say, for example, Chlorella and Spirogyra are algae. The singular form is alga so we say, for example, Chlorella is an alga. When used as an adjective, we say algal, as for example algal cells. In common with all plants, most algae contain Chlorophyll-A, as well as various other photosynthetic pigments.

This is true even though, strictly speaking, not all algae belong to the plant kingdom. Algae show a broad range of complexity, as might be expected in a group of organisms with such a mixed background. They range in complexity from tiny, microscopic forms, to very complex forms such as the kelps. Algae include both prokaryotic and eukaryotic organisms. The Cyanobacteria (sometimes called Cyanophyta or blue-green algae) and a relatively recently discovered algal division called the Prochlorophyta are both prokaryotic divisions, while all other algae are eukaryotes.

The prokaryotic algae range only from unicells and colonies, to the simplest of branched filaments. The eukaryotic algae include unicells, colonies, simple and more complex filaments, as well as the very complex parenchymatous form, which is most developed in the large kelps, such as Laminaria. Algae occur in virtually any habitat on earth as long as long as water is found there at some time, even if it is just moisture which might be present for a very short time. Algae may be found as free-floating phytoplankton, which form the base of food webs in large water bodies. There are also algae, which live, attached to rocks and other substrata at the bottom of bodies of water such as the sea. Algae may occur as epiphytes on higher plants, or on other algae. All major bodies of water have algae in abundance, including lakes, small streams, large rivers and even waterfalls.

Algae occur in fresh water, to the saline water of the sea and even in saltpans. There are also algae that thrive in the heated water of hot springs. In the sea they may occur below the range of tidal exposure—in the subtidal zone, as well as in the harsh intertidal environment of the seashore, where they may be beaten by waves. In some parts of the world, intertidal algae are even scoured by sea ice, yet they persist in living in this environment. Those algae, which live attached to the bottom of a water body, are called benthic algae and the ecosystems of which they are part are referred to as benthos. Small, microscopic algae that drift about in bodies of water, such as lakes and oceans, are called phytoplankton. Phytoplankton are important in freshwater and marine food webs and are probably responsible for producing much of the oxygen that we breathe. Some forms of algae are even able to grow in Arctic and Antarctic sea ice, where they can be quite productive and support a whole associated food web. Some algae can even grow on the seabed, beneath a thick blanket of Arctic or Antarctic sea ice, even though they are in total darkness for a considerable part of the year. Algae are found in snow too.

In some parts of the world, blooms of snow algae may paint the snow beds red in spring. Algae even occur in the driest deserts. In some areas of the Namib Desert in Namibia and the Richtersveld in South Africa, one often finds many quartz stones scattered about on the ground. Since Quartz is quite translucent, the stones permit a considerable amount of light to pass through, so there is sufficient light for photosynthesis to take place underneath the stones.

A small amount of moisture may be retained in the soil under the quartz stones; so unicellular algae are able to grow underneath them. Algae as a green coloration if you gently turn the stone over. If you do this, remember to put the stone back into position again so that the algae and other organisms that live there won't dry out and suffer damage or die. Algae are also found in the air, for there are many algae that colonize new bodies of water by simply drifting about through the air. There is even a unicellular green alga called Prototheca, which causes disease in humans, although like this specimen, you have to be very ill already to get it.

It produces skin lesions, mainly in patients whose immune systems have been damaged by other serious diseases. And of course there are the algae that enter into symbiosis with other organisms, for example, the symbiotic organisms that we call lichens. The stony corals, which construct coral reefs in warm tropical seas, are only able to build up these massive and beautiful structures because they have formed a symbiotic partnership with tiny single-celled algae called zooxanthellae.

The zooxanthellae, which live in the tissues of the coral, share with it the organic products of their photosynthesis, as well as helping the coral with the construction of its limestone skeleton. Even the chloroplast of land plants had its origin as a blue-green alga that lived within the cells of the ancestral organism. Such a special symbiotic relationship, where one organism lives inside the cells of another, is called endosymbiosis. This fascinating group of organisms forms the basis for the science of phycology—the study of algae.

Historically, algae were treated as belonging to the plant kingdom under a two-kingdom system in which all living things were considered either plants or animals. More recently, it has been recognized that algae fall into several kingdoms. Some, such as the green algae, are plants. Others fall under the kingdom Protista. Prokaryotic algae belong to the kingdom Monera together with the bacteria, but this kingdom should probably be divided into more than one.

It has even been suggested that the red algae should have a kingdom of their own. Thus it is easy to see that the "algae" constitute an artificial grouping of organisms. They are dealt with together because of historical beliefs and for the sake of convenience. Many phycologists, however, still treat the algae as though they were part of the plant kingdom, although recognizing that the prokaryotic algae (the blue-green algae) belong with the bacteria.

Seaweeds

Seaweeds are marine algae, saltwater dwelling and simple organisms that fall into the rather outdated general category of "plants." Most of them are red (6000 species), brown (2000 species) or green (1200 species) kinds shown on this page and most are attached. These plant-like organisms are found throughout the world's oceans and seas and none is known to be poisonous. Many are in fact eaten and considered to be a great delicacy. Seaweed, any of the larger, multicellular forms of algae living in fresh and salt water, especially along marine coastlines. The three main phyla, or divisions, are the brown algae, such as the kelps; the red algae, such as Irish moss and the green algae, such as the sea lettuces, all of which are commonly seen at low tide along rocky shores of northern seas.

Seaweeds differ from plants in that they lack the true stems, leaves, roots and vascular systems of higher plants.

Instead, they anchor themselves to solid objects by holdfasts and absorb nutrients directly from the water, manufacturing their food by photosynthesis. The pigments of red and brown algae mask the predominant green photosynthetic pigment, chlorophyll and probably aid in photosynthetic metabolism by absorbing and transferring light energy to the chlorophyll.

Seaweeds abound in shallow waters from the midtide line down to depths of 50 m (165 ft). Along damp cold-water shores, they are able to withstand several hours of exposure to the sun and they cover rocks high into the intertidal zone. In the Tropics, seaweeds are confined to the zone between the low-tide line and a depth of about 200 m (about 660 ft); red algae predominate, especially in lagoons and around coral reefs. The brown algae, commonly called kelp, comprise the largest seaweeds. Pacific species can reach 65 m (213 ft) in length and have structures that superficially resemble leaves and stems, as well as large air-filled bladders and strong holdfasts that anchor them against heavy surf.

Other brown algae are the common rockweed and the gulfweed, which floats in great masses in the Gulf Stream and the Sargasso Sea. Among the red algae are several species of Irish moss, which is commonly seen along northern Atlantic coasts as a matted carpet in the sublittoral zone. Red algae are abundant in clear tropical waters, where their red pigment, phycoerythrin, enables them to carry on photosynthesis at deeper levels than is possible for ordinary green algae. Seaweed is a commercially important food, especially in Japan, where it is called nori and is harvested mainly from red algae, extensively cultivated on bamboo screens submerged in estuaries. Agar, also derived from red algae, is consumed as a delicacy in Asia and is used as a laboratory medium for culturing microorganisms. Red algae are probably of little nutritive value to humans, however, other than for their limited protein, vitamin and mineral (especially iodine) content. Brown algae are used as fertilizer and as an ingredient for livestock meal.

Alginic acid, found in kelp, has wide industrial uses. It can be made into a silk-like thread or a plastic material, insoluble in water, that is used to make films, gels, rubber and linoleum and as a colloid in cosmetics, car polishes and paints. Organic derivatives of alginates are used as food gums in making ice cream, puddings and processed cheeses.

Scientific classification: Brown algae make up the phylum Phaeophyta, red algae the phylum Rhodophyta and green algae the phylum Chlorophyta. Large Pacific brown algae include those species classified in the genera Macrocystis and Nereocystis. Rockweed makes up the genus Fucus and gulfweed the genus Sargassum. Irish moss makes up the genus Chondrus. The species most commonly cultivated for food in Japan is Porphyra tenera.

Dulse

Dulse, is a common name for several edible red algae that grow on rocky marine coasts. It is used widely as a food or condiment. Purple seaweeds of the genus Porphyra, sometimes called laver, are the most widely used for this purpose. P. laciniata is grown in large quantities in Japan. Rhodymenia palmata, which is eaten in the British Isles and in other northern countries, has a purple, leathery frond. Iridaea edulis, eaten in southwestern England and Scotland, has a succulent, dull-purple frond. Dulse is rich in protein and iron and has 22% more protein than chickpeas, almonds, or whole sesame seeds. A handful gives a whole day's supply of iron. The same handful will provide more than 100% of the RDA for Vitamin B_6 and 66% of the RDA for Vitamin B_{12}. Relatively low in sodium and high in potassium. Use to add flavour and nutritional content to your food, instead of salt. Dulse is very popular in its natural original leaf form.

It has a relatively strong distinctive taste and its soft, chewy texture makes it a favourite snack food to eat right out of the bag. Kids love Dulse! It can be cooked quickly and easily in a variety of ways. Dulse chowder, dulse chips, dulse sandwiches and dulse salads are also possible. Flakes are a little coarser than granules and powder and do not "disappear" into the food as much. Mineral and vitamin content is the same.

Can be sprinkled on salads, sandwiches, in soups, stews, pasta, chowders, etc. etc. It can be used as a "salt-substitute" and "flavour-enhancer" all in one and there are unlimited ways you can add it to your diet for flavour as well as minerals and vitamin content.

Kelp

Kelp, common name for large, leafy brown algae, known as seaweed, that grow along colder coastlines. The principal genera of kelp are the true kelps, found in most cool seas and the giant kelps and bladder kelps, both of which are restricted to the northern Pacific. The giant kelps grow as long as 65 m (213 ft). The kelp plant has a rootlike holdfast that fixes to rocky surfaces; a long, slender stalk, or stipe; and long, leaflike blades that manufacture food by photosynthesis. Kelps, like ferns, reproduce by alternation of generations.

Once a major source of iodine and soda, kelp is now used to manufacture algin, a substance used to make tires and to prevent ice cream from crystallizing. Kelp is rich in vitamins and minerals and is a staple, especially in the diets of the Japanese. Scientific classification: Kelp belongs to the order Laminariales. The true kelps belong to the family Laminariaceae and are classified in the genus Laminaria. Giant kelps, classified in the genus Macrocystis and bladder kelps, classified in the genus Nereocystis, belong to the family Lessoniaceae. Kelp is exceptionally high in all major minerals, particularly calcium, potassium, magnesium and iron. Also rich in important trace minerals such as manganese, copper and zinc. It is a good source of chromium, which is instrumental in blood sugar regulation; high in iodine, which is essential to the thyroid gland.

Kelp is also a source of chlorophyll. It can be pre-soaked before cooking or added "dry" to foods, which contain liquid (soups, sauces, etc.). When adding the kelp to rice without rehydrating it, add a bit more water to allow for absorption by the kelp (kelp absorbs up to five times its weight). Uncooked kelp is chewy until soaked or marinated. To fully tenderize, soak for approx. 1 hr.

Alaria

Alaria is almost identical to Japanese wakame biologically and nutritionally, with a black or dark-green colour, yet it grows in the wild and has a more delicate taste than cultivated wakame. It also needs to cook a little longer than its cultivated Japanese counterpart. Soaking prior to cooking shortens cooking time. Keep in mind that the soak water contains many nutrients and vitamins and can be used for your water to cook with (strain for shells).

Alaria is our preferred sea vegetable for miso soup. Alaria makes a perfect calcium-rich vegetable soup and also goes well in a pot of grain, a stew, or any dish that is juicy and needs to cook for more than 20 minutes. To be eaten uncooked in salads, alaria should be presoaked or marinated in vinegar or lemon juice. Comparable to whole sesame seeds in calcium content; very high Vitamin A content (comparable to parsley, spinach or turnip greens); strong in B vitamins (B_1, B_2, B_3, B_6, B_{12}). The word alaria is Latin for "wings" since the spore-bearing leaves at the base of mature alaria plants look somewhat like wings. When cooked with rice, gives off a flavour somewhat like chicken.

Gulfweed

Any of several tropical marine seaweeds that drift in large, floating masses. Gulfweed consists of branches with leaflike blades and numerous small, airfilled sacs that keep the gulfweed afloat. One of the most widely recognized species is the common gulfweed, or sargassum weed, a kind of brown algae. The weed drifts in large masses in the Gulf Stream, which flows northeast from the Gulf of Mexico and in the Sargasso Sea, a large area of the Atlantic Ocean between the West Indies and the Azores. Scientific Classification: Gulfweeds belong to the family Fucaceae of the brown algae division, Phaeophyta. The sargassum weed is classified as Sargassum natans.

Agar

Gel-forming material of widespread commercial use,

found in the cell wall of several species of red algae, especially Oriental members of the genus Gelidium. It is used as a solidifying agent in the preparation of candies, creams and lotions and canned fish and meat; as a texturizer or emulsifier in ice cream and frozen desserts; as a clarifying agent in winemaking and brewing; and as a sizing material in fabrics. It is an excellent laboratory medium for growing bacteria, because it is not dissolved by salts or consumed by most microorganisms. Agar is extracted from seaweed by boiling and is cooled and dried and sold as flakes or cakes. Originally called agar-agar, a Malay word for a local seaweed, it was produced in the Far East but is now made in other Pacific coastal regions such as California and Australia.

CHLORELLA

Chlorella is a special algae called Pyrenoidosa that grows in fresh water and has the highest content of chlorophyll (28.9g/kg) of any known plant on earth. It is extremely high in enzymes, vitamins and minerals, including the full vitamin-B Complex. It has been shown to activate your limited number of macrophages that scavenge and digest cancer cells, foreign proteins and chemicals. This life form emerged over 2.5 billion years ago and was the first form of plant with a well-defined nucleus. There are fossils from the pre-Cambrian period that clearly indicate the presence of Chlorella. Because Chlorella is a microscopic organism, it was not discovered until the later 19th century, deriving its name from the Greek, chloros meaning green and ella, meaning small.

Chlorella belongs to the eucaryotic cell category of algae and lives in fresh water as a single celled plant. Its size is about that of a human erythrocyte (between 2-8 microns in diameter). Under favorable growth conditions; strong sunlight, pure water and clean air, chlorella multiplies at an incredible rate. The process of reproduction can generally be divided into several steps; growth-ripening-maturlty-division. Broken cell wall preparations and extracts of Chlorella pyrenoidosa, a unicellular green alga, as well as other Chlorella species, when either given orally or injected, promote growth and healing.

These preparations stimulate the immune system in such a way that the host is protected from infection and cancer.

All of the following have been associated with the consumption of Chlorella: increased production of interferon; cleansing the blood stream, liver, kidneys and bowel; stimulates production of red blood cells; increases oxygen to your body's cells and brain; aides digestion; promotes proper growth in children; stimulates tissue repair; helps raise the pH of your body to a more alkaline state; helps keep the heart functioning normally; and helps promote the production of friendly flora in your gastrointestinal tract.

Chlorella can be used by everyone. For: fatigue, high or low blood pressure, cardiovascular problems, memory loss, high cholesterol, digestive problems, obesity, headaches, infections, aged skin, toxemia, poor circulation, joint stiffness and pain, sleep disorders, allergies, injuries and overall health.

This species' proteins contain all the amino acids known to be essential for the nutrition of animals and human beings. There are also vitamins found in Chlorella pyrenoidosa including: Vitamin C, pro-vitamin A (B-carotene), thiamine (B_1), riboflavin (B_2), pyridoxine (B_6), niacin, pantothenic acid, folic acid, Vitamin B_{12}, biotin, choline, Vitamin K, lipoic acid and inositol. Minerals in Chlorella pyrenoidosa include: phosphorus, calcium, zinc, iodine, magnesium, iron and copper. Chlorella has a strong cell wall that prevents its native form from being adequately broken down and absorbed by the human digestive system and so special processing is required to break its cell wall.

In addition to amino acids, peptides, proteins, vitamins, sugars and nucleic acids, Chlorella pyrenoidosa contains a water-soluble substance known as Chlorella Growth Factor (CGF). Approximately 5% of raw Chlorella pyrenoidosa is CGF; composed of amino acids, proteins and nucleic acids believed to be derived from the nuclei of the algae. Each Chlorella pyrenoidosa microorganism is composed of a nucleus, starch grains, chloroplasts and mitochondria surrounded by a cell wall composed mainly of cellulose. Under normal conditions, Chlorella divides into four daughter cells in less than 24 hours. The length of Chlorella's life cycle depends on the strength of the sunlight, temperature and availability of nutrients.

Although the algae grow naturally in fresh water, Chlorella pyrenoidosa destined for human consumption is generally cultivated in large, fresh mineral water pools under direct sunlight. The growing process must be carefully inspected and sanitary conditions are meticulously maintained to ensure there is no contamination of the Chlorella with other microorganisms.

Once the fresh-water pools have enough Chlorella cells in them, the algae are harvested and the tough cell walls of the Chlorella must then be broken down to increase the algae's digestibility. This is accomplished with the patented process utilizing the Dyno-Mill unique method developed under the guidance of Mr. Hideo Nakayama of the Sun Chlorella Corporation.

All of the other methods, which include heating or treatment with enzymes, compromise Chlorella's digestibility, therefore eliminating full health benefits of Chlorella. The Dyno-Mill physically disintegrates the cell wall by using only natural, mechanical means and therefore there is no need for chemicals, enzymes or heating that can compromise its nutritional value, while assuring optimum assimilation and digestion. With the Dyno-Mill technique, Chlorella is more than 85% digestible. Once the cell wall has been broken, Chlorella is spray-dried, producing a powder and molded into tablets using a direct press machine.

The final results are solid tablets of pure Chlorella pyrenoidosa. A maintenance dosage of Chlorella tablets and Chlorella liquid extract for those in good health is 15 tablets (3g) and 30 ml. Those with severe medical conditions may increase the daily dosage as much as three times, depending on their specific needs. Chlorella pyrenoidosa affects the immune system by stimulating an increase in the number and activities of macrophages and polymorphonuclear leukocytes.

An acidic polysaccharide prepared from Chlorella cell wall has also been shown to induce the production of interferon in vitro and in mice and therefore, part of Chlorella pyrenoidosa's anti-cancer effect in part may be mediated through the actions of this cytokine. An acidic polysaccharide purified from the hot water extract of Chlorella pyrenoidosa possessed anti-tumor activity against five transplantable murine tumors in vivo.

Complete blood counts, differentials, cytometric determinations of natural killer (NK) cells and T-cell subsets and in vitro lymphocyte activation assays to assess level of immunosuppression, were performed on blood samples, as well as imaging studies of the brain and blood tests, performed at 3-4 month intervals, all showed Chlorella supplementation enhanced the body's immune abilities and slowed the development of tumors, while returning values to normal by around eight months. The algal cell wall of Chlorella pyreneidosa absorbs rather large amounts of toxic metals (similar to an ion exchange resin).

Either the specific combination of amino acids, the Chlorella derived growth factor, or some yet unknown other mechanism leads to the mobilization of mercury from within the cell. It definitely appears to mobilize some mercury inside the brain.

All amino acids can cross the blood brain barrier; however, there is always competition between them for cell sites. The sulfur amino acids Methionine, Cysteine and Cystine are critical for the detoxification of heavy metals and xenobiotics. L-Glutathione is a powerful antioxidant that inhibits the formation of free radicals. It can cross the blood brain barrier and can remove mercury, cadmium and other toxic metals from the brain.

The key substance for nutritional support in mercury detoxification is Chlorella. Chlorella has been shown to effectively sweep mercury out of the bowel and from the cells. Chlorella is also shown to act as an ion exchange resin in your gut, eliminating mercury from your blood.

Chlorella plays a key role in helping patients remove dangerous mercury overloads in their bodies. Chlorella can have a strengthening effect on body cells by supporting the functioning of our metabolic pathways. Chlorella can promote cell reproduction, reduce cholesterol and increase hemoglobin levels.

Since chlorella is such a broad-spectrum product, it can help to support and repair organs and tissues that have been injured by a variety of causes. Numerous research projects in the USA and Europe have indicated that Chlorella can also aid the body in the breakdown of persistent hydrocarbon and metallic toxins such as DDT, PCB, mercury, cadmium and lead as well as strengthening the immune system response.

The fibrous materials in Chlorella will also improve digestion and promote the growth of beneficial aerobic bacteria in the gut. Other research programs have indicated that regular use of chlorella can help to guard against heart disease, reduce high blood pressure and lower serum cholesterol levels. Summing up, it could be said that there is no other green plant under the Sun that is more beneficial to the human body than Chlorella. With the many positive findings of scientific researchers around the world, this food should become an indispensable part of our daily diet so that we can enjoy the many health benefits that it has to offer.

Chlorella binds strongly to cadmium and will not give it up to the body. Blood levels of cadmium were determined and demonstrated that the cadmium that was bound to the Chlorella was not absorbed into the body. Chlorella has been used to detoxify people suffering from P.C.B. (polychloro-biphenyl) exposure. Chlordecone (kepone) another very harmful chlorinated hydrocarbon insecticide, has been shown to be removed, more than twice as fast from the body, when chlorella is taken by mouth.

Chlorella given to rats speeded up the detoxification of this toxin, decreasing the half-life of the toxin from 40 days to 19 days. Chlorella walls absorb and hang onto lead. Cell components extracted from chlorella even bind uranium. Chlorella's ability to detoxify the body is very significant because of the large amount of chemicals we are exposed to in today's modern world. This ability to detoxify chemicals is also one of the important differences between chlorella and other "green" products. As a perfect food, Chlorella has no peers.

SPIRULINA PLANKTON

Spirulina, (rhymes with 'ballerina'), is a traditional food of some Mexican and African peoples. It is a planktonic blue-green algae found in warm water alkaline volcanic lakes. Wild Spirulina sustains huge flocks of flamingos in the alkaline East African Rift Valley Lakes. It possesses an amazing ability to thrive in conditions much too harsh for other algae. As might be expected, it has a highly unusual nutritional profile. Spirulina has a 62% amino acid content, is the world's richest natural source of Vitamin B_{12} and contains a whole spectrum of natural mixed carotene and xanthophyll phytopigments. Spirulina has a soft cell wall made of complex sugars and protein and is different from most other algae in that it is easily digested. Millions of people worldwide eat Spirulina cultivated in scientifically designed algaefarms. Current world production of Spirulina for human consumption is more than one thousand metric tons annually. The United States leads world production followed by Thailand, India and China. More countries are planning production as they realise it is a valuable strategic resource.

Differences between spirulina, chlorella and 'wild' blue green algae. Spirulina is not Chlorella or the blue-green algae harvested from Klamath Lake Oregon. Chlorella, a green micro-algae, is a nutritious food but does not have the same anti-viral, anti-cancer and immune stimulating properties of Spirulina. The Chlorella cell wall is made of indigestible cellulose, just like green grass, while the cell wall of Spirulina is made of complexed proteins and sugars.

The Klamath Lake blue-green algae has the scientific name Aphanizomenon flos-aquae. It is not the same as Spirulina. There are serious concerns about the safety of eating Aphanizomenon flos-aquae, because it can sometimes contain potent nerve toxins.

While the scientific literature is full of information concerning the toxicity of Aphanizomenon flos-aquae and its dangers to humans and animals, there are few, if any, peer reviewed scholarly scientific papers regarding therapeutic benefit. In contrast, the scientific literature is full of information concerning the benefits and safety of humans and animals eating Chlorella and Spirulina.

Anti-Can`cer Effects

Several studies show Spirulina or its extracts can prevent or inhibit cancers in humans and animals. Some common forms of cancer are thought to be a result of damaged cell DNA running amok, causing uncontrolled cell growth. Cellular biologists have defined a system of special enzymes called Endonuclease which repair damaged DNA to keep cells alive and healthy. When these enzymes are deactivated by radiation or toxins, errors in DNA go unrepaired and, cancer may develop. In vitro studies suggest the unique polysaccharides of Spirulina enhance cell nucleus enzyme activity and DNA repair synthesis.

This may be why several scientific studies, observing human tobacco users and experimental cancers in animals, report high levels of suppression of several important types of cancer. The subjects were fed either whole Spirulina or treated with its water extracts.

Strengthens Immune System

Spirulina is a powerful tonic for the immune system. In scientific studies of mice, hamsters, chickens, turkeys, cats and fish, Spirulina consistently improves immune system function. Medical scientists find Spirulina not only stimulates the immune system, it actually enhances the body's ability to generate new blood cells.

Important parts of the immune system, the Bone Marrow Stem Cells, Macrophages, T-cells and Natural Killer cells, exhibit enhanced activity. The Spleen and Thymus glands show enhanced function. Scientists also observe Spirulina causing macrophages to increase in number, become "activated" and more effective at killing germs.

Feeding studies show that even small amounts of

Spirulina build up both the humoral and cellular arms of the immune system. Spirulina accelerates production of the humoral system, (antibodies and cytokines), allowing it to better protect against invading germs.

The cellular immune system includes T-cells, Macrophages, B-cells and the anti-cancer Natural Killer cells. These cells circulate in the blood and are especially rich in body organs like the liver, spleen, thymus, lymph nodes, adenoids, tonsils and bone marrow.

Spirulina up-regulates these key cells and organs, improving their ability to function in spite of stresses from environmental toxins and infectious agents.

Spirulina Phycocyanin Builds Blood

Spirulina has a dark blue-green colour, because it is rich in a brilliant blue polypeptide called Phycocyanin. Studies show that Phycocyanin affects the stem cells found in bone marrow. Stem cells are "Grandmother" to both the white blood cells that make up the cellular immune system and red blood cells that oxygenate the body.

Chinese scientists document Phycocyanin stimulating hematopoiesis, (the creation of blood), emulating the affect of the hormone erythropoetin, (EPO). EPO is produced by healthy kidneys and regulates bone marrow stem cell production of red blood cells. Chinese scientists claim Phycocyanin also regulates production of white blood cells, even when bone marrow stem cells are damaged by toxic chemicals or radiation.

Based on this effect, Spirulina is approved in Russia as a "medicine food" for treating radiation sickness. The Children of Chernobyl suffer radiation poisoning from eating food grown on radioactive soil. Their bone marrow is damaged, rendering them immunodeficient. Radiation damaged bone marrow cannot produce normal red or white blood cells. The children are anemic and suffer from terrible allergic reactions. Children fed just five grams of Spirulina in tablets each day make dramatic recoveries within six weeks. Children not given Spirulina remain ill.

Other Potential Health Benefits

Spirulina is one of the most concentrated natural sources of nutrition known. It contains all the essential amino acids, rich in chlorophyll, beta-carotene and its co-factors and other natural phytochemicals. Spirulina is the only green food rich in GLA essential fatty acid. GLA stimulates growth in some animals and makes skin and hair shiny and soft yet more durable.

GLA also acts as an anti-inflammatory, sometimes alleviating symptoms of arthritic conditions. Spirulina acts as a functional food, feeding beneficial intestinal flora, especially Lactobacillus and Bifidus. Maintaining a healthy population of these bacteria in the intestine reduces potential problems from opportunistic pathogens like E. coli and Candida albicans. Studies show when Spirulina is added to the diet, beneficial intestinal flora increase.

CHARACTERISTICS

Algae are a group of protists that are very diverse in their size, shape and structure, but one thing that all algae has in common that differs from other protozoans is that they are autotrophic. They produce their own carbohydrates through photosynthesis.

These odd little organisms were once classified as plants because they contain chloroplasts and carry on photosynthesis, but they don't have tissue differentiation and therefore can't have structures like roots, stems, or leaves. Algae also have reproductive structures called gametangia. Most algae are aquatic, have flagella and contain pyrenoids.

Structure of Algae

All algae have a thallus–that is the body portion of the organism. This structure is usually haploid, but can be structured in many different ways. There are four different types of algae, based on body structure: unicellular, colonial, filamentous and multicellular. Unicellular algae have a structure that consists of a single cell. Most of this algae is aquatic and compose the phytoplankton. Colonial algae have a structure that consists of groups of cells acting in a coordinated manner. In colonial algae some cells will become specialized to allow the algae to move, feed and reproduce more efficiently. Filamentous algae have a slender, rod-shaped thallus composed of rows of cells joined end to end. Multicellular algae very often have large, complex thalluses. Most multicellular algae have large structures like leaves and stems.

Classification of Algae

Algae are classified into seven phyla based on colour, type of chlorophyll, form of food-storage substance and cell wall composition. Algae in all seven of these phyla contain chlorophyll a, but different types of algae also contain additional forms of chlorophyll. Some phyla also have pigments to give them ceratin colors.

REPRODUCTION OF ALGAE

Reproduction of Unicellular Algae

The reproduction methods of algae differ from species to species. Some species reproduce asexually and other species reproduce sexually.

Asexual Reproduction of Unicellular Algae

Many algae exist as haploid cells. In asexual reproduction these haploid cells divide up to three times by mitosis. This produces between two and eight haploid cells called zoospores. The zoospores develop inside of their parent cells. Later they will break-out of their parent cells, scatter and grow into full-sized algae.

Sexual Reproduction of Unicellular Algae

In sexual reproduction, the haploid cells again divide mitotically and produce "plus" or "minus" gamete cells. These cells look similar but have different chemical composition. To reproduce, a plus and a minus gamete will come into contact with each other, shed their cells walls and fuse to form a diploid zygote. This zygote will then develop a thick protective wall and become a zygospore. This zygospore can remain dormant through harsh environmental conditions. When conditions are favorable, the zygospore's thick wall breaks open and the zoospore emerges. The zoospore then goes though meiosis to form a number of haploid cells that will grow into full-sized algae.

Reproduction of Multicellular Algae

When dealing with multicellular algae, the methods of reproduction differ greatly among the phyla. The reproduction of the organisms in some phyla is very complex, while the reproduction methods of other organisms can be fairly simple.

Sexual Reproduction of Green Algae (Phylum Chlorophyta)

Many types of green algae have specialized cells for producing gametes. These modified calls that produce and store gametes are called unicellular gametangia. The male unicellular gametangium is called the antheridium and produces sperm. Oogonium are the female unicellular gametangium and they produce eggs. The sperm released by the antheridium swims through the surrounding water, finds an oogonium and enters through a small pore to fertilize an egg. After fertilization, a zygote is released from the oogonium and forms a thick wall to become a spore. The zygote within the spore undergoes meiosis and forms four haploid zoospores that are released into the water. Each zoospore will then settle somewhere and divide. One of the new cells becomes a holdfast to secure the algae and the rest divide to form a new filament.

Other filamentous green algae reproduce through conjugation. In this process, two filaments align side-by-side, the walls of adjacent cells dissolve and a conjugation tube forms between the two algae. A plus gamete from one algae travels through the conjugation tube to the other algae which contains a minus gamete. The gametes meet and fertilization occurs. A thick wall is developed by the new zygote and it falls from the parent filament. The resting spore will later produce a new filament. Some green algae have a sexual reproduction cycle characterized by the alternation of generations pattern.

Life cycles that involve the alternation of generations have two multicellular phases. In these phases the algae exist as gametophytes, which produce gametes and as sporophytes, which produce spores. Adult sporophytes have reproduction cells called sporangia that produce haploid zoospore by meiosis. These zoospores are divided by mitosis to create motile spores and settle on rocks to grow into gametophytes.

After developing, the gametophyte produces gametangia (single-celled structure that produces gametes), which in turn form plus and minus gametes. These gametes join to form zygotes. These zygotes are divided through mitosis to produce sporocytes and completing the cycle.

EUKARYOTE

Eukaryote, organism whose cells contain a nucleus, a saclike structure that encloses the cell's hereditary materials. The presence of a nucleus distinguishes eukaryotes from prokaryotes, those simple, one-celled organisms in which the hereditary material floats free within the cell.

Unlike prokaryotes, eukaryotes display a tremendous diversity of form, from complex, single-celled amoebas, diatoms and dinoflagellates to multicellular plants, animals and fungi.

Only the eukaryotic cell is capable of a high degree of specialization and specialization is what makes multicellular organisms possible. Just as banks, post offices and other specialized workplaces are intrinsic to a city, cells tailored for certain jobs are intrinsic to more-complex organisms. Working in concert, specialized cells can create a higher level of organization known as tissues, such as the growing shoot of a plant or the spiny skin of a sea star.

Coordinated tissues form organs and, in animals, these organs combine to form complex organ systems, such as the circulatory, digestive and respiratory systems. The orchestration of these organ systems makes up the organism.

Size and Structure

The complexity of eukaryotic cells is reflected in their size. In general, the diameter of eukaryotic cells, which range in size from 0.01 mm to 1 mm (0.000394 in to 0.0394 in), is 10 to 100 times that of typical prokaryotic cells. An average-sized animal cell measures about 0.020 mm (0.0008 in), about one-fifth the thickness of the page of a book, while a typical plant cell is slightly larger, about 0.035 mm (about 0.0014 in). The eukaryotic cell with the greatest diameter is the ostrich egg, which measures about 120 mm (4.72 in). The longest eukaryote cells on record are the nerve cells that extend 3 m (10 ft) down a giraffe's neck.

Eukaryotes house an assortment of structures, called organelles, within the cytoplasm, a gel-like substance that fills the cell. Like the unique toolbox of a carpenter or electrician, the different organelles house different molecular tools, including the specialized proteins called enzymes needed to accomplish the cell's work. With all the enzymes needed for a particular job clustered in one organelle, the eukaryotic cell can work efficiently.

The largest and most conspicuous organelle is the nucleus. The nucleus encloses and protects the cell's genetic material, deoxyribonucleic acid (DNA), so that it is not damaged by biochemical reactions in the cell. Within the eukaryotic nucleus, DNA is wrapped around specialized proteins called histones, like a thread wound around a series of spools. Each DNA strand and its histones fold back and forth several times to form a compact, stick-shaped structure called a chromosome. Depending on the organism, the nucleus contains from one to over a thousand chromosomes. Surrounding the nucleus is the nuclear envelope, a membrane with numerous pores. The pores, ringed by special protein, regulate the flow of substances into and out of the nucleus.

The most extensive organelle in the cell is the cytoskeleton, a web of protein filaments that branches extensively throughout the cytoplasm and gives the cell its shape. The cytoskeleton proteins, as well as other proteins in the cell, are made by tiny spherical organelles known as ribosomes. Several other important organelles are found in the cells. Among them are the lysosomes, membranous sacs storing enzymes that digest and recycle worn out cell parts; and the mitochondria, sacs where the cell's energy is generated.

The endoplasmic reticulum, another organelle, is an extensive network of membrane folds and tubes that serves in part as the cell's factory floor where large molecules, such as lipids, are manufactured. These large molecules are sent to another organelle, the Golgi apparatus, which consists of layers of membranes where the molecules are modified, sorted and packaged for transport.

Plants, seaweeds and microscopic algae are specialized eukaryotes that carry out photosynthesis, the process by which light is used to convert carbon dioxide and water into sugar. In addition to the organelles described above, photosynthetic eukaryotes contain chloroplasts.

Chloroplasts house the green pigment chlorophyll that is used in photosynthesis. Photosynthetic eukaryotes also may house chromoplasts, sacs of yellow, orange, or red pigments responsible for colour in flowers and fruits. Photosynthetic eukaryotes may also have organelles called amyloplasts, where energy-rich starch is stockpiled.

Passage of materials into and out of the cell is regulated by the plasma membrane. Plant cells and many algal cells are further surrounded by a tough cell wall composed of cellulose, a carbohydrate. Cells of fungi, on the other hand, are surrounded by a cell wall made of chitin, another type of carbohydrate. In addition, on the outside of the plasma membrane, many eukaryotic cells bear either cilia or flagella, slender filaments that propel cells through liquid. These slender filaments enable aquatic organisms, for example, to navigate through the water. Cilia and flagella may also move liquid past a stationary cell. For example, cilia on the cells that line the lungs and trachea push mucus and dirt particles up out of the lung.

CELL DIVISION

Eukaryotes carry out cell division to make the new cells needed for growth, to repair damaged cells and to replace worn out, dying cells. Most eukaryotic cells divide by mitosis, a process that produces two cells with thc same genetic information as the original cell. Single-celled eukaryotes, such as amoebas and diatoms, commonly reproduce by mitosis. In an adult human, mitosis spawns an estimated three million new cells every second, replacing the battered cells that line the digestive tract; dead, sloughed off skin cells; worn out red blood cells; and other cells exposed to constant abrasion or intense use.

Many eukaryotes also undergo a second type of cell division, called meiosis, which is designed for sexual reproduction, the union of male and female sex cells. In meiosis, two cell divisions occur in which the genetic material is rearranged, resulting in four genetically unique cells, each of which contains only half the number of chromosomes as the parent cell. When two cells with half the number of chromosomes unite, the new cell contains the full complement of chromosomes needed to produce the new organism.

OBTAINING NUTRIENTS

To function, eukaryotes need organic molecules: carbohydrates such as sugar and starch; proteins; lipids, which include fats and oils; and nucleic acids such as DNA. Sugars such as glucose are particularly important because eukaryotes use energy from sugar to build proteins, lipids and other organic molecules. Photosynthetic eukaryotes are known as autotrophs, a group that includes plants, seaweeds and microscopic algae, all of which can make their own sugar. Those that must take in sugar from outside sources are called heterotrophs. Among the heterotrophs are many single-celled eukaryotes and all fungi and animals.

Heterotrophic eukaryotes typically absorb the nutrients in food through the plasma membrane. To accomplish this task, they must first break down, or digest, the food. Fungi secrete digestive enzymes onto the surface of their food—often decaying leaves or branches—and then absorb the enzyme-released nutrients across the cell wall and plasma membrane. In contrast, animals first ingest their food into some sort of digestive structure such as the stomach. There, digestive enzymes break down the food and the nutrients are then absorbed into the cells.

Some single-celled eukaryotes, such as amoebas, use a process called endocytosis. In endocytosis, these organisms extrude part of the plasma membrane, scoop up a food particle and drag it into the cell, where they digest it using enzymes within the cell. In these eukaryotes, large waste molecules typically are expelled from the cell by a reverse process called exocytosis. The waste is bundled into a sac called a vesicle and transported to the plasma membrane, where it fuses with the membrane. The waste is then expelled through a hole in the fused membrane. In complex animals, cells generate wastes such as urea when nutrients are broken down within cells. These wastes are transported by blood to the kidneys. The kidneys process the waste and produce urine, which is removed from the body through the bladder. Undigested food travels through the tubelike intestines and is eliminated through the digestive system.

EVOLUTIONARY ORIGINS

Eukaryotes evolved much later than prokaryotes, whose origins date to about 3.5 billion to 3.8 billion years before present. Alga-like fossils from ancient rocks suggest that eukaryotes may have evolved about 2.1 billion years before present. Other fossil remains indicate that eukaryotes were well established 1.6 billion years before present. These fossils, called acritarchs, are hollow spheres that appear to be spores or cysts of eukaryotic algae.

Eukaryotic cells are thought to have evolved from primitive prokaryotes. Evidence for this view is found in the archaea, prokaryotes that resemble both bacteria and eukaryotes. Like bacteria, the archaea lack a nucleus and most other organelles. Like eukaryotes, they display flexible cell membranes and histone proteins and have certain segments of DNA in common. This evidence, along with other molecular studies, leads many scientists to conclude that archaea, bacteria and eukaryotes arose from a common ancestral prokaryote similar to the archaea. However, according to a theory developed by American microbiologist Carl Woese, the archaea, bacteria and eukaryotes may have arisen, not from a single common ancestor, but from a group of genetically diverse, primitive prokaryotes.

The nucleus and endoplasmic reticulum of eukaryotes probably evolved from internal folds of the plasma membrane of prokaryote ancestors. Scientists once thought that mitochondria and chloroplasts arose in the same way. But a theory proposed by American microbiologist Lynn Margulis holds that these organelles arose from certain prokaryotes, which established mutually beneficial relationships with the earliest eukaryotes. In the distant past, these free-living prokaryotes were consumed by early eukaryotes, but managed to survive within the cytoplasm. The autotrophic prokaryotes proved useful to their hosts because they produced glucose through photosynthesis. The heterotrophic prokaryotes also were useful because they could generate the energy source adenosine triphosphate (ATP) for the host. The host in turn provided protection for the engulfed prokaryotes. Over time, the autotrophic prokaryotes developed into chloroplasts and the heterotrophic prokaryotes developed into mitochondria. These organelles still contain their own DNA, with bacteria-like genes and their own ribosomes, relics handed down from their distant bacterial ancestors.

PLANKTON

Plankton, collective term for a variety of marine and freshwater organisms that drift on or near the surface of the water. Their movement depends largely on tides, currents and winds, because they are too small or weak to swim against the currents. That component of the plankton comprising photosynthetic organisms is called the phytoplankton. Important algal groups in the phytoplankton include diatoms, golden algae, green algae and cyanobacteria (formerly known as blue-green algae). The other component of the plankton, the zooplankton, comprises protozoa and small crustaceans, jellyfish, worms and mollusks, together with the eggs and larvae of the many animal species inhabiting marine and fresh waters. Important protozoan groups in the zooplankton are dinoflagellates and foraminifera.

The density of plankton varies, depending on the availability of nutrients and the stability of the water. A liter of lake water may contain more than 500 million planktonic organisms. Marine plankton occasionally becomes so numerous that the organisms colour the water; such sudden population increases are called tides. The so-called red tides are caused by billions of dinoflagellates of various species; such tides are sometimes dangerous, because they can poison both humans and fish. Red tides have occurred off the west coast of Florida and in the coastal waters of New England, southern California, Texas, Peru, eastern Australia, Chile and Japan. In 1946 such a tide killed fish, turtles, oysters and other marine organisms in the Gulf of Mexico near Fort Myers, Florida. The blue-green alga Trichodesmus sometimes, in dying off, imparts a reddish colour to water; the Red Sea is so named because of this.

An estimated 90 percent of all photosynthesis and release of free oxygen takes place in the oceans. Marine phytoplankton is the first link in the vast aquatic food chain. The zooplankton, which feeds on the phytoplankton, is consumed in turn by larger animals such as fish and even by the largest mammal, the blue whale. The high protein content of plankton has stimulated research on it as a potential food source for humans. Phytoplankton is sensitive to ocean water temperature, growing faster in cooler waters. Researchers are concerned that increased ocean temperatures due to global warming could reduce the growth of phytoplankton and thereby impact the aquatic food chain. Because phytoplankton also absorbs atmospheric carbon dioxide, which is a greenhouse gas, reduced amounts of phytoplankton in the oceans could increase the amount of carbon dioxide in the atmosphere and thereby speed up global warming.

PROKARYOTE

Prokaryote, relatively simple unicellular organism, such as a bacterium, characterized by the absence of a nucleus and other specialized cell structures. Scientists distinguish prokaryotes from eukaryotes, which are more complex organisms with cells that contain a nucleus, such as plants and animals. Scientists classify prokaryotes in different ways, depending on the classification system used.

In 1938 American biologist Herbert Copeland proposed that unicellular organisms lacking nuclei be classified in their own kingdom called Kingdom Monera, now called Kingdom Prokaryotae. All bacteria were categorized in this new kingdom. This scheme was the first to establish separate kingdoms for prokaryotes (organisms without nuclei) and eukaryotes (organisms with nuclei).

In the 1970s scientists determined that cyanobacteria, formerly known as blue-green algae, have physical features that make them more closely related to bacteria than to algae. Although the exact classification of cyanobacteria is still under debate, some scientists now classify cyanobacteria in the Kingdom Prokaryotae, while algae remains classified in the Kingdom Protista.

In 1990 American microbiologist Carl Woese proposed that bacteria be divided into two groups, archaebacteria and bacteria, based on their structural and physiological differences. Archaebacteria consist of a small group of primitive anaerobes (organisms that do not require oxygen). They are found in a narrow range of habitats—often in extreme environments such as hydrothermal vents on the deep ocean floor. In contrast, bacteria live in a wide range of environments with or without oxygen, at various temperatures and at various levels of acidity. In some classification systems, the archaebacteria are considered prokaryotes; in other systems they are classified in a category known as the Domain Archaea.

Structure

Prokaryotic cells are relatively small, ranging in size from 0.0001 to 0.003 mm (0.000004 to 0.0001 in) in diameter. With the exception of a few species, prokaryotic cells are surrounded by a protective cell wall. The cell walls of archaebacteria and bacteria contain forms of peptidoglycan, a protein-sugar molecule not present in the cell walls of fungi, plants and certain other eukaryotes.

The archaebacteria cell wall has a more diverse chemical composition than the cell wall of bacteria. Just inside the cell wall of prokaryotes is the plasma membrane, a thin structure that is both flexible and strong.

In both prokaryotes and eukaryotes, the plasma membrane is composed of two layers of phospholipid molecules interspersed with proteins and regulates the traffic that flows in and out of the cell.

The prokaryotic plasma membrane, however, carries out additional functions. It participates in replication of deoxyribonucleic acid (DNA) for cell division and synthesis of adenosine triphosphate (ATP), an energy molecule. In some prokaryotes, the plasma membrane is essential for photosynthesis, the process that uses light energy to convert carbon dioxide and water to glucose.

In the interior of the prokaryotic cell is the cytoplasm, a watery fluid that is rich in dissolved salts, nutrients, enzymes and other molecules. The great majority of the cell's biochemical reactions, which number in the thousands, take place within the cytoplasm. Prokaryotic cells typically have a single molecule of DNA in a closed loop floating free in a region of the cytoplasm called the nucleoid. Many species of prokaryotes also contain DNA in tiny ringlets known as plasmids in the cytoplasm.

Ribosomes, tiny beadlike structures that manufacture proteins, are also located in the cytoplasm. Ribosomes contain ribonucleic acid (RNA), a type of genetic material. The structure of ribosomal RNA in archaebacteria is different than the RNA structure found in bacteria and scientists often use this feature to determine whether an organism belongs to the archaebacteria group or the bacteria group.

With the exception of the ribosomes, prokaryotes lack organelles (specialized structures such as the nucleus, chloroplasts, mitochondria, lysosomes and Golgi apparatus), which are present in eukaryotes. Some photosynthetic archaebacteria and bacteria have internal membranes, extensions of the plasma membrane known as chromatophores or thylakoids, which contain the pigments for photosynthesis.

Some species of prokaryotes form endospores, thick-walled, dehydrated structures that can resist extreme dryness and very high temperatures for long periods of time. Anthrax, tetanus and botulism are diseases caused by endospore-forming bacteria.

Certain prokaryotes, such as the bacteria Salmonella, move independently by using flagella, long structures that rotate in a propeller-like fashion. Prokaryotic flagella consist of intertwined fibrils (small fibers) of the protein flagellin. A prokaryote may have a single flagellum, have a group of flagella at one or both poles of the cell, or be covered with flagella. Many species of prokaryotes also have pili (singular, pilus)—slender, hairlike extensions used for attachment to soil, rocks, teeth, or other structures.

FUNGI

Fungi are eukaryotic organisms distinct from plants and animals and members of several other smaller kingdoms. Common fungi include mushrooms, conks, corals, jellies, puffballs, stinkhorns, morels, cups, truffles, lichens, yeasts, rusts, smuts, bread molds, mildews and molds on bathroom tiles. In 1959, R.H. Whittaker introduced a five-kingdom taxonomy that granted fungi equal status with plants and animals.

The five-kingdom system has been supplanted by a multiple-kingdom classification and species traditionally treated as fungi are now distributed across several kingdoms. Those believed to form a monophyletic lineage are assigned to kingdom Eumycota (often called kingdom Fungi). Mycology, the science devoted to fungi, still covers all traditional fungi.

Characteristics of Fungi

The Eumycota consist of eukaryotic, nonchlorophyllous heterotrophs that absorb nutrients from dead or living organic matter, have cell walls composed of chitin and store excess energy as glycogen. The kingdom contains four phyla: Chytridiomycota, Zygomycota, Ascomycota and Basidiomycota.

All true fungi have a definite cell wall throughout all developmental stages. Fungal cell walls are composed of chitin, the compound also found in arthropod exoskeletons (for example, lobster shells). Most fungi produce a vegetative mycelium (filamentous thallus) composed of hyphae that branch and extend via tip elongation, although some groups (like yeasts) consist only of individual cells. Hyphae (singular, hypha) are tube-like filaments with either single multinucleate cells (coenocytes) that lack septa (cross-walls) separating nuclei, or many septate cells containing one, two, or more nuclei.

Fungal Nutrition: Saprobes, Parasites and Mutualists

A fungal thallus may be as small as a single microscopic cell (baker's yeast, Saccharomyces cerevesiae) or exceedingly large (Armillaria gallica, the several-acre-sized "humungous fungus" reported in 1992 as the world's largest organism). Sporocarp (fruit body, or "mushroom" size) also ranges from microscopic to meters in diameter (Bridgeoporus nobilissimus, an endangered bracket fungus found on noble fir trees).

Not unexpectedly, such a diverse kingdom manifests several different life cycles. Virtually all fungi produce spores. Both asexual and sexual spores may germinate to form vegetative thalli from primary and secondary mycelia. Thalli may be haploid dominant, diploid dominant, or exhibit haplo-diploid alternation of generations. Here it is important not to confuse the chromosomal state of individual nuclei (haploid versus diploid) with the number of nuclei per cell (monokartyotic versus dikaryotic). Fungi are unusual in that they often exhibit dikaryotamy, wherein hyphal cells contain two (usually haploid) nuclei that migrate, multiply and divide together.

Although superficially similar to plants, fungi are probably more closely related to animals. Like animals, fungi lack chlorophyll and do not photo-synthesize, must obtain nutrients from organic sources and store energy as glycogen instead of starch. Unlike animals, however, fungi do not engulf, but rather absorb, their nutrients after breaking them down via enzymatic action, earning them the nickname "absorbotrophs."

Fungi absorb their nutrients in three different ways:

- Saprobes decompose dead organic matter
- Parasites feed on living hosts
- mutualists live in symbiotic unions with other living organisms.

Saprophytic fungi, such as edible meadow mushrooms (Agaricus campestris), shiitake (Lentinula edodes) and oyster mushrooms (Pleurotus ostreatus), decompose dead plant and animal tissue by releasing enzymes from hyphal tips, thereby recycling organic materials back into the surrounding environment. Parasitic fungi also use enzymes to break down living tissue, usually sapping the energy of the host and frequently causing its demise.

Lichens and mycorrhizae are two important mutualistic associations. Lichens represent partnerships between a fungus (mycobiont) and one or more algae (phycobiont). Although there are a few basidiolichens, almost all

Lichen mycobionts are Ascomycota (approximately 20,000 ascolichens described thus far). Mycorrhizae are symbiotic or non to slightly pathogenic fungus-plant unions formed with approximately 85 percent of the vascular plants. Mycorrhizae are identified as ectomycorrhizal, arbuscular mycorrhizal (AM), ericoid, orchid, arbutoid and monotropoid based on anatomical form and association.

Ectomycorrhizal fungi (predominantly basidiomycetes such as boletes, amanitas and coral fungi) form thick mycelial mantles around rootlets of many trees (oaks, firs, pines, poplars) to which they transport water and minerals from the soil, receiving sugars and other organic nutrients in return. AM fungi (in the Zygomycota order Glomales) form an endo-infection by penetrating rootlets to form coils and vesicles or finely branched arbuscules. The last four mycorrhizal types are specific to individual plant groups.

ECOLOGICAL AND ECONOMIC IMPORTANCE

Fungi have a profound biological and economic impact. As decomposers, plant pathogens and symbiotic partners, their ability to grow anywhere, on anything, makes them both beneficial and harmful recyclers of carbon and nitrogen. Beneficially, they are used as food (mushrooms, truffles) and in baking and brewing (yeasts).

They are being developed to detoxify pollutants (soil fungi), control insects (pathogenic Zygomycota) and regulate plant growth (pathogenic Ascomycota). Detrimentally, rusts, smuts and molds cost billions of dollars through crop disease and spoilage while forest pathogens such as the honey mushroom (Armillaria ostoyae) and root-butt rot (Heterobasidion annosum) similarly threaten the timber industry. Some are toxic when eaten, such as the infamous destroying angel (Amanita phalloides).

Natural LSD, a hallucinogen produced by ergot (Claviceps purpurea), is associated with medieval hysterical frenzies produced by consumption of infected grain and the aflatoxin produced by Aspergillus flavus in improperly stored grain is one of the most potent carcinogens yet discovered.

As human and animal pathogens, fungi cause infections that range from the vexing (athlete's foot, yeast infections) to life threatening (histoplasmosis). Fortunately, other fungi (such as Penicillium) have been used to develop modern antibiotics and beneficial immunosuppressants.

Classification

With the introduction of deoxyribonucleic acid (DNA) sequence analyses, previous fungal evolutionary theory is undergoing rapid transformation, not surprising when one considers the extremely fragmentary fungal fossil record. Mycologists generally agree that true fungi (with animals and plants) diverged from the protozoan fungi (for example, slime molds and Oomycota) before fungi and animals diverged from plants.

The chytrids separated from the remaining phyla approximately 550 million years ago, followed by splintering of the Zygomycota from the other two phyla, which share septate hyphae and dikaryotic stages in their life cycles.

The Ascomycota split from the Basidiomycota approximately 400 million years ago, followed by an increase in fungal diversity throughout the Paleozoic. Most yeasts and "ascc-molds" are thought to have evolved in approximately the last 200 million years.

While most plants and vertebrates appear to have been described, most fungi have yet to be discovered or named. The 1995 edition of the Dictionary of Fungi counts 566,360 described species (about 100,000 in the Eumycota) but notes that there may be approximately 1.5 million species of fungi in the world.

Elias Fries (1794–1878), often regarded as the "father of mycology," classified fungi based primarily on spore print colour and sporocarp appearance. Friesian-based nomenclature, reflecting similarity of form (phenetic) rather than genetic relationships, is still used in many field guides. Now, however, taxonomists are integrating molecular and morphological characters to develop natural classifications that more adequately reflect evolutionary relationships. Realizing that fungal taxonomy and nomenclature will remain somewhat fluid until new species and data are analyzed and integrated, most mycologists generally accept the classification below.

Chytridiomycota

The fact that chytrids alone among the Eumycota produce motile zoospores explains why their phylum is sometimes assigned with the flagellate oomycetes to kingdom Chromista. Chytrids possess posteriorly uniflagellate spores, mitochondria with flattened cristae and cell walls composed of glucan and chitin. Among the simplest and smallest fungi, they live as saprobes in water and damp organic-rich habitats, or as parasites on invertebrates, plants and other fungi.

The so-called "frog chytrids" (such as Batrachochytrium dendrobatidis) are implicated in the current worldwide decline of amphibian species. Other chytrids are host-specific "Rumen fungi" (such as Piromyces communis) that thrive in anaerobic conditions in the guts of herbivores, such as cattle and sheep. There are five orders and about 800 species of chytrids recognized thus far.

Unicellular members of the order Chytridiales lack a mycelial stage and consist of a central body with a few rudimentary appendages (haustoria) that attach to and invade the host tissue. Other chytrids, such as the Blastocladiales, develop true mycelia with sporangia and male/female gametangia that produce uniflagellate zoospores.

Chytrid thalli may be either haploid or diploid and some, like the aquatic chytrid Allomyces, exhibit an isomorphic alternation of generations with similar appearing sexual and asexual zoospores.

Zygomycota.

The chytrids and this phylum are assigned the two "Bottom" branches of the fungal evolutionary tree. There are more than 1,000 species in two classes (Trichomycetes, Zygomycetes) and ten orders, representing a diverse assemblage of saprobes, soil fungi, obligate insect and fungal parasites and mycorrhizal formers.

Common representatives of the saprobic order Mucorales include the dung-inhabiting "Cap-thrower" (Pilobolus), the black bread mold (Rhizopus stolonifera) and Phycomyces blakesleeanus, sometimes referred to as the "body in the basement" because of the rapid growth of long, hairlike sporangiophores over a substrate under the right conditions.

Zygomycota are characterized by large, thick-walled, coenocytic zygospores and hyphae with relatively thin walls composed of chitin and chitosan.

Both asexual sporangiospores and sexual zygospores germinate into haploid mycelia, with the hyphae functioning as gametangia during the sexual stage. In Rhizopus, for instance, close proximity of two hyphal strands of different mating types chemically triggers each to grow branches toward the other to form septate suspensor cells and gametangia.

Eventual fusion produces a diploid zygosporangium that

undergoes meiosis to become a thick-walled zygospore with large numbers of haploid nuclei.

Ascomycota.

In addition to most lichens and so-called "imperfect fungi," about 33,000 species of unicellular yeasts, green and black molds, powdery mildews, morels, cup fungi and ascotruffles ("true" truffles) belong to this phylum. The phylum is characterized by ascospores produced within a saclike sporangium called an ascus.

Mycelia (more complex than Zygomycota mycelia) are composed of septate hyphae with chitin-glucan hyphal walls. Most species produce specialized fruiting bodies called ascocarps whose details of structure help define different species, classes, or orders. Nonascocarpic representatives (such as unicellular yeasts and mildews that reproduce primarily by budding) do not form mycelia. Both sexual and asexual reproduction are found within this phylum.

Basidiomycota.

This phylum, which also features septate hyphae and chitin-glucan cell walls, is characterized by basidiospores borne upon a club-like structure called a basidium. Approximately 22,500 species are assigned to three classes: Basidiomycetes, Teliomycetes and Ustomycetes. Basidiomycetes include mushrooms, polypores, crusts, corals, clubs, basidiolichens and jellies, which propel their spores and "gastromycetes" (or "stomach fungi") that passively release their spores (puffballs, basidiotruffles ["false truffles"], stinkhorns and birds' nests). Teliomycetes (rusts) and Ustomycetes (smuts) are obligate parasites of insects or plants. Rusts and smuts have exceedingly complex cycles involving up to five separate spore stages and multiple hosts. This ability to produce spores on different hosts in multiple ways presents a significant economic challenge to agriculture.

Oomycota

Oomycetes (kingdom Chromista) are distinguished from

true fungi by having glucan-cellulose cell walls that only occasionally incorporate small amounts of chitin. These algae-like fungi occur in aquatic or moist terrestrial habitats as single cells or mycelial mats composed of multinucleate, nonseptate hyphae.

Their life cycle generally mirrors that of plants, with a transitory haploid stage. Both resting oospores and motile zoospores are diploid, the latter propelled by two unequal flagella (tinsel type plus whiplash). This phylum contains about 700 species in nine orders, including

The gills of a mushroom. Although superficially similar to plants, fungi are members of a distinct kingdom. the generally saprophytic water molds (Saprolegniales) and the pathogenic Peronosporales and Pythiales. Most water molds are saprophytic, but there are a number of parasites that invade plants (white rusts, downy mildews, tobacco blue mold) or fish.

Among the economically significant wilts, blights and pathogens are Phytophthora infestans (responsible for the Irish potato famine in the 1850s), Plasmopara viticola (the causative agent of downy mildew of grapes) and the fish parasite Saprolegnia parasitica, a twenty-first-century threat to salmon migrating through dams in western North America.

KINGDOM

A comparison of the five kingdoms

Characteristic	Monera	Protista	Plantae	Fungi	Animalia
Internal cell membranes	Absent Present (Prokaryotes)	Present (Eukaryotes)	Present (Eukaryotes)	Present (Eukaryotes)	Present (Eukaryotes)
Cell wall	Present	Present or Absent	Present	Present	Absent
Organization	Unicellular	Unicellular or Multicellular	Multicellular	Mainly Multicellular	multicellular
Mode of nutrition	Autotrophs or Heterotrophs	Autotrophs or Heterotrophs	Autotrophs	Heterotrophs	Heterotrophs
Representative groups	Archaea, eubacteria	Protozoa, algae, slime molds	Mosses, ferns, seed plants	Molds, yeasts, mushrooms	Animals with and without backbones

Kingdom is the highest category in the hierarchical classification of organisms created by Carolus Linnaeus around 1750. Linnaeus recognized two kingdoms, plants and animals, a scheme that worked reasonably well for large multicellular organisms but failed as microscopes revealed diverse unicellular organisms.

In 1959 Robert Whittaker devised a five-kingdom system that maintained kingdoms Plantae and Animalia but added kingdoms Monera, Protista and Fungi. An autotroph is an organism that uses solar energy or energy from inorganic chemicals to make organic molecules. A heterotroph obtains organic molecules by consuming other organisms or their products.

Whittaker placed bacteria in their own kingdom, Monera, because of fundamental organizational differences between prokaryotic bacterial cells, which lack membrane-enclosed nuclei and organelles and the eukaryotic cells of other organisms that possess internal membranes. Plantae, Fungi and Animalia consist of complex, multicellular eukaryotic organisms that differ from each other in details of cell structure and in how they secure and process energy. Protista is a collection of single-celled eukaryotic organisms and simple multicellular forms, some animal-like, some plantlike.

Molecular evidence, particularly from ribosomal ribonucleic acid (RNA), suggests that the five-kingdom scheme is also too simple. Some biologists believe that Protista should be partitioned into three or more kingdoms. Similarly, kingdom Monera contains two very biochemically distinct groups of prokaryotes: archaebacteria and eubacteria.

A proposed system acknowledges this ancient evolutionary split by creating a higher level of classification, domain, above kingdom. This system distinguishes three domains: Archaea, Eubacteria and Eukarya (containing protists, plants, fungi and animals).

PROTOZOA

The word protozoa means "little animal." They are so named because many species behave like tiny animals, they hunt other microbes as food. Protozoa mainly feed on bacteria, but they also eat other protozoa,and sometimes fungi. Some protozoa absorb food through their cell tissues. Others,surround food and engulf it. Others have openings called mouth pores into which they sweep food. All protozoa digest their food in stomach-like compartments.

As they digest, they make and give off nitrogen, which is an element that plants and other higher creatures can use. Protozoa vary in size from 1/5,000 to 1/50 of an inch in diameter. They can be classified into three general groups based on their shape.

One group is the Ciliates, which are generally the largest protozoa. They have hair-like projections called cilia and they eat the other two types of protozoa as well as bacteria. The second group is the Amoebae. The third group is the Flagellates, which are usually the smallest of the protozoa and have one or several long, whip-like projections called flagella extending from their cells.

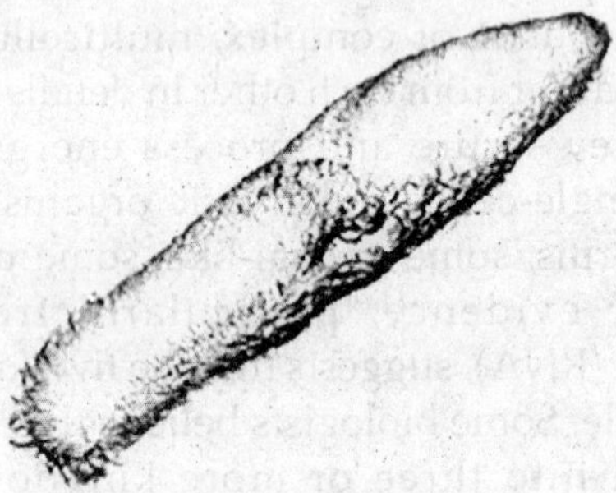

Fig. Protozoa

To hunt, protozoa have to be able to move about. Amoebas ooze about by extending parts of their cells. Amoebae have fluid cell membranes or coverings that they can stretch out, bend and curve. As the membrane moves outward, the fluid and other parts inside the cell follow, flowing into the new bulge created by the moving membrane. Many Ciliates swim along by beating their cilia in a rhythmic pattern, like so many tiny oars. Flagellates swim by waving their flagella, using them much like a fish uses its tail to push itself through water. Most protozoa do us no harm. But there are a few that cause diseases. One type of amoeba can live in human intestines. It feeds on red blood cells and causes a disease known as dysentery. Another species of protozoa can sicken hundreds of thousands of people when it gets into the tap water. Perhaps the best-known deadly protozoa causes malaria, a terrible disease that leads to about 800,000 deaths each year worldwide.

The Protozoa are considered to be a subkingdom of the kingdom Protista, although in the classical system they were placed in the kingdom Animalia. More than 50,000 species have been described, most of which are free-living organisms; protozoa are found in almost every possible habitat. The fossil record in the form of shells in sedimentary rocks shows that protozoa were present in the Pre-cambrian era. Anton van Leeuwenhoek was the first person to see protozoa, using microscopes he constructed with simple lenses.

Between 1674 and 1716, he described, in addition to free-living protozoa, several parasitic species from animals and Giardia lamblia from his own stools. Virtually all humans have protozoa living in or on their body at some time and many persons are infected with one or more species throughout their life. Some species are considered commensals, i.e., normally not harmful, whereas others are pathogens and usually produce disease.

Protozoan diseases range from very mild to life-threatening. Individuals whose defenses are able to control but not eliminate a parasitic infection become carriers and constitute a source of infection for others. In geographic areas of high prevalence, well-tolerated infections are often not treated to eradicate the parasite because eradication would lower the individual's immunity to the parasite and result in a high likelihood of reinfection.

Many protozoan infections that are inapparent or mild in normal individuals can be life-threatening in immunosuppressed patients, particularly patients with acquired immune deficiency syndrome (AIDS). Evidence suggests that many healthy persons harbor low numbers of Pneumocystis carinii in their lungs. However, this parasite produces a frequently fatal pneumonia in immunosuppressed patients such as those with AIDS. Toxoplasma gondii, a very common protozoan parasite, usually causes a rather mild initial illness followed by a long-lasting latent infection. AIDS patients, however, can develop fatal toxoplasmic encephalitis. Cryptosporidium was described in the 19th century, but widespread human infection has only recently been recognized.

Cryptosporidium is another protozoan that can produce serious complications in patients with AIDS. Microsporidiosis in humans was reported in only a few instances prior to the appearance of AIDS. It has now become a more common infection in AIDS patients. As more thorough studies of patients with AIDS are made, it is likely that other rare or unusual protozoan infections will be diagnosed. Acanthamoeba species are free-living amebas that inhabit soil and water. Cyst stages can be airborne. Serious eye-threatening corneal ulcers due to Acanthamoeba species are being reported in individuals who use contact lenses. The parasites presumably are transmitted in contaminated lens-cleaning solution. Amebas of the genus Naegleria, which inhabit bodies of fresh water, are responsible for almost all cases of the usually fatal disease primary amebic meningoencephalitis. The amebas are thought to enter the body from water that is splashed onto the upper nasal tract during swimming or diving. Human infections of this type were predicted before they were recognized and reported, based on laboratory studies of Acanthamoeba infections in cell cultures and in animals.

The lack of effective vaccines, the paucity of reliable drugs and other problems, including difficulties of vector control, prompted the World Health Organization to target six diseases for increased research and training.

Three of these were protozoan infectionsmalaria, trypanosomiasis and leishmania-sis. Although new information on these diseases has been gained, most of the problems with control persist.

Structure

Most parasitic protozoa in humans are less than 50 μm in

size. The smallest (mainly intracellular forms) are 1 to 10 μm long, but Balantidium coli may measure 150 μm. Protozoa are unicellular eukaryotes. As in all eukaryotes, the nucleus is enclosed in a membrane. In protozoa other than ciliates, the nucleus is vesicular, with scattered chromatin giving a diffuse appearance to the nucleus, all nuclei in the individual organism appear alike. One type of vesicular nucleus contains a more or less central body, called an endosome or karyosome. The endosome lacks DNA in the parasitic amebas and trypanosomes. In the phylum Apicomplexa, on the other hand, the vesicular nucleus has one or more nucleoli that contain DNA. The ciliates have both a micronucleus and macronucleus, which appear quite homogeneous in composition.

The organelles of protozoa have functions similar to the organs of higher animals. The plasma membrane enclosing the cytoplasm also covers the projecting locomotory structures such as pseudopodia, cilia and flagella. The outer surface layer of some protozoa, termed a pellicle, is sufficiently rigid to maintain a distinctive shape, as in the trypanosomes and Giardia. However, these organisms can readily twist and bend when moving through their environment. In most protozoa the cytoplasm is differentiated into ectoplasm (the outer, transparent layer) and endoplasm (the inner layer containing organelles); the structure of the cytoplasm is most easily seen in species with projecting pseudopodia, such as the amebas. Some protozoa have a cytosome or cell "mouth" for ingesting fluids or solid particles.

Contractile vacuoles for osmoregulation occur in some, such as Naegleria and Balantidium. Many protozoa have subpellicular microtubules; in the Apicomplexa, which have no external organelles for locomotion, these provide a means for slow movement. The trichomonads and trypanosomes have a distinctive undulating membrane between the body wall and a flagellum. Many other structures occur in parasitic protozoa, including the Golgi apparatus, mitochondria, lysosomes, food vacuoles, conoids in the Apicomplexa and other specialized structures. Electron microscopy is essential to visualize the details of protozoal structure. From the point of view of functional and physiologic complexity, a protozoan is more like an animal than like a single cell. The structure of the bloodstream form of a trypanosome, as determined by electron microscopy.

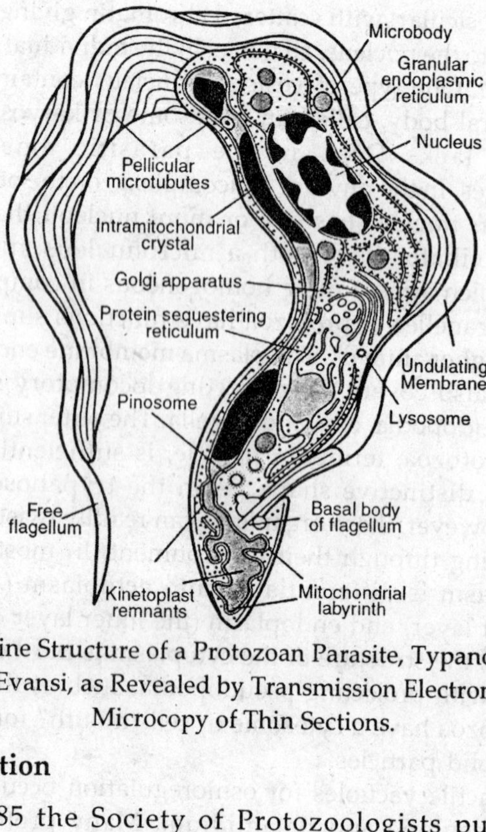

Fig. Fine Structure of a Protozoan Parasite, Typanosoma Evansi, as Revealed by Transmission Electron Microcopy of Thin Sections.

Classification

In 1985 the Society of Protozoologists published a taxonomic scheme that distributed the Protozoa into six phyla. Two of these phylathe Sarcomastigophora and the Apicomplexa—contain the most important species causing human disease. This scheme is based on morphology as revealed by light, electron and scanning microscopy. Dientamoeba fragilis, for example, had been thought to be an ameba and placed in the family Entamoebidae. However, internal structures seen by electron microscopy showed that it is properly placed in the order Trichomonadida of flagellate protozoa. In some instances, organisms that appear identical under the microscope have been assigned different species names on the basis of such criteria as geographic distribution and clinical manifestations; a good example is the genus Leishmania, for which subspecies names are often used. Biochemical methods have been employed on strains and species to determine isoenzyme patterns or to identify relevant nucleotide sequences in RNA, DNA, or both.

Extensive studies have been made on the kinetoplast, a unique mitochondrion found in the hemoflagellates and other members of the order Kinetoplastida. The DNA associated with this organelle is of great interest. Cloning is widely used in taxonomic studies, for example to study differences in virulence or disease manifestations in isolates of a single species obtained from different hosts or geographic regions. Antibodies (particularly monoclonal antibodies) to known species or to specific antigens from a species are being employed to identify unknown isolates. Eventually, molecular taxonomy may prove to be a more reliable basis than morphology for protozoan taxonomy, but the microscope is still the most practical tool for identifying a protozoan parasite.

LIFE CYCLE STAGES

During its life cycle, a protozoan generally passes through several stages that differ in structure and activity. Trophozoite (Greek for "animal that feeds") is a general term for the active, feeding, multiplying stage of most protozoa. In parasitic species this is the stage usually associated with pathogenesis. In the hemoflagellates the terms amastigote, promastigote, epimastigote and trypomastigote designate trophozoite stages that differ in the absence or presence of a flagellum and in the position of the kinetoplast associated with the flagellum.

A variety of terms are employed for stages in the Apicomplexa, such as tachyzoite and bradyzoite for Toxoplasma gondii. Other stages in the complex asexual and sexual life cycles seen in this phylum are the merozoite (the form resulting from fission of a multinucleate schizont) and sexual stages such as gametocytes and gametes. Some protozoa form cysts that contain one or more infective forms. Multiplication occurs in the cysts of some species so that excystation releases more than one organism. For example, when the trophozoite of Entamoeba histolytica first forms a cyst, it has a single nucleus. As the cyst matures nuclear division produces four nuclei and during excystation four uninucleate metacystic amebas appear.

Similarly, a freshly encysted Giardia lamblia has the same number of internal structures (organelles) as the trophozoite. However, as the cyst matures the organelles double and two trophozoites are formed. Cysts passed in stools have a protective wall, enabling the parasite to survive in the outside environment for a period ranging from days to a year, depending on the species and environmental conditions. Cysts formed in tissues do not usually have a heavy protective wall and rely upon carnivorism for transmission. Oocysts are stages resulting from sexual reproduction in the Apicomplexa. Some apicomplexan oocysts are passed in the feces of the host, but the oocysts of Plasmodium, the agent of malaria, develop in the body cavity of the mosquito vector.

REPRODUCTION

Reproduction in the Protozoa may be asexual, as in the amebas and flagellates that infect humans, or both asexual and sexual, as in the Apicomplexa of medical importance. The most common type of asexual multiplication is binary fission, in which the organelles are duplicated and the protozoan then divides into two complete organisms. Division is longitudinal in the flagellates and transverse in the ciliates; amebas have no apparent anterior-posterior axis. Endodyogeny is a form of asexual division seen in Toxoplasma and some related organisms. Two daughter cells form within the parent cell, which then ruptures, releasing the smaller progeny which grow to full size before repeating the process.

In schizogony, a common form of asexual division in the Apicomplexa, the nucleus divides a number of times and then the cytoplasm divides into smaller uninucleate merozoites. In Plasmodium, Toxoplasma and other apicomplexans, the sexual cycle involves the production of gametes (gamogony), fertilization to form the zygote, encystation of the zygote to form an oocyst and the formation of infective sporozoites (sporogony) within the oocyst.

Some protozoa have complex life cycles requiring two different host species; others require only a single host to complete the life cycle. A single infective protozoan entering a susceptible host has the potential to produce an immense population. However, reproduction is limited by events such as death of the host or by the host's defence mechanisms, which may either eliminate the parasite or balance parasite reproduction to yield a chronic infection.

For example, malaria can result when only a few sporozoites of Plasmodium falciparumperhaps ten or fewer in rare instancesare introduced by a feeding Anopheles mosquito into a person with no immunity. Repeated cycles of schizogony in the bloodstream can result in the infection of 10 percent or more of the erythrocytesabout 400 million parasites per milliliter of blood. The mode of reproduction in these organisms is the same as that of the cell generally and takes always the form of fission of some kind; that is to say, of division of the body into smaller portions, each of which repre sents a young individual.

The division of the body is preceded by that of the nucleus, if single, or of each nucleus in_ the cases where there are two different nuclei; if, however, more than one nucleus of the same kind be present, the nuclei may be simply shared amongst the daughter-individuals, this mode of division being known as plasmotomy. Other organs of the body may either, like the nucleus, undergo fission, or may be formed afresh in the daughter-individuals.

The division of the nucleus in Protozoa may take place by the direct method or by means of mitosis. Direct division, without mitosis, is of very common occurrence; the division may be simple or multiple, that is to say, into only two parts, or into a number of fragments formed simultaneously. An extreme case of multiple fission is seen in the formation of the microgametes of Coccidium schubergi, where the nucleus breaks up into a great number of chromidia, which become concentrated in patches to form the several daughter-nuclei. In some cases, on the other hand, multiple daughter-nuclei are formed by rapidly repeated simple division of the parent nucleus. The mode of division may be different in different nuclei of the same individual; thus in the Infusoria the macronucleus divides by direct division, the micronucleus by mitosis. The mitosis of the Protozoa is far from being of the uniform stereotyped pattern seen in the Metazoa, but, as might have been expected, often shows a much simpler and more primitive condition.

Centrosomes are often absent and their place may be taken, as stated above, by other bodies. The nuclear membrane may be retained throughout the mitosis. Definite chromosomes can, as a rule, be made out, but the chromosomes are often very numerous and minute, without definite form and divide irregularly. Much remains to be done in studying the mitosis of the Protozoa, but it is probable that wider knowledge will show many conditions intermediate between direct division and perfect mitosis.

The simplest method of fission in Protozoa is that termed binary, where the body divides into two halves, which may be equal and similar, so that the result is two sister-individuals impossible to distinguish as parent and offspring. In many cases of binary fission, however, the resulting daughter-individuals may be markedly unequal in size, so that one may be distinguished as the parent, the other as the offspring. If the daughterindividual be relatively very small and formed in a more or less imperfect condition at first, the process is termed gemmation or budding.

The buds formed in this way may be either external, formed on the surface of the body, or internal, that is, formed in special internal cavities, from which the offspring are later set free, as in many Acinetaria. Gemmation may be correlated with multiple nuclear fission in such a way that buds are formed over the whole body surface of the organism, which thereby undergoes a process of simultaneous multiple fission into numerous daughter-individuals. Rapid multiple fission of this kind is termed sporulation and is a form of reproduction which is of common occurrence, especially in parasitic forms. Usually the central portion of the parent body remains over as a residual body (Restharper), but sometimes the parent organism is entirely resolved into the daughter-individuals, which are termed spores I The kinetic nucleus of Trypanosoma is sometimes, but in the writer's opinion wrongly, named centrosome or blepharoplast; the bodies to which cytologists give these names are achromatic bodies; the kinetic nucleus is a true chromatic nucleus

NUTRITION

The nutrition of all protozoa is holozoic; that is, they require organic materials, which may be particulate or in solution. Amebas engulf particulate food or droplets through a sort of temporary mouth, perform digestion and absorption in a food vacuole and eject the waste substances. Many protozoa have a permanent mouth, the cytosome or micropore, through which ingested food passes to become enclosed in food vacuoles. Pinocytosis is a method of ingesting nutrient materials whereby fluid is drawn through small, temporary openings in the body wall. The ingested material becomes enclosed within a membrane to form a food vacuole.

Protozoa have metabolic pathways similar to those of higher animals and require the same types of organic and inorganic compounds. In recent years, significant advances have been made in devising chemically defined media for the in vitro cultivation of parasitic protozoa.

The resulting organisms are free of various substances that are present in organisms grown in complex media or isolated from a host and which can interfere with immunologic or biochemical studies. Research on the metabolism of parasites is of immediate interest because pathways that are essential for the parasite but not the host are potential targets for antiprotozoal compounds that would block that pathway but be safe for humans. Many antiprotozoal drugs were used empirically long before their mechanism of action was known. The sulfa drugs, which block folate synthesis in malaria parasites, are one example. The rapid multiplication rate of many parasites increases the chances for mutation; hence, changes in virulence, drug susceptibility and other characteristics may take place. Chloroquine resistance in Plasmodium falciparum and arsenic resistance in Trypanosoma rhodesiense are two examples.

Competition for nutrients is not usually an important factor in pathogenesis because the amounts utilized by parasitic protozoa are relatively small. Some parasites that inhabit the small intestine can significantly interfere with digestion and absorption and affect the nutritional status of the host; Giardia and Cryptosporidium are examples.

The destruction of the host's cells and tissues as a result of the parasites' metabolic activities increases the host's nutritional needs. This may be a major factor in the outcome of an infection in a malnourished individual. Finally, extracellular or intracellular parasites that destroy cells while feeding can lead to organ dysfunction and serious or life-threatening consequences.

MORPHOLOGY

A range of algal morphologies are exhibited, and convergence of features in unrelated groups is common. The only groups to exhibit three dimensional multicellular thalli are the reds and browns, and some chlorophytes.

Apical growth is constrained to subsets of these groups: the florideophyte reds, various browns, and the charophytes. The form of charophytes is quite different to those of reds and browns, because have distinct nodes, separated by internode 'stems'; whorls of branches reminiscent of the horsetails occur at the nodes. Conceptacles are another polyphyletic trait; they appear in the coralline algae and the Hildenbrandiales, as well as the browns.

Most of the simpler algae are unicellular flagellates or amoeboids, but colonial and non-motile forms have developed independently among several of the groups. Some of the more common organizational levels, more than one of which may occur in the life cycle of a species, are

- *Colonial*: small, regular groups of motile cells.
- *Capsoid*: individual non-motile cells embedded in mucilage.
- *Coccoid*: individual non-motile cells with cell walls.
- *Palmelloid*: non-motile cells embedded in mucilage.
- *Filamentous*: a string of non-motile cells connected together, sometimes branching.
- *Parenchymatous*: cells forming a thallus with partial differentiation of tissues.

Fig. Kelp Forest Exhibit at the Monterey Bay Aquarium

In three lines even higher levels of organization have been reached, with full tissue differentiation. These are the brown algae,—some of which may reach 50 m in length (kelps)—the red algae, and the green algae.

The most complex forms are found among the green algae, in a lineage that eventually led to the higher land plants. The point where these non-algal plants begin and algae stop is usually taken to be the presence of reproductive organs with protective cell layers, a characteristic not found in the other alga groups.

SYMBIOTIC ALGAE

Some species of Algae form symbiotic relationships with other organisms. In these symbioses, the algae supply photosynthates (organic substances) to the host organism providing protection to the algal cells. The host organism derives some or all of its energy requirements from the algae. Examples are as follows.

Lichens

Lichens are defined by the International Association for Lichenology to be "an association of a fungus and a photosynthetic symbiont resulting in a stable vegetative body having a specific structure." The fungi, or mycobionts, are from the Ascomycota with a few from the Basidiomycota. They are not found alone in nature but when they began to associate is not known.

One mycobiont associates with the same phycobiont species, rarely two, from the Green Algae, except that alternatively the mycobiont may associate with the same species of Cyanobacteria (hence "photobiont" is the more accurate term).

A photobiont may be associated with many specific mycobionts or live independently; accordingly, lichens are named and classified as fungal species. The association is termed a morphogenesis because the lichen has a form and capabilities not possessed by the symbiont species alone (they can be experimentally isolated). It is possible that the photobiont triggers otherwise latent genes in the mycobiont.

Coral Reefs

Coral reefs are accumulated from the calcareous exoskeletons of marine invertebrates of the Scleractinia order; i.e., the Stony Corals. As animals they metabolize sugar and oxygen to obtain energy for their cell-building processes, including secretion of the exoskeleton, with water and carbon dioxide as byproducts. As the reef is the result of a favorable equilibrium between construction by the corals and destruction by marine erosion, the rate at which metabolism can proceed determines the growth or deterioration of the reef.

Algae of the Dinoflagellate phylum are often endosymbionts in the cells of marine invertebrates, where they accelerate host-cell metabolism by generating immediately available sugar and oxygen through photosynthesis using incident light and the carbon dioxide produced in the host. Endosymbiont algae in the Stony Corals are described by the term zooxanthellae, with the host Stony Corals called on that account hermatypic corals, which although not a taxon are not in healthy condition without their endosymbionts. Zooxanthellae belong almost entirely to the genus *Symbiodinium*. The loss of *Symbiodinium* from the host is known as coral bleaching, a condition unless corrected leading to the deterioration and loss of the reef.

Sea Sponges

Green Algae live close to the surface of some sponges, for example, breadcrumb sponge (*Halichondria panicea*). The alga is thus protected from predators; the sponge is provided with oxygen and sugars which can account for 50 to 80% of sponge growth in some species.

LIFE-CYCLE

Rhodophyta, Chlorophyta and Heterokontophyta, the three main algal Phyla, have life-cycles which show tremendous variation with considerable complexity. In general there is an asexual phase where the seaweed's cells are diploid, a sexual phase where the cells are haploid followed by fusion of the male and female gametes.

Asexual reproduction is advantageous in that it permits efficient population increases, but less variation is possible. Sexual reproduction allows more variation but is more costly because of the waste of gametes that fail to mate, among other things. Often there is no strict alternation between the sporophyte and gametophyte phases and also because there is often an asexual phase, which could include the fragmentation of the thallus.

DISTRIBUTION

The topic of distribution of algal species has been fairly well studied since the founding of phytogeography in the mid-19th century AD. Algae spread mainly by the dispersal of spores analogously to the dispersal of Plantae by seeds and spores. Spores are everywhere in all parts of the Earth: the waters fresh and marine, the atmosphere, free-floating and in precipitation or mixed with dust, the humus and in other organisms, such as humans. Whether a spore is to grow into an organism depends on the combination of the species and the environmental conditions.

The spores of fresh-water Algae are dispersed mainly by running water and wind, as well as by living carriers. The bodies of water into which they are transported are chemically selective. Marine spores are spread by currents. Ocean water is temperature selective, resulting in phytogeographic zones, regions and provinces.

To some degree the distribution of Algae is subject to floristic discontinuities caused by geographical features, such as Antarctica, long distances of ocean or general land masses. It is therefore possible to identify species occuring by locality, such as "Pacific Algae" or "North Sea Algae". When they occur out of their localities, it is usually possible to hypothesize a transport mechanism, such as the hulls of ships. For example, *Ulva reticulata* and *Ulva fasciata* travelled from the mainland to Hawaii in this manner.

Mapping is possible for select species only: "there are many valid examples of confined distribution patterns." For example, *Clathromorphum* is an arctic genus and is not mapped far south of there. On the other hand, scientists regard the overall data as insufficient due to the "difficulties of undertaking such studies."

ECOLOGY

Algae are prominent in bodies of water, common in terrestrial environments and are found in unusual environments, such as on snow and on ice. Seaweeds grow mostly in shallow marine waters, under 100 metres (330 ft); however some have been recorded to a depth of 360 metres (1,200 ft). The various sorts of algae play significant roles in aquatic ecology. Microscopic forms that live suspended in the water column (phytoplankton) provide the food base for most marine food chains. In very high densities (algal blooms) these algae may discolor the water and outcompete, poison, or asphyxiate other life forms. Algae are variously sensitive to different factors, which has made them useful as biological indicators in the Ballantine Scale and its modifications.

Fertilizer

For centuries seaweed has been used as a fertilizer; George Owen of Henllys writing in the 16th century referring to drift weed.

This kind of ore they often gather and lay on great heapes, where it heteth and rotteth, and will have a strong and loathsome smell; when being so rotten they cast on the land, as they do their muck, and thereof springeth good corn, especially barley. After spring-tydes or great rigs of the sea, they fetch it in sacks on horse backes, and carie the same three, four, or five miles, and cast it on the lande, which doth very much better the ground for corn and grass. Today Algae are used by humans in many ways; for example, as fertilizers, soil conditioners and livestock feed. Aquatic and microscopic species are cultured in clear tanks or ponds and are either harvested or used to treat effluents pumped through the ponds. Algaculture on a large scale is an important type of aquaculture in some places. Maerl is commonly used as a soil conditioner.

Nutrition

Naturally growing seaweeds are an important source of food, especially in Asia. They provide many vitamins including: A, B1, B2, B6, niacin and C, and are rich in iodine, potassium, iron, magnesium and calcium. In addition commercially cultivated microalgae, including both Algae and Cyanobacteria, are marketed as nutritional supplements, such as Spirulina, Chlorella and the Vitamin-C supplement, Dunaliella, high in beta-carotene.

Algae are national foods of many nations: China consumes more than 70 species, including a Cyanobacterium considered a vegetable: *fat choy*; Japan, over 20 species; Ireland, dulse; Chile, cochayuyo. Laver is used to make "laver bread" in the

British Isles; in Korea, gim; in Japan, nori and aonori. It is also used along the west coast of North America from California to British Columbia, in Hawaii and by the Maoris of New Zealand. Sea lettuce and badderlocks are a salad ingredient in Scotland, Ireland, Greenland and Iceland.

The oils from some Algae have high levels of unsaturated fatty acids. For example, Arachidonic acid is very high in *Parietochloris incisa,* where it reaches up to 47% of the triglyceride pool. Some varieties of Algae favored by vegetarianism and veganism contain the long-chain, essential omega-3 fatty acids, Docosahexaenoic acid (DHA) and Eicosapentaenoic acid (EPA), in addition to vitamin B12. Fish oil contains the omega-3 fatty acids, but the original source is algae, which are eaten by marine life such as copepods and are passed up the food chain.

Pollution Control

- Sewage can be treated with algae, reducing the need for greater amounts of toxic chemicals than are already used.
- Algae can be used to capture fertilizers in runoff from farms. When subsequently harvested, the enriched algae itself can be used as fertilizer.
- Algae Bioreactors are used by some powerplants to reduce CO_2 emissions.

MOTILITY AND THE CYTOSKELETON OF PROTOZOA

The earliest observations of protozoa noted their motility. This motility resulted in their classification as 'animals', which were distinguished from the non-motile 'plants'. However, motility is not a universal feature of protozoa and different protozoa utilize different mechanisms for their movement. In fact, protozoa were initially classified based in part on their mechanism of motility.

Cilia and flagella are subcellular structures which propel protozoa through a fluid medium. Flagella are long whip-like structures which propel the organism as a result of wave-like beat which is propagated through their length. Flagellated protozoa typically have one or a few flagella per organism. In contrast, ciliated protozoa are usually covered with rows of numerous cilia.

Modes of Motility

Mechanism	Subgroup
Ameboid Movement	Amebas
Flagella	Glagellates
Cilia	Ciliates
Gliding Motility	Sporozoa

The beats of these cilia are coordinated and function like oars to propel the organism. Cilia and flagella can also assist in the procurement of food, reproduction and other functions. Cilia and flagella are made up of the same protein components and are actually equivalent structures. Both are membrane bound filamentous projections from the cell.

The filament, known as an axoneme, is composed of a series of parallel microtubules, typically exhibiting a '9 + 2' arrangement. Movement is produced when the microtubules slide past each other. The force which mediates this sliding motion is generated by a protein called dynein. Dyneins are 'motor proteins' which convert the chemical energy released by ATP hydrolysis into a mechanical energy. Microtubules are cytoskeletal elements which also play important roles in cell shape and are a major component of the mitotic spindle.

In contrast to the swimming exhibited by flagellates and ciliates, ameba are protozoa that crawl along a solid substratum in a fashion known as 'ameboid movement'. The ameba projects out a pseudopodium, or false foot, from the cell body. The pseudopodium then attaches to the substratum and then pulls the rest of the cell body forward. The force involved in this movement is generated by another cytoskeletal system, which is comprised of actin and myosin. Actin forms long filaments, also known as microfilaments, and myosin is a motor protein which moves along the microfilaments in an ATP dependent manner. Muscle contraction is another example of the force generation via actin-myosin cytoskeletal elements. In a mechanistic sense, phagocytosis is a form of ameboid movement also involving microfilaments. In this case the pseudopodia are extended to surround the particle being ingested. Fusion of the pseudopodia with the cell body results in the internalization of the particle within a vacuole.

Apicomplexa also crawl along a substratum, but by a different mechanism than the ameba. The mechanism of this so-called 'gliding motility' is just beginning to be understood and probably involves both microfilament and microtubule based cytoskeletal systems. Apicomplexa also exhibit intracellular forms and invasion of the host cell also involves this gliding motility.

Cellular motility involves force generation through either the microtubule-based cytoskeletal elements or the microfilament-based cytoskeletal elements. This is true for protozoa and other eukaryotes. The involvement of microtubules and microfilaments in both cell shape and cell movement make these subcellular structures more analogous to the musculoskeletal system.

Taxonomy [sh]

Taxonomy, or systematics, is the science of naming and classifying organisms. In addition to assigning hierarchical taxonomic classifications, systematics also attempts to place organisms into groups reflect evolutionary relationships or phylogenies. However, taxonomic criteria are often arbitrary and taxonomy is always changing to reflect new discoveries and interpretations. Furthermore, utilitarian features, such as type of disease, host range and geographical distribution, are frequently used in the systematics of pathogenic micro-organisms.

This is especially true for protozoan taxonomy. In addition, there is some debate on the overall philosophy in the classification of protozoa and the relationships between many protozoan groups are not known. The classification of protozoa is further complicated by the original definition of protozoa being unicellular heterotrophs. It is now recognized that protozoa (or protists) can utilize multiple nutritional strategies and cannot be regarded as simply either plant-like (autotroph) or animal-like (heterotroph). Thus, the term protozoa cannot be regarded as a true taxonomic group. Despite the fact that the word protozoa per se is no longer a proper taxonomic name, it is still a useful and functional term.

Historically protozoa were divided into four major groups: the ameba, the flagellates, the ciliates, and the sporozoa. The distinguishing features between the groups was based on motility (i.e., ameboid, flagella, cilia). The sporozoa were a heterogeneous group that produced spores during one stage of their life cycles and exhibited a 'gliding' motility. However, such a classification scheme is quite arbitrary and does not necessarily reflect true evolutionary relationships between organisms. One problem with using motility as a taxonomic criteria is that many protozoa utilize different types of motility during different stages of their life cycles. For example, *Naegleria* exists in an ameba form when food is plentiful and transforms into a flagellate when food is absent.

In general, the ameba are a heterogeneous group and are all probably dervived from flagellates. Among these four original protozoan groups only the ciliates are still considered a valid taxon. Beginning the 1960's the electron microscope was used to identify ultrastructural features which could serve as criteria for grouping protozoa. In many cases morphology leads to a classification which places organisims into monophyletic groups. Monophyletic means that all of the organisms in that group are probably derived from a common ancestor. For example, many of the protozoa formerly called sporozoa possess subcellular structures, collectively known as the apical organelles, and now form a monophyletic group called apicomplexa. However, subcellular structures and metabolic pathways can be lost in some lineages and placing those descendants can be problamatic.

During the 1980's and continuing until present time molecular techniques are being applied to taxonomy. Possible evolutionary histories and relationships can be derived by comparing DNA or protein sequences. Molecular sequence data has confirmed phylogenies based on other criteria, settled some debates, and led to a few surprises. For example, molecular data confirms that the apicomplexa are monophyletic, and furthermore, indicates that they are related to the ciliates and dinoflagellates. These three groups are now combined into a larger monophyletic group called alveolata. This relationship had been previously suspected and the name is in reference to morphological structures known as alveolar sacs.

But the use of single genes can be an unreliable means of determining evolutionary relationships, especially among distantly related organisms, and molecular data should be interpreted with caution. Since the mid-1980's classification of protists has been in a state of flux and afflicted with some philosophical controversies. On one hand, there is some argument for retaining elements of the Bütschlian scheme based on motility due to its familiarity and simplicity. However, these schemes are often in conflict with phylogenetic data. In addition, the hierarchies and ranks of the traditional Linnaean systematics (i.e., phylum, class, order, family, genus, species) does not always fit well with micro-organisms, and it is often difficult to decide which hierarchial level is most appropriate for any particular protozoan group.

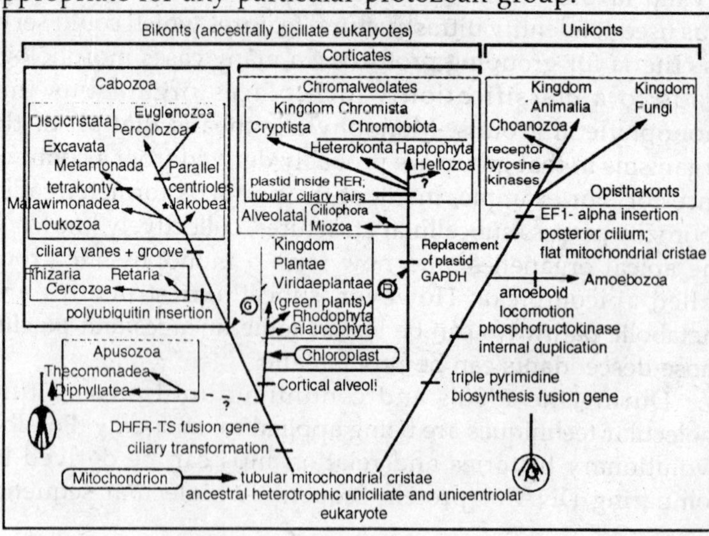

Fig. Classification of Eukaryotes into 5 Kingdoms

Previously many taxonomic schemes have defined five kingdoms of life: prokaryotes (bacteria), protists, plants, fungi, and animals. In these schemes the protozoa are part of the Protista along with unicellular algae, diatoms, oomycetes and slime molds. However, there has always been dissatisfaction with the protist group. This is in part due to protista being defined in part by negative criterion. In other words, organisms that do not fit in the other four kingdoms are defaulted into the protista. In addition, some protists are phylogenetically more closely related to the other three eukaryotic kingdoms than to other protists, and thus the protists are clearly polyphyletic.

Cavalier-Smith proposes five eukaryotic kingdoms consisting of the basal, and thus paraphyletic, kingdom Protozoa and four derived kingdoms: Animalia, Fungi, Plantae, and Chromista. Some of the former protista are now included in the derived kingdoms that they are most closely related to resulting in the kingdom Protozoa becoming monophyletic. However, the alveolates, which include many traditional protozoa such as the apicomplexa, ciliates, and dinoflagellates, form a clade with the chromista, but are nonetheless placed in the protozoan kingdom. Thus the chromista are not holophyletic. Moving the alveolates into a new kingdom called Chromalveolata would solve this problem. However, this would probably result in controversy and confusion since many of the alveolates have been long considered to be protozoa.

Chapter 4

Industrial Use of Microbes

INDUSTRIAL MICROBES

The enslavement and death of countless millions of workers may result from a newly discovered method of producing glycerine (currently produced as a byproduct of soap manufacture). Whether the process reported by Canada's National Research Council in Ottawa will be adopted commercially depends largely on the efficiency of the workers' digestive apparatus. In any case, no one will protest their exploitation. The workers are microscopic members of the clan Bacillus subtilis.

The use of microbes as minuscule chemical factories has been practiced, if not understood, since the first butter was churned, the first wine pressed, the first beer brewed. Spurred by advances in the field of biochemistry and the pressures of two wars, the employment of microbe labour has recently spread to a whole new field of chemical manufacture. Little is known of the actual metabolic process by which microbes work, but by careful control and endless experiment chemists and biologists, working together, have been able to set them hundreds of chemical tasks.

Best known of these industrial microorganisms is yeast, whose appetite for the carbohydrates in beets, sugar cane, wood and other fibrous vegetable matter made possible the production in 1944 of about 638 million gallons of alcohol—grain and wood. But the yeasts are only one group of the microbic multitude able to perform specific jobs. Bacteria resembling the bacilli of human ailments and molds like mildew have also been put to work in industry. Some microbes have done double jobs. One, fed in a certain way, yields oxalic acid, basic chemical of the blueprint industry; on a different diet it produces the gluconic acid used in medicines. The versatile Clostridium acetobutylicum, on a single diet of corn mash, produces acetone for solvents, butanol for automobile lacquers and riboflavin (Vitamin 62).

Grasshoppers and Cantaloupes. In a constant search for new microscopic workers, industrial, university and Government researchers have isolated over 200,000 varieties. Useful microbes may turn up anywhere—in the air, on the water, on forest leaf mold, in city garbage cans. A potent industrial bacillus was discovered in the intestines of a grasshopper. The best strain of the mold Penicillium notatum, which makes life-saving penicillin, was first noticed on a cantaloupe rind. Despite their humble origins, microbes are as temperamental as coloraturas.

They are fussy about temperature, about food and about the company they keep. Some like plenty of air, some like none and some govern their behaviour according to whether they get it or not. In an airless, quiet place, yeast will produce wine; in air it just reproduces itself. To keep such un reliable workers healthy, happy and productive is the responsibility of a growing new species of industrial scientist, the biological engineer. His job: to reproduce on a factory scale biological processes and conditions none too easy to control in a laboratory test tube.

Biological Engineering. Leaders in the development of this field have been technicians of such firms as Brooklyn's Charles Pfizer & Co., one of the oldest of the fermentation chemical makers and a big producer of penicillin. They have devised techniques that smack less of a factory than of the contagion ward of a hospital.

The production of penicillin, which is typical of most fermentation processes, starts from a small culture of the mold growing on a mixture of water, sugar and some nitrogenous food. A 10,000-gal. vat of nutrient may be charged with as much as 100 gal. of this mixture, which has been kept sterilized so that no impurity can possibly be present. Should another, different microbe find his way in, the batch would be spoiled. In theory, all that is necessary to start production is to introduce the microscopic microbe into the large vat. In practice, the vats are inoculated with the larger cultures, to save time, but at the end of three days the biological engineer may find the clear, brownish liquid filled with the clusters of strawlike gold that he had hoped to produce. Or he may find only an accumulation of horrible smells.

Microbe Mysteries. The biologist is often completely in the dark as to why his microbes fail so dismally, when they do. Cross contamination of one culture by another is a common cause. To prevent it at Pfizer's, no two chemicals are made in the same building. Workers with coughs or sneezes are kept away and those going in & out are thoroughly sterilized and dressed like hospital interns. Filtered air and ultra violet light are constantly at work.

In spite of all precautions, the microbes find ways to flout their keepers. A simon-pure culture may sometimes, apparently from sheer perversity, change its nature. After such a mutation, it may produce a better product, a worse product, or an entirely different product. Biologists are no more certain of the reason for this than they are of the nature of the sub-microscopic organisms which plague the microbes themselves. Absenteeism among microbe workers is no problem, but in at least one factory devoted to the manufacture of essential chemical solvents, all work had to be stopped for days because the microbes themselves got sick.

PRODUCTION AND USE OF MICROBIAL ENZYMES FOR DAIRY PROCESSING

Indians are known to be lovers of milk and its products. As a sequel to white revolution, India has surged ahead to become the largest milk producer in the world, the production figure touched to 74 million tons for the year 1998. With growing urbanization, demand for processed dairy foods has increased considerably, in particular demand for different cheese varieties and low-lactose milk due to increasing intolerance of human beings to lactose in milk and other milk products. For improving the quality of milk and milk products, a number of different enzymes from microbial as well as from nonmicrobial sources have potential applications in dairy processing.

The use of rennet in cheese manufacture was among the earliest applications of exogenous enzymes in food processing, dating back to approximately 6000 B C. In 1994, the total production of cheese was 8000 metric tons against a total demand of 9000 metric tons. The projected demand by 2000 A D is around 30,000 metric tons. The use of rennet, as an exogenous enzyme, in cheese manufacture is perhaps the largest single application of enzymes in food processing.

In recent years, proteinases have found additional applications in dairy technology, for example in acceleration of cheese ripening, modification of functional properties and preparation of dietic products. Principal among some enzymes that have important and growing applications are lipases and b-galactosidases.

Enzymes with limited applications include glucose oxidase, superoxide dismutase, sulphydryl oxidase, etc. The increasing use of enzymes to produce specific products with characteristic attributes can be emphasized by the world-wide sale of industrial enzymes approximating to US $ 1.6 billion, which is expected to reach US $ 3.0 billion by the year 2008. Almost 45% of this is shared by the food industry and the remaining is shared by detergent (34.4%), textile (11%), leather (2.8%), pulp and paper (1.2%) industries and other industries (5.6%) excluding enzymes for use in diagnostics and therapeutics. India being the highest producer of milk in the world and consequently the surplus availability of milk in our country has triggered the food and dairy industry to convert the liquid milk into value-added products using biochemical and enzymatic processes.

Microbial Rennets in Dairy Applications

Animal rennet (bovine chymosin) is conventionally used as a milk-clotting agent in dairy industry for the manufacture of quality cheeses with good flavour and texture. Owing to an increase in demand for cheese production world wide – i.e. 4% per annum over the past 20 years, approximating 13.533 million tons – coupled with reduced supply of calf rennet, has therefore led to a search for rennet substitutes, such as microbial rennets. At present, microbial rennet is used for one-third of all the cheese produced world wide.

Rennin acts on the milk protein in two stages, by enzymatic and by nonenzymatic action, resulting in coagulation of milk. In the enzymatic phase, the resultant milk becomes a gel due to the influence of calcium ions and the temperature used in the process. Many microorganisms are known to produce rennet-like proteinases which can substitute the calf rennet. Microorganisms like Rhizomucor pusillus, R. miehei, Endothia parasitica, Aspergillus oryzae and Irpex lactis are used extensively for rennet production in cheese manufacture. Extensive research that has been carried out so far on rennet substitutes has been reviewed by several authors. Different strains of species of Mucor are often used for the production of microbial rennets.

Whereas best yields of the milk-clotting protease from Rhizomucor pusillus are obtained from semisolid cultures containing 50% wheat bran, R. miehei and Endothia parasitica are well suited for submerged cultivation. Using the former, good yields of milk-clotting protease may be obtained in a medium containing 4% potato starch, 3% soybean meal and 10% barley. During growth, lipase is secreted together with the protease. Therefore, the lipase activity has to be destroyed by reducing the pH, before the preparation can be used as cheese rennet.

Microbial rennets from various microorganisms (marketed under the trade names such as Rennilase, Fromase, Marzyme, Hanilase, etc.) being marketed since the 1970s have proved satisfactory for the production of different kinds of cheese. The molecular and enzymatic properties of chymosins have been studied extensively. Although the proteolytic specificities of the three commonly used fungal rennets are considerably different from those of calf chymosin, these rennets have been used to produce acceptable cheeses. Recently Novo Nordisk has succeeded in expressing just one proteolytic enzyme from the fungus R. miehei in the well-known organism A. oryzae.

This host organism is able to produce the single protease that cleaves the casein into a glycomacropeptide and para casein by hydrolysing only at the phe105–met106 peptide bond between phenyl alanine and methionine. This monocomponent enzyme product has the trade name Novoren. One major drawback of microbial rennet use in cheese manufacture, is the development of off flavour and bitter taste in the nonripened as well as in the ripened cheeses. The rennets from microbial sources are more proteolytic in nature in comparison to rennet from animal sources, resulting in production of some bitter peptides during the process of cheese ripening. Hence, attempts have been made to clone the gene for calf chymosin and to express it in selected bacteria, yeasts and molds.

Recombinant Rennets for Cheese Manufacture

Due to shortage of calf stomachs and the economic value of cheese rennet, gene for calf chymosin was one of the first genes for mammalian enzymes that was cloned and expressed in microorganisms. Many different laboratories have cloned the gene for calf prochymosin in Escherichia coli and analysed the structure of the gene as well as the properties of the recombinant chymosin. The expressed proenzyme in E. coli, is present mainly as insoluble inclusion bodies, comprised of reduced prochymosin as well as molecules that are interlinked by disulphide bridges. After disintegration of the cells, inclusion bodies are harvested by centrifugation.

The individual laboratories have reported some differences in the procedure for renaturation of prochymosin from the inclusion bodies, but all have followed the same general scheme. The enzymatic properties of recombinant E. coli chymosin are indistinguishable from those of native calf chymosin. The enzymes were identical when observed by immunodiffusion in gels, but a slight difference was observed by enzyme linked immunosorbent assay (ELISA).

Table. Recombinal Chymosin Proparations at/or Approaching Legal and Commercial Acceptance.

Sourcc of DNA	Praducing microorganisms	Prducing company (brand namc)
Calf abomasums	klayveromyces lactis	Gist Brocades (Maxiren)
Calf abomasums	Aspergillus niger	Genccor/chr. Hansen (Chyrnoger)
Synthetic	Escherichic coli	Pfizer (Chymax)

The gene for prochymosin has also been cloned in Saccharomyces cerevisiae; the levels of expression have been reported to be 0.5 to 2.0% of total yeast protein. In yeast, about 20% of the prochymosin can be released in soluble form which can be activated directly; the remaining 80% is still associated with the cell debris.

The cloning of chymosin was carried out without the prosequence. Though the level of chymosin mRNA was similar to that of prochymosin mRNA, no milk-clotting activity was observed in clones containing the chymosin gene only. The results suggest that the prosequence is essential for correct folding of the polypeptide chain.

The zymogen for the aspartic proteases from R. pusillus, also called mucor rennin, has likewise been cloned and expressed in yeast. Studies on its conversion to active form showed that secretion of R. pusillus protease from recombinant yeast was dependent on glycosylation of the enzyme.

Compared to yeast, the filamentous fungi generally secrete larger quantities of proteins into the culture medium. Furthermore, filamentous fungi secrete the heterologous proteins with correct folding of the polypeptide chain and process correct pairing of sulphydryl groups.

The gene for R. miehei protease has been expressed in A. oryzae. Prochymosin has been expressed in Kluyveromyces lactis, A. nidulans, A. niger and Trichoderma reesei. In most of the cases, the reported yields of the model systems were about 10–40 mg of enzyme per litre of culture medium.

However, 3.3 g of enzyme per liter has also been achieved. Yeast, Kluyveromyces lactis, has recently been used as an efficient host for the secretion of recombinant chymosin, which has led to the development of large-scale production process for chymosin. If produced on an industrial scale, the yields will perhaps be of the latter magnitude. Chymosins which are approaching or have marketing and legal acceptance and generally regarded as safe (GRAS).

Most of the companies produce recombinant rennet of cattle calf origin in different microbial hosts, however in India, the major source of milk is buffalo, showing a different composition from that of cow. The rennet from buffalo source has inherent compatibility for clotting buffalo milk.

Therefore, National Dairy Research Institute, Karnal, has taken a lead in cloning gene of buffalo chymosin. The buffalo chymosin cDNA has been cloned in E. coli and partial N-terminal sequencing of the purified buffalo chymosin indicates that it is highly homologous to the cattle chymosin.

Several cheese making experiments have been carried out with recombinant chymosin and the general aspects of recombinant chymosin have been dealt with in a report. Since most of the rennet (> 90 %) added to cheese milk is lost in the whey, immobilization would considerably extend its catalytic life. Several rennets have been immobilized, but their efficiency as milk coagulants has been questioned.

So, there is a fairly general support for the view that immobilized enzymes cannot coagulate milk properly owing to inaccessability of the Phe–Met peptide bond of k-casein and that the apparent coagulating activity of immobilized rennets is due to leaching of the enzyme from the support.

Different types of conventional cheeses have been successfully made by using recombinant rennet on an experimental or pilot scale. No major differences have been detected between cheeses made with recombinant chymosins or natural enzymes, regarding cheese yield, texture, smell, taste and ripening. The recombinant chymosins are identical with calf rennet according to the report on biochemical and genetic evidences.

Lactase in Dairy Industry

Lactose, the sugar found in milk and whey and its corresponding hydrolase, lactase or b -galactosidase, have been extensively researched during the past decade. This is because of the enzyme immobilization technique which has given new and interesting possibilities for the utilization of this sugar. Because of intestinal enzyme insufficiency, some individuals and even a population, show lactose intolerance and difficulty in consuming milk and dairy products.

Hence, low-lactose or lactose-free food aid programme is essential for lactose-intolerant people to prevent severe tissue dehydration, diarrhoea and, at times, even death. Another advantage of lactase-treated milk is the increased sweetness of the resultant milk, thereby avoiding the requirement for addition of sugars in the manufacture of flavoured milk drinks.

Manufacturers of ice cream, yoghurt and frozen desserts use lactase to improve scoop and creaminess, sweetness and digestibility and to reduce sandiness due to crystallization of lactose in concentrated preparations. Cheese manufactured from hydrolysed milk ripens more quickly than the cheese manufactured from normal milk.

Technologically, lactose crystallizes easily which sets limits to certain processes in the dairy industry and the use of lactase to overcome this problem has not reached its fullest potential because of the associated high costs. Moreover, the main problem associated with discharging large quantities of cheese whey is that it pollutes the environment. But, the discharged whey could be exploited as an alternate cheap source of lactose for the production of lactic acid by fermentation. The whey permeate, which is a by-product in the manufacture of whey protein concentrates, by ultrafiltration could be fermented efficiently by Lactobacillus bulgaricus.

Lactose can be obtained from various sources like plants, animal organs, bacteria, yeasts (intracellular enzyme), or molds. Some of these sources are used for commercial enzyme preparations. Lactase preparations from A. niger, A. oryzae and Kluyveromyces lactis are considered safe because these sources already have a history of safe use and have been subjected to numerous safety tests. The most investigated E. coli lactase is not used in food processing because of its cost and toxicity problems.

Properties of Lactase

The properties of the enzyme depend on its source. Temperature and pH optima differ from source to source and with the type of particular commercial preparation. Immobilization of the enzymes, method of immobilization and type of carrier can also influence these optima values. In general, fungal lactase have pH optima in the acidic range 2.5–4.5 and yeast and bacterial lactases in the neutral region 6–7 and 6.5–7.5, respectively.

The variation in pH optima of lactases makes them suitable for specific applications, for example fungal lactases are used for acid whey hydrolysis, while yeast and bacterial lactases are suitable for milk (pH 6.6) and sweet whey (pH 6.1) hydrolysis. Product inhibition, e.g. inhibition by galactose, is another property which also depends on the source of lactase.

The enzyme from A. niger is more strongly inhibited by galactose than that from A. oryzae. This inhibition can be overcome by hydrolysing lactose at low concentrations by using immobilized enzyme systems or by recovering the enzyme using ultrafiltration after batch hydrolysis. Lactase from Bacillus species are superior with respect to thermostability, pH operation range, product inhibition and sensitivity against high-substrate concentration.

Thermostable enzymes, able to retain their activity at 60°C or above for prolonged periods, have two distinct advantages viz. they give higher conversion rate or shorter residence time for a given conversion rate and the process is less prone to microbial contamination due to higher operating temperature. Bacillus species have a pH optima of 6.8 and a temperature optima of 65°C. Its high activity for skim milk and less inhibition by galactose has made it suitable for use as a production organism for lactase.

The enzymatic hydrolysis of lactose can be achieved either by free enzymes, usually in batch fermentation process, or by immobilized enzymes or even by immobilized whole cells producing intracellular enzyme. Although numerous hydrolysis systems have been investigated, only few of them have been scaled up with success and even fewer have been applied at an industrial or semi-industrial level.

Several acid hydrolysis systems have been developed to industry-scale level. Large-scale systems which use free enzyme process have been developed for processing of UHT-milk and processing of whey, using K. lactis lactase (Maxilact, Lactozyme). Several commercial immobilized systems have been developed for commercial exploitation. Snamprogretti process of industrial-scale milk processing technology in Italy is one such working systems.

They make use of fibre-entrapped yeast lactase in a batch process and the milk used is previously sterilized by UHT. For pilot plants, there are three other processes designed and developed to handle milk:

- Gist-Brocades, Rohm GmbH (Germany);
- Sumitomo, Japan. These are continuous processes with short residence times. Processing of whey UF-permeate is accomplished by the system developed by Corning Glass, Connecticut, Lehigh, Valio and Amerace corp. The process by Corning Glass is being applied at commercial scale in the bakers yeast production using hydrolysed whey.

Microbial Enzymes in Accelerated Cheese Ripening

Cheese ripening is a complex process mediated by biochemical and biophysical changes during which a bland curd is developed into a mature cheese with characteristic flavour, texture and aroma. The desirable attributes are produced by the partial and gradual breakdown of carbohydrates, lipids and proteins during ripening, mediated by several agents, viz.

- Residual coagulants,
- Starter bacteria and their enzymes,

- Nonstarter bacteria and their enzymes,
- Indegenous milk enzymes, especially proteinases;
- Secondary inocula with their enzymes.

Proteolysis occurs in all the cheese varieties and is a prerequisite for characteristic flavour development that can be regulated by proper use of the above agents. Cheese ripening is essentially an enzymatic process which can be accelerated by augmenting activity of the key enzymes. This has the advantage of initiating more specific action for flavour development compared to use of elevated temperatures that can result in accelerating undesirable nonspecific reactions and consequently off flavour development.

Enzymes may be added to develop specific flavours in cheeses, for example lipase addition for the development of Parmesan or Blue-type cheese flavours. Attempts to accelerate the multiple secondary flavour-forming-reactions, e.g. Strecker degradation, have been scarce. The pathways leading to the formation of flavour compounds are largely unknown and therefore the use of exogenous enzymes to accelerate ripening is mostly an empirical process. Different microbial enzymes used to accelerate cheese ripening.

Studies on the hydrolysis of whole casein and isolated casein components to observe the kinetics and specificity of aspartic proteases in rennin, pepsin and four microbial rennet substitutes indicated the considerable differences in the reaction velocity and the extent of hydrolysis on the rennet curd yield. The rennet enzymes were active even in the later phases of cheese-ripening.

Proteinases and Peptidases

Proteolysis is characteristic of most cheese varieties and is indispensable for good flavour and textural development. Proteinases used in cheese processing include:

- Plasmin,
- Rennet;
- Proteinases (cell wall and/or intracellular) of the starter and nonstarter bacteria. Approximately 6% of the rennet added to cheese milk remains in the curd after

manufacture and contributes significantly to proteolysis during ripening. Combinations of individual neutral proteinases and microbial peptidases intensified cheese flavour and when used in combinations with microbial rennets reduced the intensity of bitterness caused by the latter. Acid proteases in isolation cause intense bitterness. Various animal or microbial lipases gave pronounced cheese flavour, low bitterness and strong rancidity, while lipases in combination with proteinases and/or peptidases give good cheese flavour with low levels of bitterness. In a more balanced approach to the acceleration of cheese ripening using mixtures of proteinases and peptidases, attenuated starter cells or cell-free extracts (CFE) are being favoured.

Table. Microbial Enzymes used to Ucoclcrate Chocse Ripening

Enzyme	*Source*
Microbial lipases	Aspergillus niger
	A. oryzae
	Rhizomucor miehei
Lactasce	Streptococcus lactis
	Ktuyveromyces sp.
	Escherichia coli
Microbial serline proteinases	A. niger
Neutral pmroteinases	Bacillus subtilis A. oryzae

PROTEOLYTIC ENZYMES OF LACTIC ACID BACTERIA IN FERMENTED MILK PRODUCTS

The proteolytic system of lactic acid bacteria is essential for their growth in milk and contributes significantly to flavour development in fermented milk products. The proteolytic system is composed of proteinases which initially cleaves the milk protein to peptides; peptidases which cleave the peptides to small peptides and amino acids; and transport system responsible for cellular uptake of small peptides and amino acids. Lactic acid bacteria have a complex proteolytic system capable of converting milk casein to the free amino acids and peptides necessary for their growth. These proteinases include extracellular proteinases, endopeptidases, aminopeptidases, tripeptidases and proline-specific peptidases, which are all serine proteases. Apart from lactic streptococcal proteinases, several other proteinases from nonlactostreptococcal origin have been reported.

There are also serine type of proteinases, e.g. proteinases from Lactobacillus acidophillus, L. plantarum, L. delbrueckii sp. bulgaricus, L. lactis and L. helveticus. Aminopeptidases are important for the development of flavour in fermented milk products, since they are capable of releasing single amino acid residues from oligopeptides formed by extracellular proteinase activity.

Other Dairy Enzymes

Other enzymes used for dairy food application include:

- Proteases to reduce allergic properties of cow milk products for infants;
- Lipases for development of lipolytic flavours in speciality cheeses.

The functional properties of milk proteins may be improved by limited proteolysis through the enzymatic modification of milk proteins. An acid-soluble casein, free of off flavour and suitable for incorporation into beverages and other acid foods, has been prepared by limited proteolysis. The antigenicity of casein is destroyed by proteolysis and the hydrolysate is suitable for use in milk-protein-based foods for infants allergic to cow milk.

Lipolysis makes an important contribution to swiss cheese flavours, due mainly to the lipolytic enzymes of the starter cultures. The characteristic peppery flavour of Blue cheese is due to short-chain fatty acids and methyl ketones. Most of the lipolysis in Blue cheese is catalysed by Penicillium roqueforti lipase, with a lesser contribution from indigenous milk lipase.

The NOVO process for production of enzyme modified cheese (EMC) uses medium-aged cheese which is emulsified, homogenized and pasteurized, after which 'palatase' (a lipase from R. miehei) is added, with or without a proteinase and the blend is ripened at a high temperature for one to four days. The mixture is reheated, a paste results which is suitable for inclusion in soups, dips, dressings, or snack foods. EMC technology has been developed to produce a range of characteristic cheese flavours and flavour intensities, for example swiss, blue, cheddar, provolo-nemor or romano, suitable for inclusion at low levels in many products.

The claims that exogenous enzymes are effective in accelerating ripening have not led to their wide-spread use, possibly due to their high cost, difficulties in distributing them uniformly in the curd and the possible danger of over-ripening the cheese. The other minor enzymes having limited applications in dairy processing include glucose oxidase, catalase, superoxide dismutase, sulphydryl oxidase, lactoperoxidase and lysozymes.

Glucose oxidase and catalase are often used together in selected foods for preservation. Superoxide dismutase is an antioxidant for foods and generates H_2O_2, but is more effective when catalase is present. Thermally induced generation of volatile sulphydryl groups is thought to be responsible for the cooked off-flavour in ultra high temperature (UHT) processed milk. Use of sulphydryl oxidase under aseptic conditions can eliminate this defect. The natural inhibitory mechanism in raw milk is due to the presence of low levels of lactoperoxidase (LP), which can be activated by the external addition of traces of H_2O_2 and thiocyanate.

It has been reported that the potential of LP-system and its activation enhances the keeping quality of milk. Cow milk can be provided with protective factors by the addition of lysozyme, making it suitable as an infant milk. Lysozyme acts as a preservative by reducing bacterial counts in milk without affecting the The scope of application of minor enzymes to milk and milk products has been recently reviewed.

The global market for the production of microbial enzymes for use in dairy-products manufacture is considerably large, but is being dominated only by a limited number of enzyme producers. In India the microbial dairy enzymes requirement has been very limited till now. However, with the advent of technological processes for the manufacture of different varieties of milk products, such as cheeses by the State Dairy Federations, Co-operatives and Private Dairy Product Manufacturers like Amul, Vijaya, Verka, Dynamix, Nestle, Smith Kline Beechem, etc., the markets for the sale of such products in megacities and towns has been slowly growing for the past two to three years.

Presently, many of these microbial enzymes, such as microbial rennets and other enzymes, are being imported. Hence, there is a scope for the production of enzymes such as microbial rennet, lactase, proteinases and lipases indigenously. In the near future, the requirement for these enzymes is bound to increase by leaps and bounds, basically due to requirement of value-added dairy products in the country.

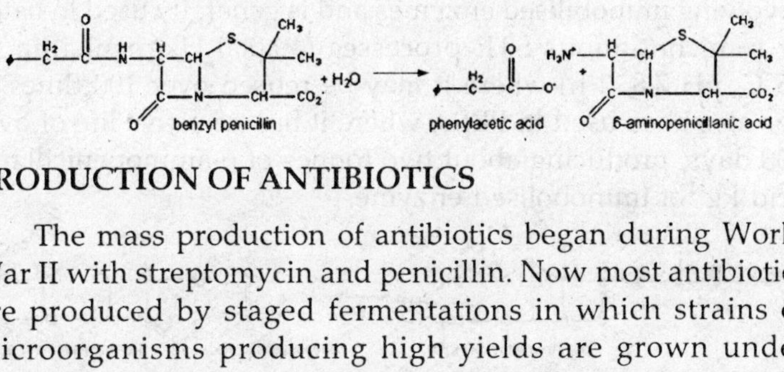

PRODUCTION OF ANTIBIOTICS

The mass production of antibiotics began during World War II with streptomycin and penicillin. Now most antibiotics are produced by staged fermentations in which strains of microorganisms producing high yields are grown under optimum conditions in nutrient media in fermentation tanks holding several thousand gallons.

The mold is strained out of the fermentation broth and then the antibiotic is removed from the broth by filtration, precipitation and other separation methods. In some cases new antibiotics are laboratory synthesized, while many antibiotics are produced by chemically modifying natural substances; many such derivatives are more effective than the natural substances against infecting organisms or are better absorbed by the body, e.g., some semisynthetic penicillins are effective against bacteria resistant to the parent substance.

Benzylpenicillins and phenoxymethylpenicillins

(penicillins G and V, respectively) are produced by fermentation and are the basic precursors of a wide range of semi-synthetic antibiotics, e.g. ampicillin. The amide link may be hydrolysed conventionally but the conditions necessary for its specific hydrolysis, whilst causing no hydrolysis of the intrinsically more labile but pharmacologically essential b-lactam ring, are difficult to attain. Such specific hydrolysis may be simply achieved by use of penicillin amidases (also called penicillin acylases).

Different enzyme preparations are generally used for the hydrolysis of the penicillins G and V, pencillin-V-amidase being much more specific than pencillin-G-amidase. Penicillin amidase may be obtained from E. coli and has been immobilised on a number of supports including cyanogen bromide-activated Sephadex G200.

It represents one of the earliest successful processes involving immobilised enzymes and is generally used in batch or semicontinuous STR processes (40,000 Ukgpenicillin G, 35°C, pH 7.8, 2 h) where it may be reused over 100 times. It has also been used in PBRs, where it has an active life of over 100 days, producing about two tonnes of 6-aminopenicillanic acid $kg^{-1}$of immobolised enzyme.

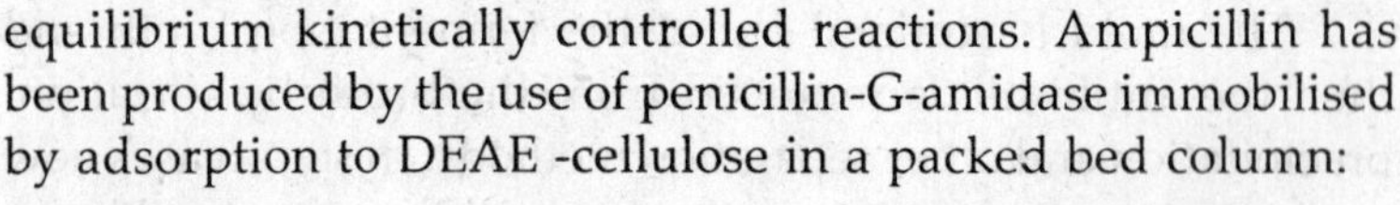

The penicillin-G-amidases may be used 'in reverse' to synthesise penicillin and cephalosporin antibiotics by non - equilibrium kinetically controlled reactions. Ampicillin has been produced by the use of penicillin-G-amidase immobilised by adsorption to DEAE -cellulose in a packed bed column:

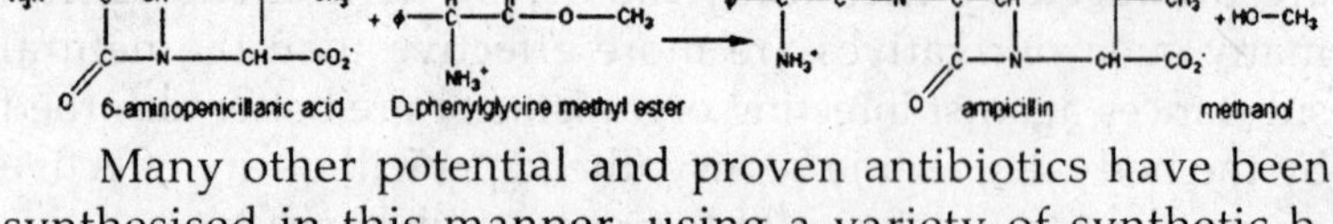

Many other potential and proven antibiotics have been synthesised in this manner, using a variety of synthetic b-lactams and activated carboxylic acids.

PREPARATION OF ACRYLAMIDE

Acrylamide is an important monomer needed for the production of a range of economically useful polymeric materials. It may be produced by the addition of water to acrylonitrile.

$$CH_2{=}CHCN + H_2O \rightarrow CH_2{=}CHCONH_2$$

This process may be achieved by the use of a reduced copper catalyst (Cu^+); however, the yield is poor, unwanted polymerisation or conversion to acrylic acid ($CH_2{=}CHCOOH$) may occur at the relatively high temperatures involved (80 - 140°C) and the catalyst is difficult to regenerate. These problems may be overcome by the use of immobilised nitrile hydratase (often erroneously called a nitrilase). The enzyme from Rhodococcus has been used by the Nitto Chemical Industry Co. Ltd, as it contains only very low amidase activity which otherwise would produce unwanted acrylic acid from the acrylamide.

Immobilised nitrile hydratase is simply prepared by entrapping the intact cells in a cross-linked 10% (w/v) polyacrylamide/dimethylaminoethylmethacrylate gel and granulating the product. It is used at 10°C and pH 8.0-8.5 in a semibatchwise process, keeping the substrate acrylonitrile concentration below 3% (w/v). Using 1% (w/v) immobilised-enzyme concentration (about 50,000 U l^{-1}) the process takes about a day. Product concentrations of up to 20% (w/v) acrylamide have been achieved, containing negligible substrate and less than 0.02% (w/w) acrylic acid. Acrylamide production using this method is about 4000 tonnes per year.

The closely related enzymes cyanidase and cyanide hydratase are used to remove cyanide from industrial waste and in the detoxification of feeds and foodstuffs containing amygdalin.

$$HCN + 2H_2O \rightarrow HCOO^- + NH_4^+$$

$$HCN + H_2O \rightarrow HCONH_2$$

IDENTIFYING USEFUL ANTIBIOTICS

Despite the wide variety of known antibiotics, less than 1% of antimicrobial agents have any medical or commerçial value. The most commonly known antibiotic, Penicillin has a highly selective toxicity and therapeutic index (as eukaryotic animal cells do not contain peptidoglycan, they are usually unaffected by it). This is not so for many antibiotics.

Others simply lack advantage over the antibiotics already in use, or have no other practical applications.In order to identify the useful antibiotics, a process of screening is often employed. Using this method, isolates of a large number of microorganisms are cultured and then tested for production of diffusible products which inhibit the growth of test organisms. However, the majority of the resulting antibiotics are already known and must therefore be disregarded.

The remainders must be tested for their selective toxicities and therapeutic activities and the best candidates can be examined and possibly modified.A more modern version of this approach is a rational design programme. This involves screening being directed towards finding new natural products that inhibit specific targets (e.g. a particular step of a metabolic pathway) on microorganisms, rather than tests to show general inhibition of a culture.

Industrial Production Techniques

Antibiotics are produced industrially by a process of fermentation, where the source microorganism is grown in large containers (100,000–150,000 liters or more) containing a liquid growth medium. Oxygen concentration, temperature, pH and nutrient levels must be optimal and are closely monitored and adjusted if necessary. As antibiotics are secondary metabolites, the population size must be controlled very carefully to ensure that maximum yield is obtained before the cells die. Once the process is complete, the antibiotic must be extracted and purified to a crystalline product. This is simpler to achieve if the antibiotic is soluble in organic solvent. Otherwise it must first be removed by ion exchange, adsorption or chemical precipitation.

Strains used for Production

Microorganisms used in fermentation are rarely identical

to the wild type. This is because species are often genetically modified to yield the maximum amounts of antibiotics. Mutation is often used and is encouraged by introducing mutagens such as ultraviolet radiation, x-rays or certain chemicals. Selection and further reproduction of the higher yielding strains over many generations can raise yields by 20-fold or more. Another technique used to increase yields is gene amplification, where copies of genes coding for proteins involved in the antibiotic production can be inserted back into a cell, via vectors such as plasmids. This process must be closely linked with retesting of antibiotic production and effectiveness.

TYPES OF ANTIBIOTICS

The great number of diverse antibiotics currently available can be classified in different ways, e.g., by their chemical structure, their microbial origin, or their mode of action. They are also frequently designated by their effective range. Tetracyclines, the most widely used broad-spectrum antibiotics, are effective against both Gram-positive and Gram-negative bacteria, as well as against rickettsias and psittacosis-causing organisms. Ciprofloxacin (Cipro) is another broad-spectrum antibiotic, effective in the treatment of mild infections of the urinary tract and sinuses.

The medium-spectrum antibiotics bacitracin, the erythromycins, penicillin and the cephalosporins are effective primarily against Gram-positive bacteria, although the streptomycin group is effective against some Gram-negative and Gram-positive bacteria. Polymixins are narrow-spectrum antibiotics effective against only a few species of bacteria.

Administration and Side Effects

Antibiotics are either injected, given orally, or applied to the skin in ointment form. Many, while potent anti-infective agents, also cause toxic side effects. Some, like penicillin, are highly allergenic and can cause skin rashes, shock and other manifestations of allergic sensitivity. Others, such as the tetracyclines, cause major changes in the intestinal bacterial population and can result in superinfection by fungi and other microorganisms. Chloramphenicol, which is now restricted in use, produces severe blood diseases and use of streptomycin can result in ear and kidney damage. Many antibiotics are less effective than formerly because antibiotic-resistant strains of microorganisms have emerged.

Nonmedical Use

Antibiotics have found wide nonmedical use. Some are used in animal husbandry, along with vitamin B_{12}, to enhance the weight gain of livestock. However, some authorities believe the addition of antibiotics to animal feeds is dangerous because continuous low exposure to the antibiotic can sensitize humans to the drug and make them unable to take the substance later for the treatment of infection.

In addition low levels of antibiotics in animal feed encourage the emergence of antibiotic-resistant strains of microorganisms. Drug resistance has been shown to be carried by a genetic particle transmissible from one strain of microorganism to another and the presence of low levels of antibiotics can actually cause an increase in the number of such particles in the bacterial population and increase the probability that such particles will be transferred to pathogenic, or disease-causing, strains. Antibiotics have also been used to treat plant diseases such as bacteria-caused infections in tomatoes, potatoes and fruit trees. The substances are also used in experimental research.

Vitamin

A vitamin is an organic compound required as a nutrient in tiny amounts by an organism. A compound is called a vitamin when it cannot be synthesized in sufficient quantities by an organism, and must be obtained from the diet. Thus, the term is conditional both on the circumstances and the particular organism. For example, ascorbic acid functions as vitamin C for some animals but not others, and vitamins D and K are required in the human diet only in certain circumstances.

Vitamins are classified by their biological and chemical activity, not their structure. Thus, each "vitamin" may refer to several *vitamer* compounds that all show the biological activity associated with a particular vitamin. Such a set of chemicals are grouped under an alphabetized vitamin "generic descriptor" title, such as "vitamin A," which includes the compounds retinal, retinol, and many carotenoids. Vitamers are often inter-converted in the body. The term *vitamin* does not include other essential nutrients such as dietary minerals, essential fatty acids, or essential amino acids, nor does it encompass the large number of other nutrients that promote health but are otherwise required less often.

Fig. Source of Vitamins

Vitamins have diverse biochemical functions, including function as hormones (e.g. vitamin D), antioxidants (e.g. vitamin E), and mediators of cell signaling and regulators of cell and tissue growth and differentiation (e.g. vitamin A). The largest number of vitamins (e.g. B complex vitamins) function as precursors for enzyme cofactor bio-molecules (coenzymes), that help act as catalysts and substrates in metabolism.

When acting as part of a catalyst, vitamins are bound to enzymes and are called prosthetic groups. For example, biotin is part of enzymes involved in making fatty acids. Vitamins also act as coenzymes to carry chemical groups between enzymes. For example, folic acid carries various forms of carbon group – methyl, formyl and methylene - in the cell. Although these roles in assisting enzyme reactions are vitamins' best-known function, the other vitamin functions are equally important.

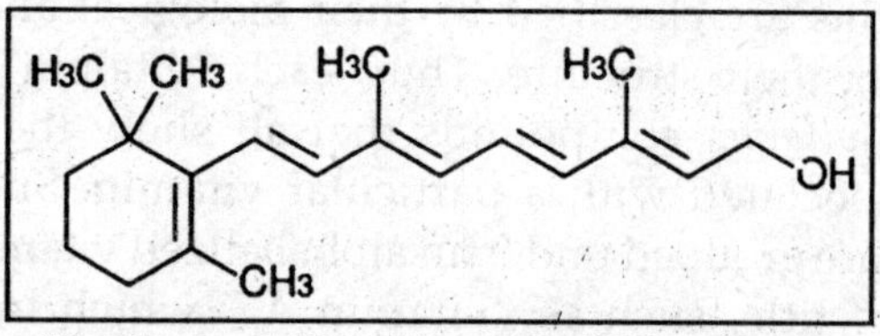

Fig. Retinol (One Vitamer of Vitamin A)

The discovery of vitamins and their sources		
Year of discovery	*Vitamin*	*Source*
1909	Vitamin A (Retinol)	Cod liver oil
1912	Vitamin B_1 (Thiamin)	Rice bran
1912	Vitamin C (Ascorbic acid)	Lemons
1918	Vitamin D (Calciferol)	Cod liver oil
1920	Vitamin B_2 (Riboflavin)	Eggs
1922	Vitamin E (Tocopherol)	Wheat germ oil, Cosmetic and Liver
1926	Vitamin B_{12} (Cyanocobalamin)	Liver
1929	Vitamin K (Phylloquinone)	Alfalfa
1931	Vitamin B_5 (Pantothenic acid)	Liver
1931	Vitamin B_7 (Biotin)	Liver
1934	Vitamin B_6 (Pyridoxine)	Rice bran
1936	Vitamin B_3 (Niacin)	Liver
1941	Vitamin B_9 (Folic acid)	Liver

IN HUMANS

Vitamins are classified as either water-soluble or fat soluble. In humans there are 13 vitamins: 4 fat-soluble (A, D, E and K) and 9 water-soluble (8 B vitamins and vitamin C).

Water-soluble

Water-soluble vitamins dissolve easily in water, and in general, are readily excreted from the body, to the degree that urinary output is a strong predictor of vitamin consumption. Because they are not readily stored, consistent daily intake is important. Many types of water-soluble vitamins are synthesized by bacteria.

Fat-soluble

Fat-soluble vitamins are absorbed through the intestinal

tract with the help of lipids (fats). Because they are more likely to accumulate in the body, they are more likely to lead to hypervitaminosis than are water-soluble vitamins. Fat-soluble vitamin regulation is of particular significance in cystic fibrosis.

NUTRITION AND DISEASES

Vitamins are essential for the normal growth and development of a multicellular organism. Using the genetic blueprint inherited from its parents, a fetus begins to develop, at the moment of conception, from the nutrients it absorbs. It requires certain vitamins and minerals to be present at certain times.

These nutrients facilitate the chemical reactions that produce among other things, skin, bone, and muscle. If there is serious deficiency in one or more of these nutrients, a child may develop a deficiency disease. Even minor deficiencies may cause permanent damage.

For the most part, vitamins are obtained with food, but a few are obtained by other means. For example, microorganisms in the intestine—commonly known as "gut flora"—produce vitamin K and biotin, while one form of vitamin D is synthesized in the skin with the help of the natural ultraviolet wavelength of sunlight. Humans can produce some vitamins from precursors they consume. Examples include vitamin A, produced from beta carotene, and niacin, from the amino acid tryptophan.

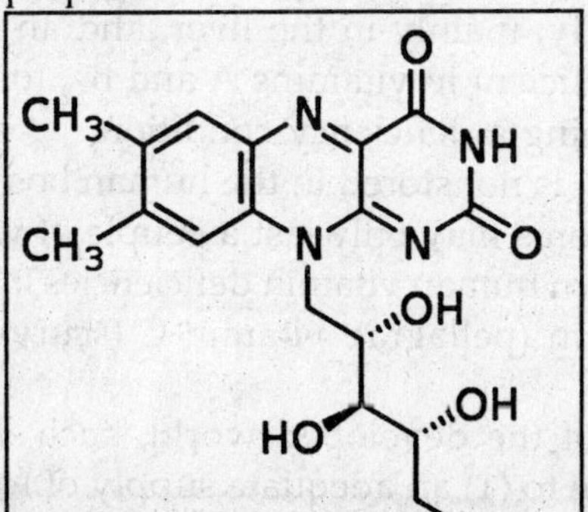

Fig. Riboflavin (Vitamin B_2)

Once growth and development are completed, vitamins remain essential nutrients for the healthy maintenance of the cells, tissues, and organs that make up a multicellular organism; they also enable a multicellular life form to efficiently use chemical energy provided by food it eats, and to help process the proteins, carbohydrates, and fats required for respiration.

Deficiencies

Deficiencies of vitamins are classified as either primary or secondary. A primary deficiency occurs when an organism does not get enough of the vitamin in its food.

A secondary deficiency may be due to an underlying disorder that prevents or limits the absorption or use of the vitamin, due to a "lifestyle factor", such as smoking, excessive alcohol consumption, or the use of medications that interfere with the absorption or use of the vitamin. People who eat a varied diet are unlikely to develop a severe primary vitamin deficiency.

In contrast, restrictive diets have the potential to cause prolonged vitamin deficits, which may result in often painful and potentially deadly diseases. Because human bodies do not store most vitamins, humans must consume them regularly to avoid deficiency.

Human bodily stores for different vitamins vary widely; vitamins A, D, and B_{12} are stored in significant amounts in the human body, mainly in the liver, and an adult human's diet may be deficient in vitamins A and B_{12} for many months before developing a deficiency condition.

Vitamin B_3 is not stored in the human body in significant amounts, so stores may only last a couple of weeks.

Well-known human vitamin deficiencies involve thiamine (beriberi), niacin (pellagra), vitamin C (scurvy) and vitamin D (rickets).

In much of the developed world, such deficiencies are rare; this is due to (1) an adequate supply of food; and (2) the addition of vitamins and minerals to common foods, often called fortification.

Some evidence also suggests that there is a link between vitamin deficiency and mental disorders.

Side Effects and Overdose

In large doses, some vitamins have documented side effects that tend to be more severe with a larger dosage. The likelihood of consuming too much of any vitamin from food is remote, but overdosing from vitamin supplementation does occur. At high enough dosages some vitamins cause side effects such as nausea, diarrhea, and vomiting.

When side effects emerge, recovery is often accomplished by reducing the dosage. The concentrations of vitamins an individual can tolerate vary widely, and appear to be related to age and state of health.

In the United States, overdose exposure to all formulations of vitamins was reported by 62,562 individuals in 2004 (nearly 80% of these exposures were in children under the age of 6), leading to 53 "major" life-threatening outcomes and 3 deaths;a small number in comparison to the 19,250 people who died of unintentional poisoning of all kinds in the U.S. in the same year.

NAMES IN CURRENT AND PREVIOUS NOMENCLATURES

The reason the set of vitamins seems to skip directly from E to K is that the vitamins corresponding to "letters" F-J were either reclassified over time, discarded as false leads, or renamed because of their relationship to "vitamin B", which became a "complex" of vitamins.

The German-speaking scientists who isolated and described vitamin K (in addition to naming it as such) did so because the vitamin is intimately involved in the *Koagulation* of blood following wounding. At the time, most (but not all) of the letters from F through to J were already designated, so the use of the letter K was considered quite reasonable.

The following table lists chemicals that had previously been classified as vitamins, as well as the earlier names of vitamins that later became part of the B-complex:

Previous name	*Chemical name*	*Reason for Name Change*
Vitamin B_4	Adenine	DNA metabolite
Vitamin B_8	Adenylic acid	DNA metabolite
Vitamin F	Essential fatty acids	Needed in large quantities (does not fit the definition of a vitamin).
Vitamin G	Riboflavin	Reclassified as Vitamin B_2
Vitamin H	Biotin	Reclassified as Vitamin B_7
Vitamin J	Catechol, Flavin	Protein metabolite
Vitamin L_1	Anthranilic acid	Protein metabolite
Vitamin L_2	Adenylthiomethylpentose	RNA metabolite
Vitamin M	Folic acid	Reclassified as Vitamin B_9
Vitamin O	Carnitine	Protein metabolite
Vitamin P	Flavonoids	No longer classified as a vitamin
Vitamin PP	Niacin	Reclassified as Vitamin B_3
Vitamin U	S-Methylmethionine	Protein metabolite

VACCINE

A vaccine is a biological preparation that establishes or improves immunity to a particular disease. Vaccines can be prophylactic (e.g. to prevent or ameliorate the effects of a future infection by any natural or "wild" pathogen), or therapeutic. The term *vaccine* derives from Edward Jenner's 1796 use of the term *cow pox* which, when administered to humans, provided them protection against smallpox.

The early vaccines were inspired by the concept of variolation originating in China, in which a person is deliberately infected with a weak form of smallpox as a form of inoculation.

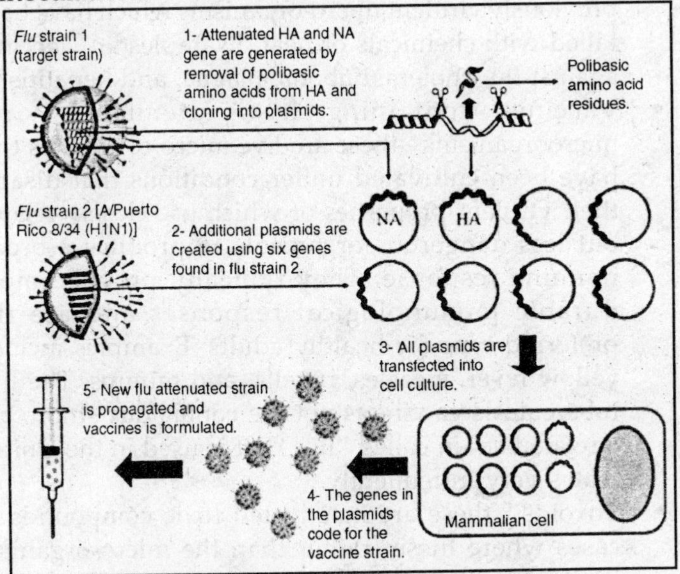

Fig. Vian Flu Vaccine Development by Reverse Genetics Techniques

Jenner observed that milkmaids who had contact with cowpox did not get smallpox. He discovered that deliberate vaccination with cowpox (which has very mild effect in humans) would prevent smallpox (which is often fatal). Jenner's work was continued by Louis Pasteur and others in the 19th century.

Since vaccination against smallpox was much safer than

smallpox inoculation, the latter fell into disuse and was eventually banned in England in 1849. The 19th and 20th centuries saw the introduction of several successful vaccines against a number of infectious diseases. These included bacterial and viral diseases, but not (to date) any parasitic diseases.

Types

Vaccines may be dead or inactivated organisms or purified products derived from them. There are four types of traditional vaccines:

- Vaccines containing killed microorganisms - these are previously virulent micro-organisms which have been killed with chemicals or heat. Examples are vaccines against flu, cholera, bubonic plague, and hepatitis A.
- Vaccines containing live, attenuated virus microorganisms - these are live micro-organisms that have been cultivated under conditions that disable their virulent properties or which use closely-related but less dangerous organisms to produce a broad immune response. They typically provoke more durable immunological responses and are the preferred type for healthy adults. Examples include yellow fever, measles, rubella, and mumps. The live tuberculosis vaccine is not the contagious strain, but a related strain called "BCG"; it is used in the United States very infrequently.
- Toxoids - these are inactivated toxic compounds in cases where these (rather than the micro-organism itself) cause illness. Examples of toxoid-based vaccines include tetanus and diphtheria. Not all toxoids are for micro-organisms; for example, Crotalis atrox toxoid is used to vaccinate dogs against rattlesnake bites.
- Protein subunit - rather than introducing an inactivated or attenuated micro-organism to an immune system (which would constitute a "whole-agent" vaccine), a fragment of it can create an immune response. Characteristic examples include the subunit

vaccine against Hepatitis B virus that is composed of only the surface proteins of the virus (produced in yeast) and the virus-like particle (VLP) vaccine against human papillomavirus (HPV) that is composed of the viral major capsid protein.

A number of innovative vaccines are also in development and in use:

- Conjugate - certain bacteria have polysaccharide outer coats that are poorly immunogenic. By linking these outer coats to proteins (e.g. toxins), the immune system can be led to recognize the polysaccharide as if it were a protein antigen. This approach is used in the *Haemophilus influenzae* type B vaccine.
- Recombinant Vector - by combining the physiology of one micro-organism and the DNA of the other, immunity can be created against diseases that have complex infection processes
- DNA vaccination - in recent years a new type of vaccine, created from an infectious agent's DNA called *DNA vaccination*, has been developed. It works by insertion (and expression, triggering immune system recognition) into human or animal cells, of viral or bacterial DNA. Some cells of the immune system that recognize the proteins expressed will mount an attack against these proteins and cells expressing them. Because these cells live for a very long time, if the pathogen that normally expresses these proteins is encountered at a later time, they will be attacked instantly by the immune system. One advantage of DNA vaccines is that they are very easy to produce and store. As of 2006, DNA vaccination is still experimental.

While most vaccines are created using inactivated or attenuated compounds from micro-organisms, synthetic vaccines are composed mainly or wholly of synthetic peptides, carbohydrates or antigens. Vaccines may be *monovalent* (also called *univalent*) or *multivalent* (also called *polyvalent*). A monovalent vaccine is designed to immunize against a single antigen or single microorganism. A multivalent or polyvalent vaccine is designed to immunize against two or more strains of the same microorganism, or against two or more microorganisms. In certain cases a monovalent vaccine may be preferable for rapidly developing a strong immune response.

EFFECTIVENESS

Vaccines do not guarantee complete protection from a disease. Sometimes this is because the host's immune system simply doesn't respond adequately or at all. This may be due to a lowered immunity in general (diabetes, steroid use, HIV infection) or because the host's immune system does not have a B cell capable of generating antibodies to that antigen. Even if the host develops antibodies, the human immune system is not perfect and in any case the immune system might still not be able to defeat the infection.

Adjuvants are typically used to boost immune response. Adjuvants are sometimes called the *dirty little secret* of vaccines [2] in the scientific community, as not much is known about how adjuvants work. Most often aluminium adjuvants are used, but adjuvants like squalene are also used in some vaccines and more vaccines with squalene and phosphate adjuvants are being tested. The efficacy or performance of the vaccine is dependent on a number of factors:

- The disease itself (for some diseases vaccination performs better than for other diseases).
- The strain of vaccine.
- Whether one kept to the timetable for the vaccinations.
- Some individuals are 'non-responders' to certain vaccines, meaning that they do not generate antibodies even after being vaccinated correctly.
- Other factors such as ethnicity or genetic predisposition.

When a vaccinated individual does develop the disease vaccinated against, the disease is likely to be milder than without vaccination. The following are important considerations in the effectiveness of a vaccination program:

1. Careful modelling to anticipate the impact that an immunisation campaign will have on the epidemiology of the disease in the medium to long term.
2. Ongoing surveillance for the relevant disease following introduction of a new vaccine.
3. Maintaining high immunisation rates, even when a disease has become rare.

Types of Vaccine Developments

- *Rational Attenuation.* Specific modifications or deletions of genes that confer virulence reduce or remove the pathogenicity of the microbe while still allowing an immune response to be generated. This type of rational attenuation can be viewed as creating a "live" attenuated vaccine.
- *Vector-mediated Subunit Delivery.* Introducing a non-infectious, non-pathogenic subunit into a live vector can prompt an immune response without presence of the pathogen. This is called *vector-mediated subunit delivery*. For example, rabies surface protein gene has been inserted into vaccinia virus.
- *Virus-like Particles.* Capsid proteins of icosahedral viruses assemble without the presence of a genome. These virus-like particles are antigenically authentic, but non-infectious. This has been used for HPV-16 and HPV-18 vaccines.
- *InVacc-platform* is another advance on the original DNA vaccines and generates new vaccines with greatly enhanced properties. The platform consists of a chain of amino acids inserted into the adenovirus, which ensures that more of the antigens are shown to the immune system.

ECONOMICS OF DEVELOPMENT

One challenge in vaccine development is economic: many of the diseases most demanding a vaccine, including HIV, malaria and tuberculosis, exist principally in poor countries.

Pharmaceutical firms and biotechnology companies have little incentive to develop vaccines for these diseases, because there is little revenue potential. Even in more affluent countries, financial returns are usually minimal and the financial and other risks are great.

Most vaccine development to date has relied on 'push' funding by government, universities and non-profit organizations.

Many vaccines have been highly cost effective and beneficial for public health. The number of vaccines actually administered has risen dramatically in recent decades. This increase, particularly in the number of different vaccines administered to children before entry into schools may be due to government mandates and support, rather than economic incentive.

Many researchers and policymakers are calling for a different approach, using 'pull' mechanisms to motivate industry. Mechanisms such as prizes, tax credits, or advance market commitments could ensure a financial return to firms that successfully developed a HIV vaccine. If the policy were well-designed, it might also ensure people have access to a vaccine if and when it is developed.

INTELLECTUAL PROPERTY

Intellectual property can also be viewed as an obstacle to the development of new vaccines. Because of the weak protection offered through the patent of the final product, the protection of the innovation regarding vaccines is often made through the patent of processes used on the development of new vaccines as well as the protection of secrecy.

Preservatives

Many vaccines need preservatives to prevent serious adverse effects such as the *Staphylococcus* infection that, in one 1928 incident, killed 12 of 21 children inoculated with a diphtheria vaccine that lacked a preservative.

Several preservatives are available, including thiomersal, phenoxyethanol, and formaldehyde. Thiomersal is more effective against bacteria, has better shelf life, and improves vaccine stability, potency, and safety, but in the U.S., the European Union, and a few other affluent countries, it is no longer used as a preservative in childhood vaccines, as a precautionary measure due to its mercury content. Controversial claims have been made that thiomersal contributes to autism; no convincing scientific evidence supports these claims.

DELIVERY SYSTEMS

There are several new delivery systems in development, which will hopefully make vaccines more efficient to deliver. Possible methods include liposomes and *ISCOM*.The latest developments in vaccine delivery technologies have resulted in oral vaccines. A polio vaccine was developed and tested by volunteer vaccinations with no formal training; the results were very positive in that the ease of the vaccines increased dramatically.

With an oral vaccine, there is no risk of blood contamination. Oral vaccines are likely to be solid which have proven to be more stable and less likely to freeze; this stability eliminates the need for a "cold chain": the resources required to keep vaccines within a restricted temperature range from the manufacturing stage to the point of administration, which, in turn, will decrease costs of vaccines.

Finally, a microneedle approach, which is still in stages of development, seems to be the vaccine of the future, the microneedle, which is "pointed projections fabricated into arrays that can create vaccine delivery pathways through the skin".

Plasmids

The use of plasmids has been validated in preclinical studies as a protective vaccine strategy for cancer and infectious diseases. However, the crossover application into human studies has been met with poor results based on the inability to provide clinically relevant benefit. The overall efficacy of plasmid DNA immunization depends on increasing the plasmid's immunogenicity while also correcting for factors involved in the specific activation of immune effector cells.

USE IN NONHUMANS

Vaccinations of animals are used both to prevent their contracting diseases and to prevent transmission of disease to humans. Both animals kept as pets and animals raised as livestock are routinely vaccinated. In some instances, wild populations may be vaccinated. This is sometimes accomplished with vaccine-laced food spread in a disease-prone area and has been used to attempt to control rabies in raccoons.

Where rabies occurs, rabies vaccination of dogs may be required by law. Other canine vaccines include canine distemper, canine parvovirus, infectious canine hepatitis, adenovirus-2, leptospirosis, bordatella, canine parainfluenza virus, and Lyme disease among others.

Trends

Vaccine development has several trends:

- Until now, most vaccines have been aimed at infants and children, but adolescents and adults are increasingly being targeted.
- Combinations of vaccines are becoming more common; vaccines containing five or more components are used in many parts of the world.
- New methods of administering vaccines are being developed, such as skin patches, aerosols via inhalation devices, and eating genetically engineered plants.
- Vaccines are being designed to stimulate innate immune responses, as well as adaptive.
- Attempts are being made to develop vaccines to help cure chronic infections, as opposed to preventing disease.
- Vaccines are being developed to defend against bioterrorist attacks such as anthrax, plague, and smallpox.

Principles that govern the immune response can now be used in tailor-made vaccines against many noninfectious human diseases, such as cancers and autoimmune disorders. For example, the experimental vaccine CYT006-AngQb has been investigated as a possible treatment for high blood pressure.

SINGLE CELL PROTEIN

In essence a single cell protein is a protein extracted from cultured algae, yeasts, or bacteria and used as a substitute for protein-rich foods, especially in animal feeds. Many types of animal feeds contain single cell proteins. Single cell proteins develop when microbes ferment their waste materials. Examples are wood and straw. The problem with extracting single cell proteins from the wastes is the dilution and cost. They are found in very low concentrations, usually less than 5%. Luckily, scientists have developed a way to increase the concentrations. These include: centrifugation, precipitation, coagulation and filtration; these methods can be combined with semi-permeable membranes.

The next problem to be addressed is the cost. The methods to increase the concentrations to adequate levels, and de-watering process require equipment that is expensive and is not feasible for small-scale operations. For most instances the de-watering process that is required for storage and stabilization of single cell proteins is not economical. The single cell protein needs to be dehydrated to approximately 10% moisture content. It can also be condensed and acidified to prevent spoilage. Perhaps the most economical method and the simplest is to feed the product shortly after it is produced.

Microorganisms which are allowed to grow on waste products from agro based industries produce a large amount of proteins and store them in their cell bodies. These organisms are called as single cell proteins. Description: Single cell protein has the potential to be developed into a very large source of supplemental protein that could be used in livestock feeding. In some regions single cell protein could become the principal protein source that is used for domestic livestock, depending upon the population growth and the availability of plant feed protein sources.

This could develop because microbes can be used to ferment some of the vast amounts of waste materials, such as straws; wood and wood processing wastes; food, cannery and food processing wastes; and residues from alcohol production or from human and animal excreta. Producing and harvesting microbial proteins is not without costs, unfortunately. In nearly all instances where a high rate of production would be achieved, the single cell protein will be found in rather dilute solutions, usually less than 5% solids.

Methods available for concentrating include, filtration, precipitation, coagulation, centrifugation, and the use of semi-permeable membranes. These de-watering methods require equipment that is quite expensive and would not be suitable for most small-scale operations. Removal of the amount of water necessary to stabilize the material for storage, in most instances, is not currently economical. Single cell protein must be dried to about 10% moisture, or condensed and acidified to prevent spoilage from occurring, or fed shortly after being produced.

Uses

A single cell protein has the capability to be used and transformed into one of the most immense sources of supplemental proteins in animal, mainly livestock feeds. In come locations the single cell protein could become the main protein source to feed domestic livestock, this however will depend on the population growth and the other protein sources available.

BIOFERTILIZER

After the introduction of chemical fertilizers in the last century, farmers were happy of getting increased yield in agriculture in the beginning. But slowly chemical fertilizers started displaying their ill-effects such as leaching out, and polluting water basins, destroying micro-organisms and friendly insects, making the crop more susceptible to the attack of diseases, reducing the soil fertility and thus causing irreparable damage to the overall system. A number of intellectuals throughout the world started working on the alternatives and found that biofertilizers can help in increasing the yield without causing the damage associated with chemical fertilizers.

The name itself is self explanatory. The fertilizers are used to improve the fertility of the land using biological wastes, hence the term biofertilizers, and biological wastes do not contain any chemicals which are detrimental to the living soil. They are extremely beneficial in enriching the soil with those micro-organisms, which produce organic nutrients for the soil and help combat diseases. The farm produce does not contain traces of hazardous and poisonous materials. Thus those products are accepted across the world as Organic ones. Hence for organic farming the use of biofertilizers is mandatory.

Galaxy of Biofertilizers

Phospho: It releases insoluble phosphorus in soil and fix this phosphorus in clay minerals which is of great significance in agriculture.

Rhizo: Rhizo Bacterial plays a very important role in agriculture by inducing nitrogen fixings nodules on the root of legumes such as peas,beans clove and alfalfa.

Azotobactor: Atmosphere contains 78% nitrogen which is a very important nutrient for plant growth. Azotobactor fixes the atmospheric nitrogen in the soil and make it available to the plants. It protects the roots from other pathogens present in the soil

Trichoderma: It is a non- pathogenic and eco-friendly product. The product is antagonistic hyper parasitic against different pathogens in the field and economically well established biocontrol agent

Composter: (Decomposing Culture): Composter breaks down any organic matter such as dead plants farm yard waste, cattle waste etc. thereby increasing the soil productivity.

Tricho-Card: Trichogramma is an efficient destroyer of eggs of many leaf and flower eaters, stems, fruit, shoot borers etc. It can be used in a variety of crops as well as in horticultural and ornamental plants,such as sugarcane, cotton, brinjal, tomato, corn, jawar, vegetables, citrus, paddy apple etc.

Vermi Compost: It is 100% pure eco-friendly organic fertilizer. This organic fertilizer has nitrogen phosphorus, potassium,organic carbon, sulphur, hormones, vitamins, enzymes and antibiotics which helps to improve the quality and quantity of yield.

It is observed that due to continuous misuse of chemical fertiliser soil losses its fertility and gets salty day by day. To overcome such problems natural farming is the only remedy and Vermi compost is the best solution.

Biocompost: It is eco-friendly organic fertilizer which is prepared from the sugar industry waste material which is decomposed and enriched of with various plants and human friendly bacteria and fungi. Biocompost consists of nitrogen, phosphate solubilizing bacteria and various useful fungi like decomposing fungi, trichoderma viridea which protects the plants from various soil borne disease and also help to increase soil fertility which results to a good quality product to the farmers.

NITROGEN FIXATION

Nitrogen fixation is the process by which nitrogen is taken from its natural, relatively inert molecular form (N_2) in the atmosphere and converted into nitrogen compounds (such as ammonia, nitrate and nitrogen dioxide).Nitrogen fixation is performed naturally by a number of different prokaryotes, including bacteria, actinobacteria, and certain types of anaerobic bacteria. Microorganisms that fix nitrogen are called diazotrophs. Some higher plants, and some animals (termites), have formed associations with diazotrophs. Nitrogen fixation also occurs as a result of non-biological processes. These include lightning, industrially through the Haber-Bosch Process, and combustion.

Biological Nitrogen Fixation

Biological Nitrogen Fixation (BNF) occurs when

atmospheric nitrogen is converted to ammonia by a pair of bacterial enzymes called nitrogenase. The formula for BNF is:

$N_2 + 8H^+ + 8e^{''} + 16\ ATP$ '! $2NH_3 + H_2 + 16ADP + 16\ P_i$

Although ammonia (NH_3) is the direct product of this reaction, it is quickly protonated into ammonium (NH_4^+). In free-living diazotrophs, the nitrogenase-generated ammonium is assimilated into glutamate through the glutamine synthetase/glutamate synthase pathway. In most bacteria, the nitrogenase enzymes are very susceptible to destruction by oxygen (and many bacteria cease production of the enzyme in the presence of oxygen).

Low oxygen tension is achieved by different bacteria by: living in anaerobic conditions, respiring to draw down oxygen levels, or binding the oxygen with a protein such as Leghemoglobin - also spelled leghaemoglobin.

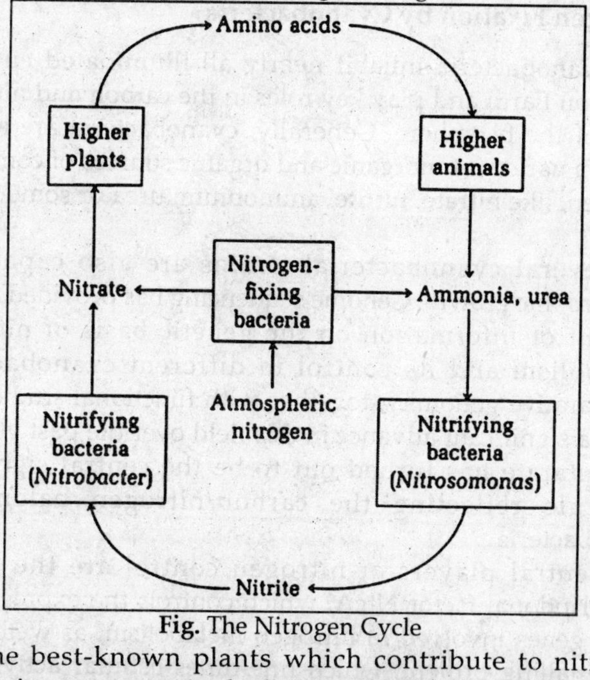

Fig. The Nitrogen Cycle

The best-known plants which contribute to nitrogen fixation in nature, are in the legume family - Fabaceae, which includes such taxa as clover, beans, alfalfa, lupines and peanuts. They contain symbiotic bacteria called rhizobia within nodules in their root systems, producing nitrogen compounds that help the plant to grow and compete with other plants. When the plant dies, the fixed nitrogen is released, making it available to other plants and this helps to fertilize the soil The great majority of legumes have this association, but a few genera (e.g., *Styphnolobium*) do not.

In many traditional and organic farming practices, fields are rotated through various types of crops, which usually includes one consisting mainly or entirely of clover or buckwheat (family Polygonaceae), which were often referred to as "green manure", since the other natural way of adding nitrogen to the soil is via animal waste products. The entire plant is often ploughed back into the field, thus not only adding more nitrogen, but also improving the soil's organic content and volume.

Nitrogen Fixation by Cyanobacteria

Cyanobacteria inhabit nearly all illuminated environments on Earth and play key roles in the carbon and nitrogen cycle of the biosphere. Generally, cyanobacteria are able to utilize a variety of inorganic and organic sources of combined nitrogen, like nitrate, nitrite, ammonium, urea or some amino acids.

Several cyanobacterial strains are also capable of diazotrophic growth. Genome sequencing has provided a large amount of information on the genetic basis of nitrogen metabolism and its control in different cyanobacteria. Comparative genomics, together with functional studies, has led to a significant advance in this field over the past years. 2-oxoglutarate has turned out to be the central signalling molecule reflecting the carbon/nitrogen balance of cyanobacteria.

Central players of nitrogen control are the global transcriptional factor NtcA, which controls the expression of many genes involved in nitrogen metabolism, as well as the P_{II} signalling protein, which fine-tunes cellular activities in response to changing C/N conditions. These two proteins are sensors of the cellular 2-oxoglutarate level and have been conserved in all cyanobacteria. In contrast, the adaptation to nitrogen starvation involves heterogeneous responses in different strains.

CHEMICAL NITROGEN FIXATION

Nitrogen can also be artificially fixed for use in fertilizer, explosives, or in other products. The most popular method is by the Haber process. This artificial fertilizer production has achieved such scale that it is now the largest source of fixed nitrogen in the Earth's ecosystem.

The Haber process requires high pressures and very high temperatures and active research is committed to the development of catalyst systems that convert nitrogen to ammonia at ambient temperatures. Many compounds can react with atmospheric nitrogen under ambient conditions (eg lithium makes lithium nitride if left exposed), but the products of such reactions are not easily converted into biologically accessible nitrogen sources. After the first dinitrogen complex was discovered in 1965 based on ammonia coordinated to ruthenium ($[Ru(NH_3)_5(N_2)]^{2+}$), research in chemical fixation focused on transition metal complexes. Since that time a large number of transition metal compounds that contain dinitrogen as ligand have been discovered. The dinitrogen ligand can be bound either to a single metal or bridge two (or more) metals. The coordination chemistry of dinitrogen is rich and under intense study. This research may lead to new ways of using dinitrogen in synthesis and on an industrial scale.

The first example of homolytic cleavage of dinitrogen under mild conditions was published in 1995. Two equivalents of a molybdenum complex reacted with one equivalent of dinitrogen, creating a triple bonded MoN complex. Since this triple bounded complex has been used to make nitriles . The first catalytic system converting nitrogen to ammonia at room temperature and 1 atmosphere was discovered in 2003 and is based on another molybdenum compound, a proton source and a strong reducing agent. Unfortunately, the catalytic reduction only undergoes a few turnovers before the catalyst dies.

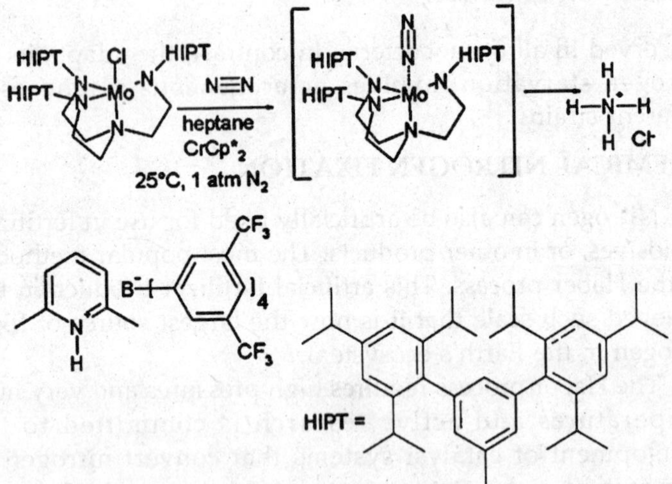

In contrast to the graphic shown above, the major product of this reaction is ammonia (NH_3) and not an ammonium salt ($[NH_4][X]$). In fact, approximately 75% of the ammonia produced can be distilled away from the reaction vessel (suggesting the ammonia is not protonated) into a vessel containing HCl as a trap.

This method of trapping the NH3 was doubtlessly chosen because it makes the product easier to handle. Also, note that because only 1 equiv of Cl anion is available under catalytic conditions therefore it is unlikely that the product ammonium salt would always have this counterion.

Vermiculture

Vermiculture means artificial rearing or cultivation of worms (Earthworms) and the technology is the scientific process of using them for the betterment of human beings. Vermicompost is the excreta of earthworm, which is rich in humus. Earthworms eat cow dung or farm yard manure along with other farm wastes and pass it through their body and in the process convert it into vermicompost.

The municipal wastes; non-toxic solid and liquid waste of the industries and household garbage's can also be converted into vermicompost in the same manner. Earthworms not only convert garbage into valuable manure but keep the environment healthy. Conversion of garbage by earthworms into compost and the multiplication of earthworms are simple process and can be easily handled by the farmers. The goal is to continually increase the number of worms in order to obtain a sustainable harvest. The worms are either used to expand a vermicomposting operation or sold to customers who use them for the same or other purposes.

VERMICOMPOST

Vermicompost (also called compost, vermicast, worm castings, worm humus or worm manure) is the end-product of the breakdown of organic matter by some species of earthworm. Vermicompost is the best nutrient-rich, organic fertilizer and soil conditioner. The process of producing vermicompost is called *vermicomposting*. The earthworm species (or composting worms) most often used are Red Wigglers (*Eisenia foetida*).

These species are commonly found in organic rich soils throughout Europe and North America and especially prefer the special conditions in rotting vegetation, compost and manure piles. Composting worms are available from nursery mail-order suppliers or angling (fishing) shops where they are sold as bait.

Small-scale vermicomposting is well-suited to turn kitchen waste into high-quality soil, where space is limited. Together with bacteria, earthworms are the major catalyst for decomposition in a healthy vermicomposting system, although other soil species also play a contributing role: these include insects, other worms and molds.

Bins

Vermicomposting bins (also known as worm bins) vary drastically depending on the desired size of the system. Garden-scale containers can be made out of bricks arranged in the shape of a box.

Bins for an apartment or similar dwelling can be anything from reused plastic buckets to purpose-built commercial containers.

Small Scale

Small-scale systems may use a wide variety of bins. Often, small-scale composters build their own bins. Companies also sell such bins. Commonly, bins are made of old plastic containers, wood, Styrofoam containers, or metal containers.

Some materials are less desirable than others in bin construction. Styrofoam is believed to release toxins into the earthworms' environment. Metal containers often conduct heat too readily, are prone to rusting, and may release heavy metals into compost.

Bins should have holes in the sides to allow air to flow, and a spout that can be opened or closed or holes in the bottom to drain into a collection tray. Plastic bins require more drainage than wooden ones because they are non-absorbent. The design of a small bin usually depends on where an individual wishes to store the bin and how they wish to feed the worms. Most small bins can be grouped into three categories:

- Non-continuous – an undivided container. A layer of bedding materials is placed in the bin, lining the bottom. Worms are added and organic matter for composting is added in a layer above the bedding. Another layer of bedding is added on top of the organic matter and the worms will start to compost the organic matter and bedding. This type of bin is often used because it is small and easy to build. But it is relatively difficult to harvest because all the materials and worms must be emptied out when harvesting.
- Continuous vertical flow – a series of trays stacked vertically. The bottom-most tray is filled first, in a similar fashion to any other bin, but is not harvested when it is full. Instead, a thick layer of bedding is added on top and the tray above is used for adding organic material. Worms finish composting the materials in the bottom tray and then migrate to the one above. When a sufficient number of worms have migrated, the vermicompost in the bottom tray can

be collected and should be relatively free of worms. These bins provide an easier method of harvesting.

- Continuous horizontal flow – a serics of trays lined horizontally. This method too relies on the earthworms migrating towards a food source in order to ease the process of harvesting. The bin is usually constructed to be similar to a non-continuous bin but longer horizontally. It is divided in half, usually by a large gauge screen of chicken wire. One half is used until it becomes full, then the other half is filled with bedding and organic matter. In time, the worms migrate to the side with the food and the compost can be collected. These bins are larger than a non-continuous system but still small enough to be convenient.

Large Scale

There are two main methods of large-scale vermiculture. Some systems use a windrow, which consists of bedding materials for the earthworms to live in and acts as a large bin; organic material is added to it. Although the windrow has no physical barriers to prevent worms from escaping, in theory they should not due to an abundance of organic matter for them to feed on. Often windrows are used on a concrete surface to prevent predators from killing the worm population. When large scale windrows are fed on one side consistently, a wave motion is generated over time.

The second type of large-scale vermicomposting system is the raised bed or flow-through system. Here the worms are fed an inch of "worm chow" across the top of the bed, and an inch of castings are harvested from below by pulling a breaker bar across the large mesh screen which forms the base of the bed. Because red worms are surface dwellers and are constantly moving towards the new food source, the flow-through system eliminates the need to separate worms from the castings before packaging. Flow-through systems are well suited to indoor facilities, making them the preferred choice for operations in colder climates.

Starting off

When beginning a vermicomposting bin, moist bedding is put into the bin and the worms are added. In hot climates, the bin is placed away from direct sunlight. Appropriate waste can be added daily or weekly. At first, the worms are fed at most half their body weight per day. After they have established themselves, they can be fed up to their entire body weight. It is best not to add new food on top of old food until the old food has been processed by the worms. However, new food can be added in a different location in the bin.

Bedding

Bedding is the living medium and also a food source for the worms. It should be material high in carbon and made to mimic decaying dried leaves on the forest floor, the worms' natural habitat. The bedding should be moist (similar to the consistency of a wrung-out sponge) and loose to enable the worms to breathe and to facilitate aerobic decomposition of the food that is buried in it.

A wide variety of bedding materials can be used, including shredded newspaper, sawdust, hay, cardboard, coir, burlap coffee sacks, peat moss, pre-composted (aged) manure, and dried leaves. Cat litter, and pet and human waste should not be used, because they may carry disease.

Most vermicomposters avoid using glossy paper from newspapers and magazines, junk mail, and shredded paper from offices, because they may contain toxins which may disrupt the system. Also, coated cardboard that contains wax or plastic, such as milk boxes, should not be used. Newspaper and phone books printed on regular, non-glossy paper with non-toxic soy ink are safe for use, and decompose relatively quickly.

Climate and Temperature

Worms used in composting systems prefer temperatures of 55 to 70 degrees Fahrenheit (12-21 degrees Celsius). The temperature of the bedding should not drop below freezing or rise above 89.6 °F (32 °C). This temperature range means that indoor vermicomposting is suitable for homes in all but tropical climates.

Food

Worms and other composting organisms have a preferred ratio of carbon to nitrogen (C:N), approximately 30:1. As some waste is richer in carbon and others in nitrogen, waste must be mixed to approximate the ideal ratio. "Brown matter", or wood products such as shredded papers, is rich in carbon. "Green matter", such as food scraps, has more nitrogen, which is related to the amount of protein in the waste.If the waste is mostly vegetable and fruit scraps, and does not regularly include animal products or high-protein vegetable foods like beans, the resulting vermicompost and waste liquid will be low in nitrogen.

Suitable

Kitchen waste suitable for worms includes coffee grounds and paper filters, tea bags, plate scrapings, rotting fruit (including citrus fruit but NOT citrus peel), vegetable peels, leftovers, moldy bread, etc. These materials can be raw or cooked. They do not have to be ground up, as the micro-organisms in the bin will gradually soften them. However, chopping fruit and vegetable scraps into smaller pieces will speed the composting process. If a large quantity of dry food (e.g., moldy bread) is added and covered with bedding, pour a little purified water over the bedding to moisten the mixture.

If too much kitchen waste is added, the bin mixture putrifies before the worms can process it and becomes harmful to the worms. High-protein foods like beans are particularly susceptible. For these reasons many vermicomposters limit their mix to only include fruit and vegetable matter, avoiding grains, proteins or prepared food scraps altogether. Check the bin at least once a week, give the materials a stir to oxygenate, and add bedding if the bin appears too moist

Soft garden wastes such as carrot tops and tomato leaves are suitable foods. An occasional sprinkling of garden soil in the bin gives the worms the grit they need to digest food. It's not harmful to throw in an entire plant, but the worms will not process the woody parts or large roots and these will have to be hand-removed later from the finished vermicompost.

Compostable plates, cups, etc. are also suitable, but in small bins they should be torn first into smaller pieces so as not to block oxygen flow.

Unsuitable

Grass clippings and other products sprayed with pesticides should be avoided.

Although worms can digest proteins and fats in meat scraps, these materials can attract scavengers. Too much oil or fat can hinder the breathing of the worms, as they breathe through their skin. Worms cannot break down bone and are said to dislike highly spiced foods and salt. Adding meat and dairy products increases the difficulty of maintaining a healthy, low-odor vermicomposting mix, and is usually not recommended.

Acidic foods (tomatoes, citrus), starchy foods (bread, rice), garlic and onions should only be added in moderation. Large amounts of these materials can change the balance of the system.

BIN MAINTENANCE

Worms and composting microorganisms require oxygen, so the bin must "breathe". This can be accomplished by regularly removing the composted material, adding holes to the bin, or using a continuous-flow bin. If insufficient oxygen is available, the decay becomes anaerobic, like that in swamps and bogs, producing a strong odor and creating a toxic environment for the worms.

The moisture level and oxygen flow in a home worm bin should be checked at least once a week.

Over the long term, care should be taken to maintain optimum moisture levels. In a non-continuous-flow vermicomposting bin, excess liquid can be drained via a tap and used as plant food. A continuous flow bin does not retain excess liquid and, depending on the foods used, may require sprinklings of water to keep the bedding moist. The pH should be slightly alkaline. Alkalinity can be increased by occasionally adding a handful of calcium carbonate, sold as "garden lime." Overly acidic compost can also be easily corrected by adding crushed eggshells, which many kitchen-scrap vermicomposters include along with vegetable matter as part of a balanced mix.

Adding many citrus peels can hinder the worms, but probably due not to acidity but to d-limonene, a fragrant chemical present in the rind of citrus fruits. Coffee grounds have sometimes been blamed for acidity, but analysis shows they are only mildly acidic, with a pH of 6.2.

Feeding

There are two methods of adding matter to the bin.

- Top feeding — organic matter is placed directly on top of the existing layer of bedding in a bin and then covered with another layer of bedding. This is repeated every time the bin is fed.
- Pocket feeding — a top layer of bedding is maintained and food is buried beneath. The location of the food is changed each time, rotating around the bin to give the worms time to decompose the food in the previously fed pockets. The top layer of bedding is replenished as necessary.

Vermicomposters often use a combination of both methods. Sometimes unburied food can attract fruit flies, so food should be buried under at least one inch of bedding material.

HARVESTING

Vermicompost is ready for harvest when it contains few to no scraps of uneaten food or bedding. Even a properly composted mixture will contain visible large items that should be discarded, such as peach pits, glassine-like sheets from melon skins, and twigs.

There are several methods of harvesting, depending on the purpose for which the vermicompost will be used, and whether or not the composter wishes to salvage as many worms and worm eggs as possible from the vermicompost.

Vermicompost Properties

Vermicompost is richer in many nutrients than compost produced by other composting methods. It is also rich in microbial life which helps break down nutrients already present in the soil into plant-available forms. Unlike other compost, worm castings also contain worm mucus which keeps nutrients from washing away with the first watering and holds moisture better than plain soil.

For this reason, some fruit and seed pits are reported to germinate in vermicompost easily. Vermicompost made from ordinary kitchen scraps will contain small seeds, especially those of tomatoes, peppers, and eggplants, that may sprout weeks later.

Vermicompost benefits soil by

- improving its physical structure;
- enriching soil in micro-organisms, adding plant hormones such as auxins and gibberellic acid, and adding enzymes such as phosphatase and cellulase;
- attracting deep-burrowing earthworms already present in the soil;
- improving water holding capacity;
- enhancing germination, plant growth, and crop yield; and
- improving root growth and structure.

Vermicompost can be used to make compost tea (worm tea), by mixing some vermicompost in water and steeping for a number of hours or days. The resulting liquid is used as a fertilizer.

The dark brown waste liquid that drains into the bottom of some vermicomposting systems, as water-rich foods break down, is also excellent as fertilizer. However, the pH and nutrient contents of these liquids varies, depending on the food fed to the worms and whether or not lime has been added to the system. pH and nitrogen, phosphorus, and potassium (NPK) measurements should be taken periodically to determine the fertilizer composition before use. Home kits for testing are sold in hardware stores and nurseries.

Problems

Odor, usually due to overabundance of "greens" in the bin, results from too much nitrogen combining with hydrogen to form ammonia. To neutralize the odors, add a fair amount of shredded newspaper or other "browns" to the mix to absorb excess moisture, remove the smelly waste, and stop adding food to the bin until a substantial portion of the uneaten food has been turned into compost.

The carbon will balance the nitrogen and form a compound that is not smelly. The higher level of carbon means that decomposition will be slower. Also, always add new material deep in the bin to disallow access to would-be pests. Consistently doing so will greatly reduce any nuisance from odor and undesirable organisms.

Pests such as rodents and flies may be attracted by certain materials and odors, especially lots of kitchen waste and especially meat.

This problem is largely avoided if a sealed bin is used where the pests cannot access the material, although many proponents recommend having ample access to air. This promotes natural decomposition, as worms and beneficial bacteria require oxygen. Fruit flies are common in warm weather if the food is not thoroughly covered with bedding.

COMPOSTING

Composting is the aerobic decomposition of biodegradable organic matter, producing compost. The decomposition is performed primarily by facultative and obligate aerobic bacteria, yeasts and fungi, helped in the cooler initial and ending phases by a number of larger organisms, such as springtails, ants, nematodes and oligochaete worms.

Composting can be divided into home composting and industrial composting. Essentially the same biological processes are involved in both scales of composting, however techniques and different factors must be taken into account.

Importance

Composting recycles or "downcycles" organic household and yard waste and manures into an extremely useful humus-like, soil end-product called compost. Examples are fruits, vegetables and yard clippings. Ultimately this permits the return of needed organic matter and nutrients into the foodchain and reduces the amount of "green" waste going into landfills.

Composting is widely believed to considerably speed up the natural process of decomposition as a result of the higher temperatures generated. The elevated heat results from exothermic processes, and the heat in turn reduces the generational time of microorganisms, and thereby speeds up the energy and nutrient exchanges taking place. It is a popular misconception that composting is a "controlled" process; if the right environmental circumstances are present the process virtually runs itself.

Hence the popular expression: "compost happens". It is, however, important to engineer the best possible circumstances for large amounts of organic waste to break down properly. This is especially so when it is accompanied by heating, since at elevated temperatures oxygen within the piles is consumed more rapidly, and if not controlled, will lead to malodor.

Decomposition similar to composting occurs throughout nature as garbage dissolves in the absence of all the conditions, and weather patterns, that modern composters talk about; however, the process can be slow. For example, in the forest bark, wood and leaves break down into humus over 3-7 years. In restricted environments, for example, vegetables in a plastic trash container, decomposition with a lack of air encourages growth of anaerobic microbes, which produce disagreeable odors.

Another form of degradation practiced deliberately in absence of oxygen is called anaerobic digestion- an increasingly popular companion to composting as it enables capture of residual energy in the form of biogas, whereas composting releases the majority of bound carbon-energy as excess heat (which helps sanitize the material) as well as copious amounts of biogenic CO_2 to the atmosphere.

It is important to distinguish between terms such as "biodegradable", "compostable", and "compost-compatible". Another term for composting is yeesting, it is literally translated into changing garden while making gestures.

- A biodegradable material is capable of being broken down completely under the action of microorganisms into carbon dioxide, water and biomass. It may take a very long time for a material to biodegrade depending on its environment (e.g. hardwood in an arid area), but it ultimately breaks down completely.
- A compostable material biodegrades substantially under composting conditions, into carbon dioxide, methane, water and compost biomass. Compost biomass refers to the portion of the material that is metabolized by the microorganisms and which is incorporated into the cellular structure of the organisms or converted into humic acids etc. Compost biomass residues from a compostable material are fully biodegradable. "Compostable" is thus a subset of "biodegradable". The size of the material is a factor in determining compostability because it affects the rate of degradation. Large pieces of hardwood may not be compostable under a specific set of composting conditions, whereas sawdust of the same type of wood may be.
- A compost-compatible material does not have to be compostable or even biodegradable. It may biodegrade too slowly to be itself compostable, or it may not biodegrade at all. However, it is not readily distinguishable from the compost on a macroscopic scale and does not have a deleterious effect on the compost (e.g. it is not a biocide). Compost-compatible materials are generally inert and are present in compost at relatively low levels. Examples of compost-compatible materials include sand particles and inert particles of plastic.

Reducing Waste

Although composting has historically focused on creating garden-ready soil, it is becoming more important as a tool for reducing solid waste. More than 60 percent of household waste in the United States is recyclable or compostable. But Americans only compost 8 percent of their waste. The decomposition of material sent to landfill is a principal cause of methane, an important greenhouse gas, so reducing the amount of waste sent to landfill is a key element of the fight against climate change.

Surveys have shown that the #1 reason Americans don't compost their waste is because they feel the process is complicated, time-consuming or requires special equipment. However, especially in rural areas, much of the solid waste could be removed from the waste stream by promoting "extremely passive composting" where consumers simply discard their yard waste and kitchen scraps on their own land, regardless of whether the material is ever re-used as "compost".

HOME COMPOSTING

Home composters use a range of techniques, varying from extremely passive (throw everything in a pile and leave it for a year or two) to extremely active (monitor the temperature, turn the pile regularly, and adjust the ingredients over time). Some composters use mineral powders to absorb smells, although a well-maintained pile seldom has bad odors. It is usually located in the back garden.

Moisture and heat

An effective compost pile is about as damp as a well wrung-out sponge. This provides the moisture that all life requires. Microorganisms vary by their ideal temperature and the heat they generate as they digest. Mesophilic bacteria survive best at temperatures of 20 to 44 °C (70 to 120 °F). thermophilic (heat-surviving) bacteria grow optimally at around 55°C (130 °F), and can attain the fastest decomposition, since metabolic processes proceed more rapidly under higher temperatures. Elevated temperature is also preferred since it causes the most rapid pathogen reduction, and is more destructive of weed seeds. To minimally achieve it, the heap should be about 1 m (3 ft) wide, 1 m (3 ft) tall, and as long as is practicable. This provides enough insulating mass to build up heat but also allows aeration. The center of the pile heats up the most.

If the pile does not heat up, common reasons include that:

- The heap is too wet, limiting the oxygen which bacteria require
- The heap is too dry for the bacteria to survive and reproduce
- There is insufficient protein (nitrogen-rich material)

The necessary material should be added, or the pile should be turned to aerate it and bring the outer layers inside and vice versa.

One guideline is to turn the pile when the high temperature has begun to drop, indicating that the food source for the fastest-acting bacteria (in the center of the pile) has been largely consumed. When turning the pile does not cause a temperature rise, it brings no further advantage. When all the material has turned into dark brown crumbly matter, it is ready to use.

Worm Composting

Worm composting or vermicomposting is a method of composting using Red Wiggler worms in a container. Food waste and moistened bedding are added, and the worms and micro-organisms eventually convert them to rich compost. The worms excrete a soil-nutrient material called worm castings. Worm composting can be done indoors, allowing year-round composting, and providing apartment dwellers with a means of composting.

Worms are low in the food chain, and so are critical to healthy soil. This is why farmers have historically wanted healthy worm populations to live in their fields. The nutrients and micro-organisms can be concentrated in liquid form called worm tea, made by running distilled water through worm castings. When it is poured into the soil, the micro-organisms multiply, creating a healthy growing environment for plants.

INDUSTRIAL COMPOSTING

Industrial composting systems are increasingly being installed as a waste management alternative to landfills, along with other advanced waste processing systems. Industrial composting or anaerobic digestion combined with mechanical sorting of mixed waste streams is called mechanical biological treatment increasingly used in Europe due to stringent new regulations controlling the amount of organic matter allowed in landfills.

Treating biodegradable waste before it enters a landfill reduces global warming from fugitive methane; untreated waste breaks down anaerobically in a landfill, producing landfill gas that contains methane, a greenhouse gas even more potent than carbon dioxide.

Most commercial and industrial composting operations use active composting techniques. These ensure that the process does not get out of control especially with the high through-put demand imposed by contracted, incoming waste. This means that as short as possible a processing time must be maintained to keep the facility properly functioning. Partly for this reason composters have declined to support compost maturity standards if it would increase the required holding time. The greatest amount of technological control of composting is seen in systems using an enclosed vessel and controlling its temperature, air flow, moisture and other parameters.

HERBICIDE

A herbicide is used to kill unwanted plants. Selective herbicides kill specific targets while leaving the desired crop relatively unharmed. Some of these act by interfering with the growth of the weed and are often synthetic "imitations" of plant hormones. Herbicides used to clear waste ground, industrial sites, railways and railway embankments are non-selective and kill all plant material with which they come into contact. Some plants produce natural herbicides, such as the genus *Juglans* (walnuts).

They are applied in total vegetation control (TVC) programs for maintenance of highways and railroads. Smaller quantities are used in forestry, pasture systems, and management of areas set aside as wildlife habitat. Herbicides are widely used in agriculture and in landscape turf management. In the U.S., they account for about 70% of all agricultural pesticide use.Classification of herbicides

Herbicides can be grouped by activity, use, chemical family, mode of action, or type of vegetation controlled.

By activity:

- Contact herbicides destroy only the plant tissue in contact with the chemical. Generally, these are the fastest acting herbicides. They are less effective on perennial plants, which are able to regrow from rhizhomes, roots or tubers.
- Systemic herbicides are translocated through the plant, either from foliar application down to the roots, or from soil application up to the leaves. They are capable of controlling perennial plants and may be slower acting but ultimately more effective than contact herbicides.

By use:

- Soil-applied herbicides are applied to the soil and are taken up by the roots and/or hypocotyl of the target plant. The following two categories are examples of soil-applied herbicides.
- Pre-plant incorporated herbicides are soil applied prior to planting and mechanically incorporated into the soil. The objective for incorporation is to prevent dissipation through photo decomposition and/or volatility.
- Preemergent herbicides are applied to the soil before the crop emerges and prevent germination or early growth of weed seeds.
- Post-emergent herbicides are applied after the crop has emerged.

Their classification by mechanism of action (MOA) indicates the first enzyme, protein, or biochemical step affected in the plant following application. The main mechanisms of action are:

- ACCase inhibitors are compounds that kill grasses. Acetyl coenzyme A carboxylase (ACCase) is part of the first step of lipid synthesis. Thus, ACCase inhibitors affect cell membrane production in the meristems of the grass plant. The ACCases of grasses are sensitive to these herbicides, whereas the ACCases of dicot plants are not.
- ALS inhibitors: the acetolactate synthase (ALS) enzyme (also known as acetohydroxyacid synthase, or AHAS) is the first step in the synthesis of the branched-chain amino acids (valine, leucine, and isoleucine). These herbicides slowly starve affected plants of these amino acids which eventually leads to inhibition of DNA synthesis. They affect grasses and dicots alike. The ALS inhibitor family includes sulfonylureas (SUs), imidazolinones (IMIs), triazolopyrimidines (TPs), pyrimidinyl oxybenzoates (POBs), and sulfonylamino carbonyl triazolinones (SCTs). ALS is a biological pathway that exists only in plants and not in animals thus making the ALS-inhibitors among the safest herbicides.
- EPSPS inhibitors: The enolpyruvylshikimate 3-phosphate synthase enzyme EPSPS is used in the synthesis of the amino acids tryptophan, phenylalanine and tyrosine. They affect grasses and dicots alike. Glyphosate (Roundup) is a systemic EPSPS inhibitor but inactivated by soil contact.
- Synthetic auxin inaugurated the era of organic herbicides. They were discovered in the 1940s after a long study of the plant growth regulator auxin. Synthetic auxins mimic this plant hormone. They have several points of action on the cell membrane, and are effective in the control of dicot plants. 2,4-D is a synthetic auxin herbicide.

- Photosystem II inhibitors reduce electron flow from water to NADPH2+ at the photochemical step in photosynthesis. They bind to the Qb site on the D1 protein, and prevent quinone from binding to this site. Therefore, this group of compounds cause electrons to accumulate on chlorophyll molecules. As a consequence, oxidation reactions in excess of those normally tolerated by the cell occur, and the plant dies. The triazine herbicides (including atrazine) and urea derivatives (diuron) are photosystem II inhibitors.

Organic Herbicides

Almost all herbicides in use today are considered "organic" herbicides in that they contain carbon as a primary molecular component. A notable exception would be the arsenical class of herbicides. Sometimes they are referred to as synthetic organic herbicides. Recently the term "organic" has come to imply products used in organic farming. Under this definition an organic herbicide is one that can be used in a farming enterprise that has been classified as organic.

Organic herbicides are expensive and may not be affordable for commercial production. They are much less effective than synthetic herbicides and are generally used along with cultural and mechanical weed control practices.

Organic herbicides include:

- Spices are now effectively used in patented herbicides.
- Vinegar is effective for 5-20% solutions of acetic acid with higher concentrations most effective but mainly destroys surface growth and so respraying to treat regrowth is needed. Resistant plants generally succumb when weakened by respraying.
- Steam has been applied commercially but is now considered uneconomic and inadequate. It kills surface growth but not underground growth and so respraying to treat regrowth of perennials is needed.
- Flame is considered more effective than steam but suffers from the same difficulties.
- D-limonene (citrus oil), which is the active ingredient

in Nature's Avenger Organic Herbicide. D-limonene (citrus oil) is a natural degreasing agent that strips the waxy skin or cuticle from weeds, causing dehydration and ultimately death.

Application

Most herbicides are applied as water-based sprays using ground equipment. Ground equipment varies in design, but large areas can be sprayed using self-propelled sprayers equipped with a long boom, of 60 to 80 feet (20 to 25 m) with flat fan nozzles spaced about every 20 in (500 mm). Towed, handheld, and even horse-drawn sprayers are also used.

BIOPESTICIDE

The term biopesticide is used for microbial biological pest control agents that are applied in a similar manner to chemical pesticides. Commonly these are bacterial, but there are also examples of fungal control agents, including *Trichoderma* spp. and *Ampelomyces quisqualis* (a control agent for grape powdery mildew). *Bacillus subtilis* are used to control plant pathogens. Weeds and rodents have also been controlled with microbial agents.

One well-known insecticide example is *Bacillus thuringiensis*, a bacterial disease of Lepidoptera, Coleoptera and Diptera. Because it has little effect on other organisms, it is considered more environmentally friendly than synthetic pesticides. The toxin from *Bacillus thuringiensis* (Bt toxin) has been incorporated directly into plants through the use of genetic engineering.

Biological insecticides include products based on:

- Entomopathogenic fungi (*e.g. Metarhizium anisopliae*),
- Entomopathogenic nematodes (*e.g. Steinernema feltiae*) and
- Entomopathogenic viruses (*e.g. Cydia pomonella* granulovirus).

The Manual of Biocontrol Agents gives a review of the available biological insecticide (and other biology-based control) products. In order to implement these environmentally-friendly pest control agents, it is often especially important to pay attention to their formulation and application.

Biopesticide Production

Biological control is the most important alternative to chemical pesticides in protecting crops from pests, pathogens, and weeds. Major breakthroughs in molecular biology and biotechnology since the early 1980s indicate that quick improvement in the competitive ability of biological control methods is possible, and that biopesticides can play a major role in crop protection in the future. It has become possible to improve some of the critical properties that earlier hampered the usefulness of many biocontrol agents. Valuable genes from completely unrelated organisms can now be utilized for biological control purposes.

Biological control using recombinant DNA (genetic engineering) technology can be achieved in several different ways: control agents may be improved; crop plants can be engineered to carry better resistance genes; or organisms associated with the plant may be modified to provide protection. All these approaches have successfully been used in several different ways experimentally.

Product development has been very active in the area of incorporating resistance genes-mainly from *Bt*-directly into plants. Successes include potato, tomato, tobacco, and cotton. General root colorizing bacteria of plants have also been engineered to produce insecticidal toxins, which protect against pests such as the corn rootworm. Another bacterium living in the vascular tissues of corn has also been modified to give protection against the corn borer. None of these modified plants or associated organisms is available commercially yet. Similar approaches are used for the biological control of plant pathogens and weeds, but research has been most active in the area of insect control.

Engineering Crop Plants

The first published reports of successful engineering of

crop plants to produce insecticidal or antifeedant proteins appeared in 1987. The crop plants were tobacco and tomato, producing the delta endotoxin of *Bacillus thuringiensis* to make them resistant against caterpillars. To date, transgenic crop plants have been produced of at least 27 different species, including potato, cabbage, sugarbeet, rice, soybeans, corn, rapeseed, sunflower, walnut, and poplar. Within only two years of the first reports, at least 53 field trials in seven countries were conducted, involving eight plant species.

Instead of being inserted directly into the crop plant genome, the protective insecticidal genes can be engineered into associated organisms. Two bacteria have been successfully tested for this purpose. *Pseudomonas fluorescens*, which colonizes the root systems of crops, has been engineered to express *Bacillus thuringiensis* (Bt) toxins, and thus provide continuous protection against such pests as corn rootworm.

The genes for all the major proteins that account for the insecticidal properties of *Bt* have been cloned and sequenced. Now we have nucleotide sequences for more than 20 *Bt* genes that encode proteins active against leopidopterans, eight genes encoding proteins active against dipterans, and two genes encoding proteins active against coleopterans. To increase the environmental stability and effectiveness of the various *Bt* toxins in the field, genes encoding proteins active against beetles and caterpillars have also been cloned into the rhizobacterium *Pseudomonas fluorescens*. After fermentation, the bacteria are killed and the cell walls hardened chemically. The endotoxins are thereby microencapsulated, resulting in insecticides with greatly enhanced residual activity.

Similar strategies have been employed to develop mosquitocidal species of algae. Even more significantly, several major crop species, including cotton, tobacco, and soybeans, have been transformed with *Bt* genes, becoming resistant to attack by caterpillars and beetles. Through genetic-engineering techniques, the *Autographa californica multinucleocapsid nucleopolyhedrosis virus* (AcMNPV) has been engineered to kill insects more quickly by expressing either enzymes or toxins soon after host invasion.

Of particular interest is the possibility of making viruses produces insect neurohormones, which can cause rapid physiological disruptions in minutely defined target hosts. This strategy is in its early stages of development, but there is little doubt that within the very near future we will have viruses with extended or specifically designed host ranges, capable of killing insects within 24 to 48 hours. These genetically engineered viruses should have an advantage for use against hosts that are not easily controlled by *Bt*.

Very little is known about the genetics of entomopathogenic fungi. The first transformation system for an entomopathogenic fungus was developed using *Metarhizium anisopliae* protoplasts mixed with a fungicide-resistant plasmid. A benomyl-resistant strain of *M.anisoplieae* has thus been obtained. Fungal enzymes involved in the penetration of the insect cuticle have now been identified. Knowledge of these genes and gene products will eventually lead to the possibility of genetic alteration of fungal pathogens that possess those genes. Transformation systems for some fungi exist already and may soon be applied to the entomopathogenic species. Parasexual recombination not only facilitates genetic analysis in asexually reproducing fungi, but also provides an important tool in strain improvement of bioprotectant fungi. Entomopathogenic fungi, which are facultative parasites, are subjected to many environmental factors, and the host insect can exert a selective pressure by favoring one or a few genotypes. Pathogenicity tests showed that some strains are selective hosts for virulence.

BIOTRANSFORMATION

Biotransformation is the chemical modification (or modifications) made by an organism on a chemical compound. If this modification ends in mineral compounds like CO_2, NH_3^+ or H_2O, the biotransformation is called mineralisation. Biotransformation means chemical alteration of the drug in the body. It is needed to render nonpolar compounds polar so that they are not reabsorbed in renal tubules and are excreted.

Microbial Biotransformation

Biotransformation of various pollutants is a sustainable way to clean up contaminated environments. These bioremediation and biotransformation methods harness the naturally occurring, microbial catabolic diversity to degrade, transform or accumulate a huge range of compounds including hydrocarbons (e.g. oil), polychlorinated biphenyls (PCBs), polyaromatic hydrocarbons (PAHs), pharmaceutical substances, radionuclides and metals. Major methodological breakthroughs in recent years have enabled detailed genomic, metagenomic, proteomic, bioinformatic and other high-throughput analyses of environmentally relevant micro-organisms providing unprecedented insights into biotransformation and biodegradative pathways and the ability of organisms to adapt to changing environmental conditions.

Biological processes play a major role in the removal of contaminants and pollutants from the environment. Some microorganisms possess an astonishing catabolic versatility to degrade or transform such compounds. New methodological breakthroughs in sequencing, genomics, proteomics, bioinformatics and imaging are producing vast amounts of information. In the field of Environmental Microbiology, genome-based global studies open a new era providing unprecedented *in silico* views of metabolic and regulatory networks, as well as clues to the evolution of biochemical pathways relevant to biotransformation and to the molecular adaptation strategies to changing environmental conditions.

Functional genomic and metagenomic approaches are increasing our understanding of the relative importance of different pathways and regulatory networks to carbon flux in particular environments and for particular compounds and they are accelerating the development of bioremediation technologies and biotransformation processes. Also there is other approach of biotransformation called enzymatic biotransformation.

ENVIORNMENTAL BIOTECHNOLOGY

The Chambers Science and Technology Dictionary defines

biotechnology as 'the use of organisms or their components in industrial or commercial processes, which can be aided by the techniques of genetic manipulation in developing e.g. novel plants for agriculture or industry.' Despite the inclusiveness of this definition, the biotechnology sector is still often seen as largely medical or pharmaceutical in nature, particularly amongst the general public. While to some extent the huge research budgets of the drug companies and the widespread familiarity of their products makes this understandable, it does distort the full picture and somewhat unfairly so. However, while therapeutic instruments form, in many respects, the 'acceptable' face of biotechnology, elsewhere the science is all too frequently linked with unnatural interference.

While the agricultural, industrial and environmental applications of biotechnology are potentially very great, the shadow of Frankenstein has often been cast across them. Genetic engineering may be relatively commonplace in pharmaceutical thinking and yet in other spheres, like agriculture for example, society can so readily and thoroughly demonise it. The history of human achievement has always been episodic. For a while, one particular field of endeavour seems to hold sway as the preserve of genius and development, before the focus shifts and development forges ahead in dizzy exponential rush in an entirely new direction.

So it was with art in the renaissance, music in the 18th century, engineering in the 19th and physics in the 20th. Now it is the age of the biological, possibly best viewed almost as a rebirth, after the great heyday of the Victorian naturalists, who provided so much input into the developing science.

It is then, perhaps, no surprise that the European Federation of Biotechnology begins its 'Brief History' of the science in the year 1859, with the publication of On the Origin of Species by Means of Natural Selection by Charles Darwin. Though his famous voyage aboard HMS Beagle, which led directly to the formulation of his (then) revolutionary ideas, took place when he was a young man, he had delayed making them known until 1858, when he made a joint presentation before the Linnaean Society with Alfred Russell Wallace, who had, himself, independently come to very similar conclusions. Their contribution was to view evolution as the driving force of life, with successive selective pressures over time endowing living beings with optimised characteristics for survival. Neo-Darwinian thought sees the interplay of mutation and natural selection as fundamental. The irony is that Darwin himself rejected mutation as too deleterious to be of value, seeing such organisms, in the language of the times, as 'sports' – oddities of no species benefit. Indeed, there is considerable evidence to suggest that he seems to have espoused a more Lamarckist view of biological progression, in which physical changes in an organism's lifetime were thought to shape future generations.

Darwin died in 1882. Ninety-nine years after his death, the first patent for a genetically modified organism was granted to Ananda Chakrabarty of the US General Electric, relating to a strain of Pseudomonas aeruginosa engineered to express the genes for certain enzymes in order to metabolise crude oil. Twenty years later still, in the year that saw the first working draft of the human genome sequence published and the announcement of the full genetic blueprint of the fruit fly, Drosophila melanogaster, that archetype of eukaryotic genetics research, biotechnology has become a major growth industry with increasing numbers of companies listed on the world's stock exchanges.

Thus, at the other end of the biotech timeline, a century and a half on from Origin of Species, the principles it first set out remain of direct relevance for what has been termed the 'chemical evolution' of biologically active substances and are commonly used in laboratories for in vitro production of desired qualities in biomolecules.

The Role of Environmental Biotechnology

While pharmaceutical biotechnology represents the glamorous end of the market, environmental applications are decidedly more in the Cinderella mould. The reasons for this are fairly obvious. The prospect of a cure for the many diseases and conditions currently promised by gene therapy and other biotech-oriented medical miracles can potentially touch us all. Our lives may, quite literally, be changed.

Environmental biotechnology, by contrast, deals with far less apparently dramatic topics and, though their importance, albeit different, may be every bit as great, their direct relevance is far less readily appreciated by the bulk of the population. Cleaning up contamination and dealing rationally with wastes is, of course, in everybody's best interests, but for most people, this is simply addressing a problem which they would rather had not existed in the first place.

Even for industry, though the benefits may be noticeable on the balance sheet, the likes of effluent treatment or pollution control are more of an inevitable obligation than a primary goal in themselves. In general, such activities are typically funded on a distinctly limited budget and have traditionally been viewed as a necessary inconvenience. This is in no way intended to be disparaging to industry; it simply represents commercial reality. In many respects, there is a logical fit between this thinking and the aims of environmental biotechnology. For all the media circus surrounding the grand questions of our age, it is easy to forget that not all forms of biotechnology involve xenotransplantation, genetic modification, the use of stem cells or cloning.

Some of the potentially most beneficial uses of biological engineering and which may touch the lives of the majority of people, however indirectly, involve much simpler approaches. Less radical and showy, certainly, but powerful tools, just the same. Environmental biotechnology is fundamentally rooted in waste, in its various guises, typically being concerned with the remediation of contamination caused by previous use, the impact reduction of current activity or the control of pollution.

Thus, the principal aims of this field are the manufacture of products in environmentally harmonious ways, which allow for the minimisation of harmful solids, liquids or gaseous outputs or the clean-up of the residual effects of earlier human occupation. The means by which this may be achieved are essentially two-fold.

Environmental biotechnologists may enhance or optimise

conditions for existing biological systems to make their activities happen faster or more efficiently, or they resort to some form of alteration to bring about the desired outcome.

The variety of organisms which may play a part in environmental applications of biotechnology is huge, ranging from microbes through to trees and all are utilised on one of the same three fundamental bases – accept, acclimatise or alter. For the vast majority of cases, it is the former approach, accepting and making use of existing species in their natural, unmodified form, which predominates.

The Scope for Use

There are three key points for environmental biotechnology interventions, namely in the manufacturing process, waste management or pollution control. Accordingly, the range of businesses to which environmental biotechnology has potential relevance is almost limitless. One area where this is most apparent is with regard to waste.

All commercial operations generate waste of one form or another and for many, a proportion of what is produced is biodegradable. With disposal costs rising steadily across the world, dealing with refuse constitutes an increasingly high contribution to overheads. Thus, there is a clear incentive for all businesses to identify potentially cost-cutting approaches to waste and employ them where possible.

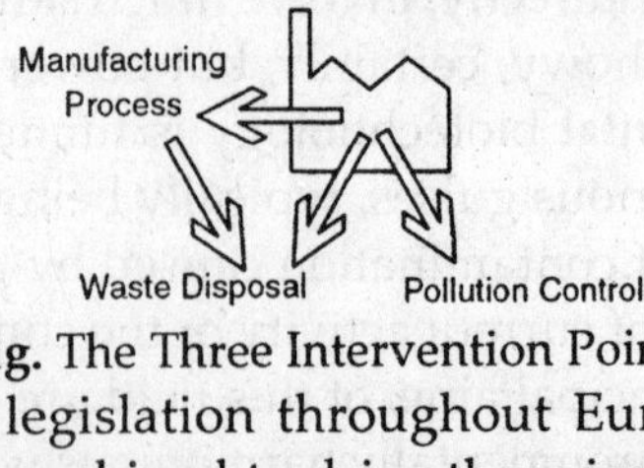

Fig. The Three Intervention Points

Changes in legislation throughout Europe, the US and elsewhere, have combined to drive these issues higher up the political agenda and biological methods of waste treatment have gained far greater acceptance as a result. For those industries with particularly high biowaste production, the various available treatment biotechnologies can offer considerable savings. Manufacturing industries can benefit from the applications of whole organisms or isolated biocomponents. Compared with conventional chemical processes, microbes and enzymes typically function at lower temperatures and pressures.

The lower energy demands this makes leads to reduced costs, but also has clear benefits in terms of both the environment and workplace safety. Additionally, biotechnology can be of further commercial significance by converting low-cost organic feedstocks into high value products or, since enzymatic reactions are more highly specific than their chemical counterparts, by deriving final substances of high relative purity.

Almost inevitably, manufacturing companies produce wastewaters or effluents, many of which contain biodegradable contaminants, in varying degrees. Though traditional permitted discharges to sewer or watercourses may be adequate for some, other industries, particularly those with recalcitrant or highly concentrated effluents, have found significant benefits to be gained from using biological treatment methods themselves on site. Though careful monitoring and process control are essential, biotechnology stands as a particularly cost-effective means of reducing the pollution potential of wastewater, leading to enhanced public relations, compliance with environmental legislation and quantifiable cost-savings to the business.

Those involved in processing organic matter, for example, or with drying, printing, painting or coating processes, may give rise to the release of volatile organic compounds (VOCs) or odours, both of which represent environmental nuisances, though the former is more damaging than the latter. For many, it is not possible to avoid producing these emissions altogether, which leaves treating them to remove the offending contaminants the only practical solution. Especially for relatively low concentrations of readily water-soluble VOCs or odorous chemicals, biological technologies can offer an economic and effective alternative to conventional methods.

The use of biological cleaning agents is another area of potential benefit, especially where there is a need to remove oils and fats from process equipment, work surfaces or drains. Aside from typically reducing energy costs, this may also obviate the need for toxic or dangerous chemical agents.

The pharmaceutical and brewing industries, for example, both have a long history of employing enzyme-based cleaners to remove organic residues from their process equipment. In addition, the development of effective biosensors, powerful tools which rely on biochemical reactions to detect specific substances, has brought benefits to a wide range of sectors, including the manufacturing, engineering, chemical, water, food and beverage industries.

With their ability to detect even small amounts of their particular target chemicals, quickly, easily and accurately, they have been enthusiastically adopted for a variety of process monitoring applications, particularly in respect of pollution assessment and control.

Contaminated land is a growing concern for the construction industry, as it seeks to balance the need for more houses and offices with wider social and environmental goals. The reuse of former industrial sites, many of which occupy prime locations, may typically have associated planning conditions attached which demand that the land be cleaned up as part of the development process.

With urban regeneration and the reclamation of 'brown-field' sites increasingly favoured in many countries over the use of virgin land, remediation has come to play a significant role and the industry has an ongoing interest in identifying cost-effective methods of achieving it. Historically, much of this has involved simply digging up the contaminated soil and removing it to landfill elsewhere.

Bioremediation technologies provide a competitive and sustainable alternative and in many cases, the lower disturbance allows the overall scheme to make faster progress. As the previous brief examples show, the range of those which may benefit from the application of biotechnology is lengthy and includes the chemical, pharmaceutical, water, waste management and leisure industries, as well as manufacturing, the military, energy generation, agriculture and horticulture.

Clearly, then, this may have relevance to the viability of these ventures and, as was mentioned at the outset, biotechnology is an essentially commercial activity. Environmental biotechnology must compete in a world governed by the best practicable environmental option (BPEO) and the best available techniques not entailing excessive cost (Batneec). Consequently, the economic aspect will always have a large influence on the uptake of all initiatives in environmental biotechnology and, most particularly, in the selection of methods to be used in any given situation. It is impossible to divorce this context from the decisionmaking process. By the same token, the sector itself has its own implications for the wider economy.

THE MARKET FOR ENVIRONMENTAL BIOTECHNOLOGY

The UK's Department of Trade and Industry estimated that 15–20% of the global environmental market in 2001 was biotech-based, which amounted to about 250–300 billion US dollars and the industry is projected to grow by as much as ten-fold over the following five years. This expected growth is due to greater acceptance of biotechnology for clean manufacturing applications and energy production, together with increased landfill charges and legislative changes in waste management which also alter the UK financial base favourably with respect to bioremediation.

Biotechnology-based methods are seen as essential to help meet European Union (EU) targets for biowaste diversion from landfill and reductions in pollutants. Across the world the existing regulations on environmental pollution are predicted to be more rigorously enforced, with more stringent compliance standards implemented. All of this is expected to stimulate the sales of biotechnology-based environmental processing methods significantly and, in particular, the global market share is projected to grow faster than the general biotech sector trend, in part due to the anticipated large-scale EU aid for environmental clean-up in the new accession countries of Eastern Europe.

Other sources paint a broadly similar picture. The BioIndustry Association (BIA) survey, Industrial Markets for UK Biotechnology – Trends and Issues, published in 1999 does not quote any monetary sector values per year, but gives the size of the UK sector as employing 40 000 people in 1998 with an average yearly growth over 1995–98 of 20%. Environmental biotech is reported as representing around 10% of this sector. An Arthur Anderson report of 1997 gives the turnover of the UK biotech sector as 702 million pounds sterling in 1995/96, with a 50% growth over three years. A 1998 Ernst and Young report on the European Life Sciences Sector says that the market for biotechnology products has the potential to reach 100 billion pounds sterling worldwide by 2005.

The Organisation for Economic Cooperation and Development (OECD) estimates that the global market for environmental biotechnology products and services alone will rise to some US$75 billion by the year 2000, accounting for some 15 to 25% of the overall environmental technology market, which has a growth rate estimated at 5.5% per annum. The UK potential market for environmental biotechnology products and services is estimated at between 1.65 and 2.75 billion US dollars and the growth of the sector stands at 25% per annum, according to the Bio- Commerce Data European Biotechnology Handbook.

An unsourced quote found on a Korean University website says that the world market size of biotechnology products and services was estimated to be approximately 390 billion US dollars in the year 2000. The benefits are not, however, confined to the balance sheet. The Organisation for Economic Cooperation and Development (OECD 2001) concluded that the industrial use of biotechnology commonly leads to increasingly environmentally harmonious processes and additionally results in lowered operating and/or capital costs. For years, industry has appeared locked into a seemingly unbreakable cycle of growth achieved at the cost of environmental damage.

The OECD investigation provides what is probably the first hard evidence to support the reality of biotechnology's long-heralded promise of alternative production methods, which are ecologically sound and economically efficient. A variety of industrial sectors including pharmaceuticals, chemicals, textiles, food and energy were examined, with a particular emphasis on biomass renewable resources, enzymes and bio-catalysis.

While such approaches may have to be used in tandem with other processes for maximum effectiveness, it seems that their use invariably leads to reduction in operating or capital costs, or both. Moreover, the research also concludes that it is clearly in the interests of governments of the developed and developing worlds alike to promote the use of biotechnology for the substantial reductions in resource and energy consumption, emissions, pollution and waste production it offers. The potential contribution to be made by the appropriate use of biotechnology to environmental and economic sustainability would seem to be clear.

The upshot of this is that few biotech companies in the environmental sector perceive problems for their own business development models, principally as a result of the wide range of businesses for which their services are applicable, the relatively low market penetration to date and the large potential for growth.

Competition within the sector is not seen as a major issue either, since the field is still largely open and unsaturated. Moreover, there has been a discernible tendency in recent years towards niche specificity, with companies operating in more specialised subarenas within the environmental biotechnology umbrella. Given the number and diversity of such possible slots, coupled with the fact that new opportunities and the technologies to capitalise on them, are developing apace, this trend seems likely to continue.

It is not without some irony that companies basing their commercial activities on biological organisms should themselves come to behave in such a Darwinian fashion. However, the picture is not entirely rosy. Typically the sector comprises a number of relatively small, specialist companies and the market is, as a consequence, inevitably fragmented.

Often the complexities of individual projects make the application of 'standard' off-theshelf approaches very difficult, the upshot being that much of what is done must be significantly customised.

While this, of course, is a strength and of great potential environmental benefit, it also has hard commercial implications which must be taken into account.

A sizeable proportion of companies active in this sphere, have no products or services which might reasonably be termed suitable for generalised use, though they may have enough expertise, experience or sufficiently perfected techniques to deal with a large number of possible scenarios. The fact remains that one of the major barriers to the wider uptake of biological approaches is the high perceived cost of these applications.

Part of the reason for this lies in historical experience. For many years, the solutions to all environmental problems were seen as expensive and for many, particularly those unfamiliar with the multiplicity of varied technologies available, this has remained the prevalent view.

Generally, there is often a lack of financial resource allocation available for this kind of work and biotech providers have sometimes come under pressure to reduce the prices for their services as a result.

Greater awareness of the benefits of biotechnology, both as a means to boost existing markets and for the opening up of new ones, is an important area to be addressed. Many providers, particularly in the UK, have cited a lack of marketing expertise as one of the principal barriers to their exploitation of novel opportunities.

In addition, a lack of technical understanding of biotech approaches amongst target industries and, in some cases, downright scepticism regarding their efficacy, can also prove problematic.

Good education, in the widest sense, of customers and potential users of biological solutions will be one major factor in any future upswing in the acceptance and utilisation of these technologies.

Modalities and Local Influences

Another of the key factors affecting the practical uptake of environmental biotechnology is the effect of local circumstances. Contextual sensitivity is almost certainly the single most important factor in technology selection and represents a major influence on the likely penetration of biotech processes into the marketplace.

Neither the nature of the biological system, nor of the application method itself, play anything like so relevant a role. This may seem somewhat unexpected at first sight, but the reasons for it are obvious on further inspection. While the character of both the specific organisms and the engineering remain essentially the same irrespective of location, external modalities of economics, legislation and custom vary on exactly this basis.

Accordingly, what may make abundant sense as a biotech intervention in one region or country, may be totally unsuited to use in another. In as much as it is impossible to discount the wider global economic aspects, disassociating political, fiscal and social conditions equally cannot be done, as the following example illustrates.

In 1994, the expense of bioremediating contaminated soil in the United Kingdom greatly exceeded the cost of removing it to landfill. Six years later, with successive changes of legislation and the imposition of a landfill tax, the situation has almost completely reversed. In those other countries where landfill has always been an expensive option, remediation has been embraced far more readily.

While environmental biotechnology must, inevitably, be viewed as contextually dependent, as the previous example shows, contexts can change. In the final analysis, it is often fiscal instruments, rather than the technologies, which provide the driving force and sometimes seemingly minor modifications in apparently unrelated sectors can have major ramifications for the application of biotechnology.

The legal framework is another aspect of undeniable importance in this respect. Increasingly tough environmental law makes a significant contribution to the sector and changes in regulatory legislation are often enormously influential in boosting existing markets or creating new ones.

When legislation and economic pressure combine, as, for example, they have begun to do in the European Landfill Directive, the impetus towards a fundamental paradigm shift becomes overwhelming and the implications for relevant biological applications can be immense. There is a natural tendency to delineate, seeking to characterise technologies into particular categories or divisions. However, the essence of environmental biotechnology is such that there are many more similarities than differences. Though it is, of course, often helpful to view individual technology uses as distinct, particularly when considering treatment options for a given environmental problem, there are inevitably recurrent themes which feature throughout the whole topic. Moreover, this is a truly applied science. While the importance of the laboratory bench cannot be denied, the controlled world of research translates imperfectly into the harsh realities of commercial implementation. Thus, there can often be a dichotomy between theory and application and it is precisely this fertile ground which is explored in the present work.

In addition, the principal underlying approach of specifically environmental biotechnology, as distinct from other kinds, is the reliance on existing natural cycles, often directly and in an entirely unmodified form. Thus, this science stands on a foundation of fundamental biology and biochemistry. To understand the application, the biotechnologist must simply examine the essential elements of life, living systems and ecological circulation sequences. However engineered the approach, this fact remains true. In many respects, environmental biotechnology stands as the purest example of the newly emergent bioindustry, since it is the least refined, at least in terms of the basis of its action.

In essence, all of its applications simply encourage the natural propensity of the organisms involved, while seeking to enhance or accelerate their action. Hence, optimisation, rather than modification, is the typical route by which the particular desired end result, whatever it may be, is achieved and, consequently, a number of issues feature as common threads of individual technologies.

Integrated Approach

ntegration is an important aspect for environmental biotechnology. One theme that will be developed throughout this book is the potential for different biological approaches to be combined within treatment trains, thereby producing an overall effect which would be impossible for any single technology alone to achieve. However, the wider goal of integration is not, of necessity, confined solely to the specific methods used. It applies equally to the underpinning knowledge that enables them to function in the first place and an understanding of this is central to the rationale behind this book. In some spheres, traditional biology has become rather unfashionable and the emphasis has shifted to more exciting sounding aspects of life science.

While the new-found concentration on 'Ecological processes', or whatever, sounds distinctly more 'Environmental', in many ways and somewhat paradoxically, it sometimes serves the needs of environmental biotechnology rather less well. The fundamentals of living systems are the stuff of this branch of science and, complex though the whole picture may be, at its simplest the environmental biotechnologist is principally concerned with a relatively small number of basic cycles. In this respect, a good working knowledge of biological processes like respiration, fermentation and photosynthesis, a grasp of the major cycles by which carbon, nitrogen and water are recycled and an appreciation of the flow of energy through the biosphere must be viewed as prerequisites.

The intent here has been neither to insult the readership by parading what is already well known, nor gloss over aspects which, if left unexplained, at least in reasonable detail, might only serve to confuse. However, this is expressly not designed to be a substitute for much more specific texts on these subjects, nor an entire alternative to a cohesive course on biology or biochemistry. The intention is to introduce and explain the necessary aspects and elements of various metabolic pathways, reactions and abilities as required to advance the reader's understanding of this particular branch of biotechnology.

A large part of the reasons for approaching the subject in this way is the fact that there really is no such thing as a 'typical' environmental biotechnologist. Practitioners come into the profession from a wide variety of disciplines and by many different routes. Thus, amongst their ranks are agronomists, biochemists, biologists, botanists, enzymologists, geneticists, microbiologists, molecular biologists, process engineers and protein technologists, all of whom bring their own particular skills, knowledge base and experiences. The applied nature of environmental biotechnology is obvious. While the science underlying the processes themselves may be as pure as any other, what distinguishes this branch of biological technology are the distinctly real-life purposes to which it is put. Hence, part of the intended function of this book is to attempt to elucidate the former in order to establish the basis of the latter.

At the same time, as any applied scientist will confirm, what happens in the field under operational conditions represents a distinct compromise between the theoretical and the practically achievable. At times, anything more than an approximation to the expected results may be counted as something of a triumph of environmental engineering.

Closing Remarks

The celebrated astronomer and biologist, Sir Fred Hoyle, said that the solutions to major unresolved problems should be sought by the exploration of radical hypotheses, while simultaneously adhering to well-tried and tested scientific tools and methods. This approach is particularly valid for environmental biotechnology. With new developments in treatment technologies appearing all the time, the list of what can be processed or remediated by biological means is ever changing. By the same token, the applications for which biotechnological solutions are sought are also subject to alteration. For the biotech sector to keep abreast of these new demands it may be necessary to examine some truly 'radical hypotheses' and possibly make use of organisms or their derivatives in ways previously unimagined. This is the basis of innovation; the inventiveness of an industry is often a good measure of its adaptability and commercial robustness.

MICROBIAL BIOMASS

THE AQUATIC NITROGEN CYCLE

The aquatic nitrogen cycle is a complex system consisting of several different forms of nitrogen with different oxidation states that are interconnected by numerous conversion reactions. The different components of the aquatic nitrogen cycle can be separated from each other in space and/or time. The nitrogen cycle plays an important role in aquatic systems since nitrogen is usually the element limiting microbial production in these systems. However, due to its complexity, various aspects of the nitrogen cycle are still poorly understood.

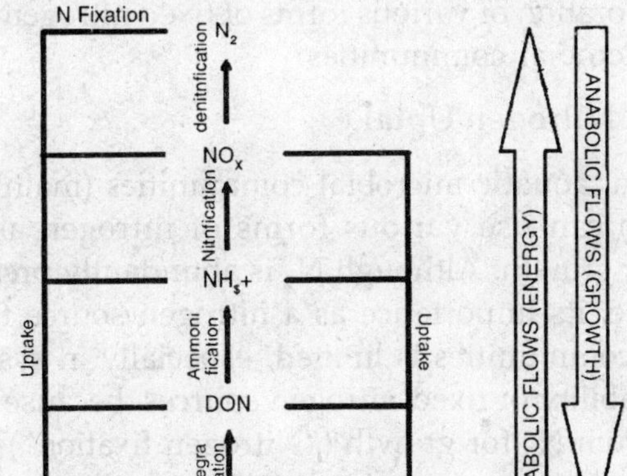

Fig. General Overview of Nitrogen Transformation Processes within the Nitrogen Cycle, Including Nitrogen flows Through Bacteria and algae.

Bioavailable nitrogen is present in the environment as "free" nitrogen (atmospheric N_2) and "fixed" nitrogen. The latter can roughly be divided into particulate nitrogen (functionally defined as all nitrogenous material that can be captured by filtration) and dissolved nitrogen (all nitrogenous material that passes through a filter, typically GF/F).

The dissolved fixed nitrogen pool can be subdivided into dissolved inorganic nitrogen (DIN) and dissolved organic nitrogen (DON). The DIN pool includes ammonium (NH^{4+}), nitrate (NO^{3-}) and nitrite (NO^{2-}) of which the last two can be taken together as NOx-. The DON pool is a more complex pool of which only a small fraction (< 20%) has been chemically identified (Benner 2002, Bronk 2002). The identifiable fraction mainly consists of urea, dissolved free amino acids (DFAAs) and dissolved combined amino acids (DCAAs, mainly proteins and peptides). The conversion reactions within the nitrogen cycle can have two "Directions". In catabolic (or "Dissimilatory") reactions, nitrogenous substrates are used as a source of energy, these reactions are generally mediated by bacteria. In anabolic (or "assimilatory") reactions, nitrogenous substrates are used as a source of nitrogen for growth (i.e. production of biomass), which requires energy. The uptake and incorporation of various forms of fixed nitrogen by natural aquatic microbial communities.

Microbial Nitrogen Uptake

Natural aquatic microbial communities (mainly bacteria and algae) can use various forms of nitrogen as nitrogen sources for growth. Although N_2 is abundantly present in the atmosphere, its importance as a nitrogen source for aquatic microbial communities is limited, especially in systems with high availability of fixed nitrogen sources, because the use of nitrogen from N_2 for growth ("Nitrogen fixation") requires a high energy investment and only a limited group of organisms (cyanobacteria and some heterotrophic bacteria) is capable of nitrogen fixation Therefore, the main nitrogen sources for microbial communities in coastal waters and sediments are DIN and DON.

Traditionally, studies on aquatic microbial nitrogen uptake focused on uptake of DIN by algae while bacteria were only considered as remineralizers of organic matter. However, the present view of microbial nitrogen uptake is considerably more complex since DON (especially amino acids and urea) has been found to be a potentially important nitrogen source as well and because it is now clear that bacteria are also important players in aquatic microbial nitrogen uptake. In addition to bacteria and algae, aquatic microbial communities can also include Archaea. However, at present their role in total microbial nitrogen uptake is unknown.

The pathways for uptake of dissolved nitrogen by algae and bacteria are schematized. The uptake process consists of two steps that are not necessarily coupled. The first step is the uptake of dissolved nitrogenous compounds from the surrounding water into the cells. Small molecules like NH4+, NOx- and urea can easily be taken up into cells while larger compounds like proteins and peptides first need to be broken down into individual amino acids that can either be directly taken up into the cells or be broken down to NH4+ which is then taken up into the cells.

The second step is incorporation of the nitrogen in new biomass. For this, intracellular NO^{3-} and urea are first converted to NH^{4+} which is used to synthesize amino acids that form the building blocks for proteinaceous biomass components. Since amino acids can directly be incorporated into biomass after uptake (i.e. little or no conversion is required), these are a highly preferred nitrogen source. Conversely, the microbial preference for NO^{x-} as a nitrogen source is low because of the relatively high energy investment required for its conversion to NH^{4+}. Preferences for NH4+ and urea are generally in between those for amino acids and NO^{x-}.

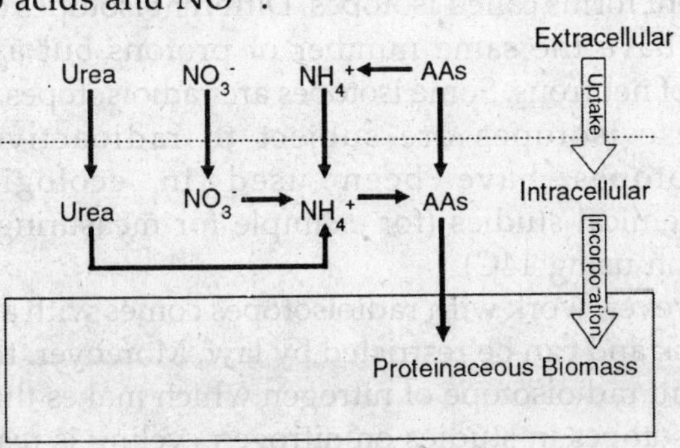

Fate of Microbial Nitrogen

An interesting, yet poorly studied, aspect of the aquatic nitrogen cycle is the fate of microbial nitrogen (i.e., what happens with the nitrogen after it has been incorporated into microbial biomass?) which is coupled with the fate of other elements in microbial biomass such as carbon. Although nitrogen can be retained in microbial biomass during continued growth (by cell division), microbial biomass (especially that of bacteria) is often relatively constant over time, meaning that production of new biomass is balanced by loss of biomass due to cell death. Death of microorganisms can be due to various causes such as viral lysis, predation and deteriorating environmental conditions.

After cell death, the leftovers (remnants) of microbial biomass are subject to degradation. During degradation, nitrogen is released back into the system as DON and/or DIN that can be used as nitrogen source for production of new microbial biomass or be lost from the system.

Some microbial biomass components are more resistant to degradation ("Refractory") than others and may accumulate in the system where they serve as a long term sink for microbial-derived nitrogen. Another potential fate of microbial nitrogen is transfer to higher trophic levels of the food web when microbial biomass is consumed by larger organisms.

Stable Isotopes in Ecological and Biogeochemical

Elements such as carbon (C) and nitrogen (N) can occur in different forms called isotopes. Different isotopes of a single element have the same number of protons but a different number of neutrons. Some isotopes are radioisotopes, meaning that these isotopes are subject to radioactive decay. Radioisotopes have been used in ecological and biogeochemical studies (for example for measuring primary production using 14C).

However, work with radioisotopes comes with a potential health risk and can be restricted by law. Moreover, there is no convenient radioisotope of nitrogen which makes that the use of radioisotopes in studies on nitrogen cycling is restricted to work with the short-lived radioisotope 13N (e.g., Zehr et al. 1988). Isotopes that are not subject to radioactive decay are stable isotopes. A single element can have two or more stable isotopes, of which one is by far the most abundant (e.g. 12C and 14N) while the other forms are rare (for example: 99.6% of all nitrogen in nature is 14N and only 0.4% is 15N). The abundance of these rare isotopes is measured by isotope ratio mass spectrometry (IRMS), a technique that uses a magnetic field to separate gas molecules (CO_2 in case of 13C analysis and N2 in case of 15N analysis) of different molecular weights resulting from different stable isotope compositions. The relative abundance of these molecules with different molecular weights can be used to determine the relative abundance of the rare (or "Heavy") isotope.

This abundance is typically expressed by the delta notation (E) which is the ratio between the rare ("Heavy") isotope and the common (or "Light") isotope relative to that of a standard in permille (%°) units.

Table. Carbon and Nitrogen Isotopes

		Type of Isotope	**Natural Abundance**
Carbon	10_C	Radio (½ life 17 Sec)	
	12_C	Stable	98.9%
	13_C	Stable	1.1%
	13_C	Radio (½ life 5730 yr)	
Nitrogen	13_N	Radio (½ life 10 min)	
	14_N	Stable	99.6%
	15_N	Stable	0.4%
	16_N	Radio (½ life 7 Sec)	

Stable isotopes can be used in ecological and biogeochemical studies in two different ways. The first way is to study the natural abundance of the rare, heavy isotope. Here, the relevant information lies in small differences in relative abundance of the heavy isotope between different compartments of a system.

These differences are the results of isotopic fractionation, which involves changes in the natural abundance of the heavy isotope during reactions caused by small differences in mass-dependent reaction rates of heavy and light isotopes. The second way to use stable isotopes is by stable isotope labeling. In this approach, a substrate that is highly enriched in the heavy isotope is deliberately added to a system after which the abundance of the heavy isotope (the "Label") is measured in various components of the system which allows to trace and quantify flows of the label trough these components. The material presented in this thesis all concerns stable isotope labeling work with the nitrogen isotope 15N and the carbon isotope 13C.

BIOMARKERS

Biomarkers are compounds that are unique to a specific group of organisms. Ideally, biomarker compounds are part of the organism's biomass and make up a relatively constant fraction of that biomass so that biomarker concentrations can be converted to total biomass of the group it represents. A variety of compounds have been used as biomarkers for various microbial groups.

Some of these compounds are rapidly degraded after death of the organism, which makes them specific for living biomass ("ecological biomarkers"), for example phospholipid-derived fatty acids (PLFAs). Other components are more resistant to degradation ("biogeochemical biomarkers"), for example D-amino acids from bacterial cell walls, which makes that these components can accumulate in a system, particularly in sediment where the abundance of these biomarkers can provide information on the contribution of material from the corresponding group of organisms to the total organic matter pool in the sediment. A third group of biomarkers are highly refractory molecules ("geological biomarkers").

The concentrations and characteristics of these "molecular fossils" in sediments can be used for paleoenvironmental reconstructions (e.g., Sinninghe Damsté et al. 2004). A valuable application of biomarkers in ecological- and biogeochemical studies is analysis of their stable isotope composition, which requires compound-specific stable isotope analysis.

This can be done at natural abundance level or in combination with stable isotope labeling where the latter provides the unique opportunity to directly link microbial identity (biomarker), biomass (biomarker concentration) and activity (label incorporation).

Selection of bacterial biomarkers that have been used in combination with stable isotope analysis. Biomarkers and stable isotope analysis of biomarkers, including some examples of applications in microbial ecology, can be found in Boschker and Middelburg while technical aspects of compound-specific stable isotope analysis.

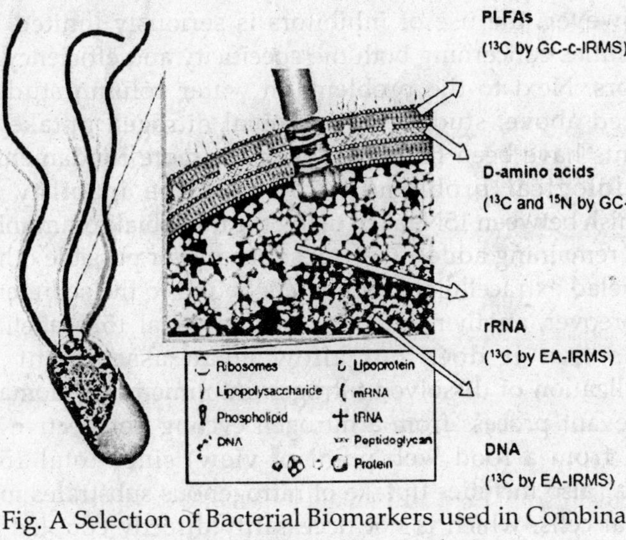

Fig. A Selection of Bacterial Biomarkers used in Combination with Stable Isotope Analysis.

The Problem

Microbial uptake of nitrogen from various nitrogenous substrates is generally studied by analysis of uptake of 15N from 15N-labeled substrates. This approach has frequently been used in water column studies where 15N enrichment of total particulate material in the water is considered to represent total microbial 15N uptake. However, more detailed assessment of 15N uptake by specific microbial groups (bacteria versus algae) has long been hampered by methodological problems. Until recently, two different approaches have been used to discriminate 15N uptake by bacteria versus algae. The first approach is size fractionation, which concerns physical separation of relatively small bacteria from relatively large algae by filtration or flow cytometry.

Although this approach may work in clear waters when a size difference between bacteria and algae is present, size fractionation is not suitable for studies in more turbid waters where bacteria and algae are closely associated with particulate matter. The second approach involves the use of inhibitors to eliminate activity of a specific microbial group, for example antibiotics to inhibit bacterial activity.

However, the use of inhibitors is seriously limited by uncertainties concerning both the specificity and efficiency of inhibitors. Next to the problems in water column studies presented above, studies on microbial nitrogen uptake in sediments have been hampered by even more fundamental methodological problems because of the inability to distinguish between 15N taken up by the microbial community and the remaining added 15N-substrate and/or possible other 15N-labeled extracellular compounds bound to the sediment.

Moreover, another limitation of traditional 15N-labeling work is that it does not allow clean assessment of immobilization of dissolved nitrogen into microbial biomass (the relevant process from a nitrogen cycling perspective as well as from a food web point of view) since total 15N "Uptake" also includes uptake of nitrogenous substrates into microbial cells, which is not necessarily directly coupled to incorporation into biomass. The method presented in this thesis is a new approach that has the potential to overcome the problems mentioned above and therefore may help to further clarify nitrogen flows through bacteria and algae in natural microbial communities.

The Solution

The method presented in this thesis involves the use of compound-specific stable isotope analysis of biomarkers. This approach has already been used at NIOO-CEME for some years in 13C-labeling studies where analysis of 13C incorporation into various group-specific PLFAs provided valuable information on 13C flows through different microbial groups, including bacteria and algae. However, since PLFAs do not contain nitrogen, another type of compounds was required for 15Nlabeling studies.

This thesis investigates the potential of 15N analysis of hydrolysable amino acids (HAAs), including the bacterial biomarker D-alanine (D-Ala). HAAs are amino acids that can be released from proteinaceous biomass (proteins and protein-like material) by hydrolysis in hot acid. HAAs make up a large fraction (50- 80%) of total microbial biomass meaning that analysis of 15N incorporation into total HAAs (THAAs) should provide a good indication of 15N incorporation into total microbial biomass. Amino acids can occur as D- and L-stereoisomers. Most HAAs are L-amino acids that are common biomass constituents of all organisms and therefore have very limited biomarker potential

Conversely, hydrolysable D-amino acids are rare and only occur in bacterial biomass, more specifically in peptidoglycan, a structural component of bacterial cell walls. Therefore, these D-amino acids can be used as bacterial biomarkers. Although different D-amino acids can be present in peptidoglycan, our method focuses on D-Ala because this is the only D-amino acid that is present in all bacteria and because it makes up a relatively stable fraction of total bacterial biomass. Moreover, D-Ala has been shown to be most suitable for stable isotope analysis by GC-c-IRMS.

The method can also be used to measure incorporation of the carbon isotope 13C into D-Ala and other HAAs, which allows assessment of carbon flows through bacteria and algae. This is particularly interesting in direct combination with analysis of nitrogen flows.

BIOREMEDIATION

Bioremediation can be defined as any process that uses microorganisms, fungi, green plants or their enzymes to return the natural environment altered by contaminants to its original condition. Bioremediation may be employed to attack specific soil contaminants, such as degradation of chlorinated hydrocarbons by bacteria. An example of a more general approach is the cleanup of oil spills by the addition of nitrate and/or sulfate fertilisers to facilitate the decomposition of crude oil by indigenous or exogenous bacteria.

Applications

Naturally occurring bioremediation and phytoremediation have been used for centuries. For example, desalination of agricultural land by phytoextraction has a long tradition. Bioremediation technology using microorganisms was reportedly invented by George M. Robinson. He was the assistant county petroleum engineer for Santa Maria, California. During the 1960's, he spent his spare time experimenting with dirty jars and various mixes of microbes

Bioremediation technologies can be generally classified as *in situ* or *ex situ*. *In situ* bioremediation involves treating the contaminated material at the site while *ex situ* involves the removal of the contaminated material to be treated elsewhere. Some examples of bioremediation technologies are bioventing, landfarming, bioreactor, composting, bioaugmentation, rhizofiltration, and biostimulation.

Not all contaminants, however, are easily treated by bioremediation using microorganisms. For example, heavy metals such as cadmium and lead are not readily absorbed or captured by organisms. The assimilation of metals such as mercury into the food chain may worsen matters. Phytoremediation is useful in these circumstances, because natural plants or transgenic plants are able to bioaccumulate these toxins in their above-ground parts, which are then harvested for removal.

The heavy metals in the harvested biomass may be further concentrated by incineration or even recycled for industrial use. The elimination of a wide range of pollutants and wastes from the environment is an absolute require increasing our understanding of the relative importance of different pathways and regulatory networks to carbon flux in particular environments and for particular compounds and they will certainly accelerate the development of bioremediation technologies and biotransformation processes.

Advantages

There are a number of cost/efficiency advantages to bioremediation, which can be employed in areas that are inaccessible without excavation. For example, hydrocarbon spills (specifically, petrol spills) or certain chlorinated solvents may contaminate groundwater, and introducing the appropriate electron acceptor or electron donor amendment, as appropriate, may significantly reduce contaminant concentrations after a lag time allowing for acclimation. This is typically much less expensive than excavation followed by disposal elsewhere, incineration or other *ex situ* treatment strategies, and reduces or eliminates the need for "pump and treat", a common practice at sites where hydrocarbons have contaminated clean groundwater.

Chapter 5

Infectious Disease Transmission

INFECTIOUS DISEASE

An infectious disease is a clinically evident disease resulting from the presence of pathogenic microbial agents, including pathogenic viruses, pathogenic bacteria, fungi, protozoa, multicellular parasites and aberrant proteins known as prions. These pathogens are able to cause disease in animals and/or plants.Infectious pathologies are usually qualified as contagious diseases (also called communicable diseases) due to their potentiality of transmission from one person or species to another. Transmission of an infectious disease may occur through one or more of diverse pathways including physical contact with infected individuals. These infecting agents may also be transmitted through liquids, food, body fluids, contaminated objects, airborne inhalation, or through vector-borne spread.

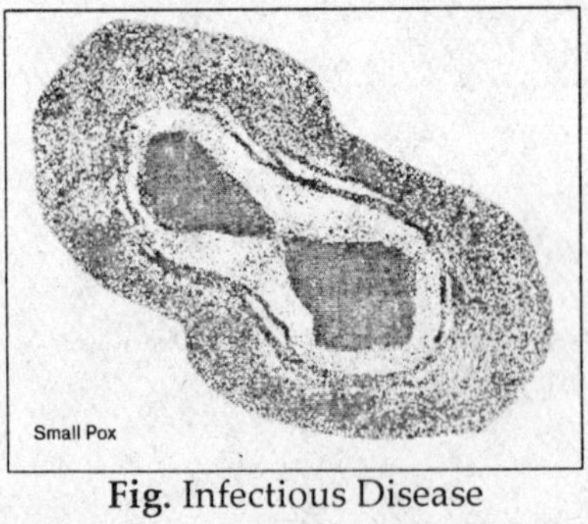

Fig. Infectious Disease

The term infectivity describes the ability of an organism to enter, survive and multiply in the host, while the infectiousness of a disease indicates the comparative ease with which the disease is transmitted to other hosts. An infection however, is not synonymous with an infectious disease, as an infection may not cause important clinical symptoms or impair host function.

Classification

Among the almost infinite varieties of microorganisms, relatively few cause disease in otherwise healthy individuals. Infectious disease results from the interplay between those few pathogens and the defenses of the hosts they infect. The appearance and severity of disease resulting from any pathogen depends upon the ability of that pathogen to damage the host as well as the ability of the host to resist the pathogen. Infectious microorganisms, or microbes, are therefore classified as either primary pathogens or as opportunistic pathogens according to the status of host defenses.

Primary pathogens cause disease as a result of their presence or activity within the normal, healthy host and their intrinsic virulence (the severity of the disease they cause) is, in part, a necessary consequence of their need to reproduce and spread. Many of the most common primary pathogens of humans only infect humans, however many serious diseases are caused by organisms acquired from the environment or which infect non-human hosts.

Organisms which cause an infectious disease in a host with depressed resistance are classified as opportunistic pathogens. Opportunistic disease may be caused by microbes that are ordinarily in contact with the host, such as pathogenic bacteria or fungi in the gastrointestinal or the upper respiratory tract and they may also result from (otherwise innocuous) microbes acquired from other hosts (as in Clostridium difficile colitis) or from the environment as a result of traumatic introduction (as in surgical wound infections or compound fractures). An opportunistic disease requires impairment of host defenses, which may occur as a result of genetic defects (such as Chronic granulomatous disease), exposure to antimicrobial drugs or immunosuppressive chemicals (as might occur following poisoning or cancer chemotherapy), exposure to ionizing radiation, or as a result of an infectious disease with immunosuppressive activity (such as with measles, malaria or HIV disease). Primary pathogens may also cause more severe disease in a host with depressed resistance than would normally occur in an immunosufficient host.

One way of proving that a given disease is "infectious", is to satisfy Koch's postulates, which demands that the infectious agent be identified only in patients and not in healthy controls and that patients who contract the agent also develop the disease. These postulates were first used in the discovery that Mycobacteria species cause tuberculosis. Koch's postulates cannot be met ethically for many human diseases because they require experimental infection of a healthy individual with a pathogen produced as a pure culture. Often, even diseases that are quite clearly infectious do not meet the infectious criteria.

For example, Treponema pallidum, the causative spirochete of syphilis, cannot be cultured in vitro - however the organism can be cultured in rabbit testes. It is less clear that a pure culture comes from an animal source serving as host than it is when derived from microbes derived from plate culture. Epidemiology is another important tool used to study disease in a population. For infectious diseases it helps to determine if a disease outbreak is sporadic (occasional occurrence), endemic (regular cases often occurring in a region), epidemic (an unusually high number of cases in a region), or pandemic (a global epidemic).

TRANSMISSION

An infectious disease is transmitted from some source. Defining the means of transmission plays an important part in understanding the biology of an infectious agent and in addressing the disease it causes. Transmission may occur through several different mechanisms. Respiratory diseases and meningitis are commonly acquired by contact with aerosolized droplets, spread by sneezing, coughing, talking, kissing or even singing. Gastrointestinal diseases are often acquired by ingesting contaminated food and water. Sexually transmitted diseases are acquired through contact with bodily fluids, generally as a result of sexual activity. Some infectious agents may be spread as a result of contact with a contaminated, inanimate object (known as a fomite), such as a coin passed from one person to another, while other diseases penetrate the skin directly.

Transmission of infectious diseases may also involve a "Vector". Vectors may be mechanical or biological. A mechanical vector picks up an infectious agent on the outside of its body and transmits it in a passive manner. An example of a mechanical vector is a housefly, which lands on cow dung, contaminating its appendages with bacteria from the feces and then lands on food prior to consumption. The pathogen never enters the body of the fly.

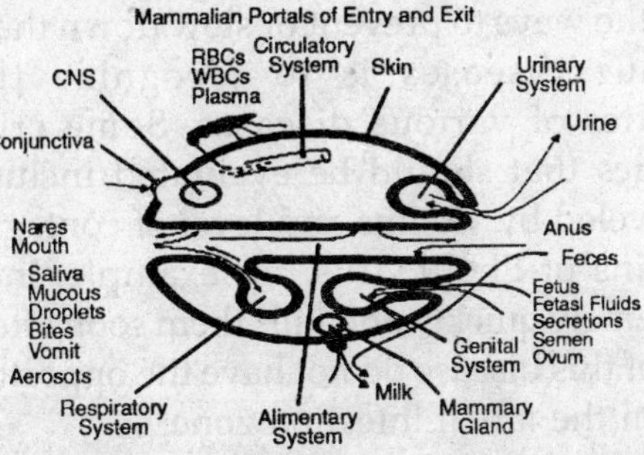

Fig. Infectious Disease of Transmission

In contrast, biological vectors harbor pathogens within their bodies and deliver pathogens to new hosts in an active manner, usually a bite. Biological vectors are often responsible for serious blood-borne diseases, such as malaria, viral encephalitis, Chagas disease, Lyme disease and African sleeping sickness. Biological vectors are usually, though not exclusively, arthropods, such as mosquitoes, ticks, fleas and lice. Vectors are often required in the life cycle of a pathogen. A common strategy, used to control vector borne infectious diseases, is to interrupt the life cycle of a pathogen, by killing the vector.

The relationship between virulence and transmission is

complex and has important consequences for the long term evolution of a pathogen. Since it takes many generations for a microbe and a new host species to co-evolve, an emerging pathogen may hit its earliest victims especially hard.

It is usually in the first wave of a new disease that death rates are highest. If a disease is rapidly fatal, the host may die before the microbe can get passed along to another host. However, this cost may be overwhelmed by the short term benefit of higher infectiousness if transmission is linked to virulence, as it is for instance in the case of cholera (the explosive diarrhea aids the bacterium in finding new hosts) or many respiratory infections (sneezing and coughing create infectious aerosols).

PREVENTING TRANSMISSION

One of the ways to prevent or slow down the transmission of infectious diseases is to recognize the different characteristics of various diseases. Some critical disease characteristics that should be evaluated include virulence, distance traveled by victims and level of contagiousness. The human strains of Ebola virus, for example, incapacitate its victims extremely quickly and kills them soon after. As a result, the victims of this disease do not have the opportunity to travel very far from the initial infection zone.

Also, this virus must spread through skin lesions or permeable membranes such as the eye. Thus, the initial stage of Ebola is not very contagious since its victims experience only internal hemorrhaging. As a result of the above features, the spread of Ebola is very rapid and usually stays within a relatively confined geographical area. In contrast, Human Immunodeficiency Virus (HIV) kills its victims very slowly by attacking their immune system. As a result, a lot of its victims transmit the virus to many others before even realizing that they are carrying the disease. Also, the relatively low virulence allows its victims to travel long distances, increasing the likelihood of an epidemic.

Another effective way to decrease the transmission rate of infectious diseases is to recognize the effects of small-world networks. In epidemics, there are often extensive interactions within hubs or groups of infected individuals and other interactions within discrete hubs of susceptible individuals. Despite the low interaction between discrete hubs, the disease can jump to and spread in a susceptible hub via a single or few interactions with an infected hub.

4Thus, infection rates in small-world networks can be reduced somewhat if interactions between individuals within infected hubs are eliminated. However, infection rates can be drastically reduced if the main focus is on the prevention of transmission jumps between hubs. The use of needle exchange programs in areas with a high density of drug users with HIV is an example of the successful implementation of this treatment method. Another example is the use of ring culling or vaccination of potentially susceptible livestock in adjacent farms to prevent the spread of the foot-and-mouth virus in 2001.

DIAGNOSIS AND THERAPY

Diagnosis of infectious disease sometimes involves identifying an infectious agent either directly or indirectly. In practice most minor infectious diseases such as warts, cutaneous abscesses, respiratory system infections and diarrheal diseases are diagnosed by their clinical presentation. Conclusions about the cause of the disease are based upon the likelihood that a patient came in contact with a particular agent, the presence of a microbe in a community and other epidemiological considerations.

Given sufficient effort, all known infectious agents can be specifically identified. The benefits of identification, however, are often greatly outweighed by the cost, as often there is no specific treatment, the cause is obvious, or the outcome of an infection is benign.

Specific identification of an infectious agent is usually only determined when such identification can aid in the treatment or prevention of the disease, or to advance knowledge of the course of an illness prior to the development of effective therapeutic or preventative measures. For example, in the early 1980s, prior to the appearance of AZT for the treatment of

AIDS, the course of the disease was closely followed by monitoring the composition of patient blood samples, even though the outcome would not offer the patient any further treatment options. In part, these st idies on the appearance of HIV in specific communities permitted the advancement of hypotheses as to the route of transmission of the virus.

By understanding how the disease was transmitted, resources could be targeted to the communities at greatest risk in campaigns aimed at reducing the number of new infections. The specific serological diagnostic identification and later genotypic or molecular identification, of HIV also enabled the development of hypotheses as to the temporal and geographical origins of the virus, as well as a myriad of other hypothesis. The development of molecular diagnostic tools have enabled physicians and researchers to monitor the efficacy of treatment with anti-retroviral drugs.

Molecular diagnostics are now commonly used to identify HIV in healthy people long before the onset of illness and have been used to demonstrate the existence of people who are genetically resistant to HIV infection. Thus, while there still is no cure for AIDS, there is great therapeutic and predictive benefit to identifying the virus and monitoring the virus levels within the blood of infected individuals, both for the patient and for the community at large.

Methods of Diagnosis

Diagnosis of infectious disease is nearly always initiated by medical history and physical examination. More detailed identification techniques involve the culture of infectious agents isolated from a patient. Culture allows identification of infectious organisms by examining their microscopic features, by detecting the presence of substances produced by pathogens and by directly identifying an organism by its genotype. Other techniques (such as X-rays, CAT scans, PET scans or NMR) are used to produce images of internal abnormalities resulting from the growth of an infectious agent. The images are useful in detection of, for example, a bone abscess or a spongiform encephalopathy produced by a prion.

Microbial Culture

Microbiological culture is a principal tool used to diagnose infectious disease. In a microbial culture, a growth medium is provided for a specific agent. A sample taken from potentially diseased tissue or fluid is then tested for the presence of an infectious agent able to grow within that medium. Most pathogenic bacteria are easily grown on nutrient agar, a form of solid medium that supplies carbohydrates and proteins necessary for growth of a bacterium, along with copious amounts of water.

A single bacterium will grow into a visible mound on the surface of the plate called a colony, which may be separated from other colonies or melded together into a "Lawn". The size, colour, shape and form of a colony is characteristic of the bacterial species, its specific genetic makeup (its strain) and the environment which supports its growth. Other ingredients are often added to the plate to aid in identification.

Plates may contain substances that permit the growth of some bacteria and not others, or that change colour in response to certain bacteria and not others. Bacteriological plates such as these are commonly used in the clinical identification of infectious bacteria. Microbial culture may also be used in the identification of viruses: the medium in this case being cells grown in culture that the virus can infect and then alter or kill. In the case of viral identification, a region of dead cells results from viral growth and is called a "plaque". Eukaryotic parasites may also be grown in culture as a means of identifying a particular agent.

In the absence of suitable plate culture techniques, some microbes require culture within live animals. Bacteria such as Mycobacterium leprae and T. pallidum can be grown in animals, although serological and microscopic techniques make the use of live animals unnecessary. Viruses are also usually identified using alternatives to growth in culture or animals. Some viruses may be grown in embryonated eggs. Another useful identification method is Xenodiagnosis, or the use of a vector to support the growth of an infectious agent. Chaga's disease is the most significant example, because it is difficult to directly demonstrate the presence of the causative agent, Trypanosoma cruzi in a patient, which therefore makes it difficult to definitively make a diagnosis. In this case, xenodiagnosis involves the use of the vector of the Chaga's agent T. cruzi, an uninfected triatomine bug (subfamily Triatominae), which takes a blood meal from a person suspected of having been infected. The bug is later inspected for growth of T. cruzi within its gut.

MICROSCOPY

Another principle tool in the diagnosis of infectious disease is microscopy. Virtually all of the culture techniques at some point, on microscopic examination for definitive identification of the infectious agent. Microscopy may be carried out with simple instruments, such as the compound light microscope, or with instruments as complex as an electron microscope. Samples obtained from patients may be viewed directly under the light microscope and can often rapidly lead to identification. Microscopy is often also used in conjunction with biochemical staining techniques and can be made exquisitely specific when used in combination with antibody based techniques.

For example, the use of antibodies made artificially fluorescent (fluorescently labeled antibodies) can be directed to bind to and identify a specific antigens present on a pathogen. A fluorescence microscope is then used to detect fluorescently labeled antibodies bound to internalized antigens within clinical samples or cultured cells. This technique is especially useful in the diagnosis of viral diseases, where the light microscope is incapable of identifying a virus directly.

Other microscopic procedures may also aid in identifying infectious agents. Almost all cells readily stain with a number of basic dyes due to the electrostatic attraction between negatively charged cellular molecules and the positive charge on the dye. A cell is normally transparent under a microscope and using a stain increases the contrast of a cell with its background. Staining a cell with a dye such as Giemsa stain or crystal violet allows a microscopist to describe its size, shape, internal and external components and its associations with other cells. The response of bacteria to different staining procedures is used in the taxonomic classification of microbes as well. Two methods, the Gram stain and the acid-fast stain, are the standard approaches used to classify bacteria and to diagnosis of disease.

The Gram stain identifies the bacterial groups Firmicutes and Actinobacteria, both of which contain many significant human pathogens. The acid-fast staining procedure identifies the Actinobacterial genera Mycobacterium and Nocardia.

BIOCHEMICAL TESTS

Biochemical tests used in the identification of infectious agents include the detection of metabolic or enzymatic products characteristic of a particular infectious agent. Since bacteria ferment carbohydrates in patterns characteristic of their genus and species, the detection of fermentation products is commonly used in bacterial identification. Acids, alcohols and gases are usually detected in these tests when bacteria are grown in selective liquid or solid media.

The isolation of enzymes from infected tissue can also provide the basis of a biochemical diagnosis of an infectious disease. For example, humans can make neither RNA replicases nor reverse transcriptase and the presence of these enzymes are characteristic of specific types of viral infections. The ability of the viral protein hemagglutinin to bind red blood cells together into a detectable matrix may also be characterized as a biochemical test for viral infection, although strictly speaking hemagglutinin is not an enzyme and has no metabolic function.

Serological methods are highly sensitive, specific and often extremely rapid tests used to identify microorganisms. These tests are based upon the ability of an antibody to bind specifically to an antigen.

The antigen, usually a protein or carbohydrate made by an infectious agent, is bound by the antibody. This binding then sets off a chain of events that can be visibly obvious in various ways, dependent upon the test. For example, "Strep throat" is often diagnosed within minutes and is based on the appearance of antigens made by the causative agent, S. pyogenes, that is retrieved from a patients throat with a cotton swab. Serological tests, if available, are usually the preferred route of identification, however the tests are costly to develop and the reagents used in the test often require refrigeration. Some serological methods are extremely costly, although when commonly used, such as with the "strep test", they can be inexpensive.

Molecular Diagnostics

Technologies based upon the polymerase chain reaction (PCR) method will become nearly ubiquitous gold standards of diagnostics of the near future, for several reasons. First, the catalog of infectious agents has grown to the point that virtually all of the significant infectious agents of the human population have been identified.

Second, an infectious agent must grow within the human body to cause discase; essentially it must amplify its own nucleic acids in order to cause a disease. This amplification of nucleic acid in infected tissue offers an opportunity to detect the infectious agent by using PCR. Third, the essential tools for directing PCR, primers, are derived from the genomes of infectious agents and with time those genomes will be known, if they are not already.

Thus, the technological ability to detect any infectious agent rapidly and specifically are currently available. The only remaining blockades to the use of PCR as a standard tool of diagnosis are in its cost and application, neither of which is insurmountable.

The diagnosis of a few diseases will not benefit from the development of PCR methods, such as some of the clostridial diseases (tetanus and botulism). These diseases are fundamentally biological poisonings by relatively small numbers of infectious bacteria that produce extremely potent neurotoxins. A significant proliferation of the infectious agent does not occur, this limits the ability of PCR to detect the presence of any bacteria.

Clearance and Immunity

Infection with most pathogens does not result in death of the host and the offending organism is ultimately cleared after the symptoms of the disease have waned. This process requires immune mechanisms to kill or inactivate the inoculum of the pathogen. Specific acquired immunity against infectious diseases may be mediated by antibodies and/or T lymphocytes. Immunity mediated by these two factors may be manifested by:

- A direct effect upon a pathogen, such as antibody-initiated complement-dependent bacteriolysis, opsonoization, phagocytosis and killing, as occurs for some bacteria,
- Neutralization of viruses so that these organisms cannot enter cells,
- Or by T lymphocytes which will kill a cell parasitized by a microorganism.

The immune response to a microorganism often causes symptoms such as a high fever and inflammation and has the potential to be more devastating than direct damage caused by a microbe. Resistance to infection (immunity) may be acquired following a disease, by asymptomatic carriage of the pathogen, by harboring an organism with a similar structure (crossreacting), or by vaccination. Knowledge of the protective antigens and specific acquired host immune factors is more complete for primary pathogens than for opportunistic pathogens.

Immune resistance to an infectious disease requires a critical level of either antigen-specific antibodies and/or T cells when the host encounters the pathogen. Some individuals develop natural serum antibodies to the surface polysaccharides of some agents although they have had little or no contact with the agent, these natural antibodies confer specific protection to adults and are passively transmitted to newborns.

MORTALITY FROM INFECTIOUS DISEASES

The World Health Organization collects information on global deaths by International Classification of Disease (ICD) code categories. The top infectious disease killers which caused more than 100,000 deaths in 2002 (estimated). 1993 data is included for comparison.

Worldwide Mortality due to Infectious Diseases

Rank	Cause of death	Deaths 2002	Percent age of all deaths	Deaths 1993	1993 Rank
N/A	All infectious diseases	14.7 million	25.9%	16.4 million	32.2%
1	Lower respiratory infections	3.9 million	6.9%	4.1 million	1
2	HIV/AIDS	2.8 million	4.9%	0.7 million	7
3	Diarrheal diseases	1.8 million	3.2%	3.0 million	2
4	Tuberculosis (TB)	1.6 million	2.7%	2.7 million	3
5	Malaria	1.3 million	2.2%	2.0 million	4
6	Measles	0.6 million	1.1%	1.1 million	5
7	Pertussis	0.29 million	0.5%	0.36 million	7
8	Tetanus	0.21 million	0.4%	0.15 million	12
9	Meningitis	0.17 million	0.3%	0.25 million	8
10	Syphilis	0.16 million	0.3%	0.19 million	11
11	Hepatitis B	0.10 million	0.2%	0.93 million	6
12-17	Tropical diseases(6)	0.13 million	0.2%	0.53 million	9, 10, 16-18

The top three single agent/disease killers are HIV/AIDS, TB and malaria. While the number of deaths due to nearly every disease have decreased, deaths due to HIV/AIDS have increased fourfold. Childhood diseases include pertussis, poliomyelitis, diphtheria, measles and tetanus. Children also make up a large percentage of lower respiratory and diarrheal deaths.

Historic Pandemics

A pandemic (or global epidemic) is a disease that affects people over an extensive geographical area.

- Plague of Justinian, from 541 to 750, killed between 50 and 60 percent of Europe's population.
- The Black Death of 1347 to 1352 killed 25 million in Europe over 5 years (estimated to be between 25 and 50% of the populations of Europe, Asia and Africa - the world population at the time was 500 million).
- The introduction of smallpox, measles and typhus to the areas of Central and South America by European explorers during the 15th and 16th centuries caused pandemics among the native inhabitants. Between 1518 and 1568 disease pandemics are said to have caused the population of Mexico to fall from 20 million to 3 million.
- The first European influenza epidemic occurred between 1556 and 1560, with an estimated mortality rate of 20%.
- Smallpox killed an estimated 60 million Europeans in the 18th century alone. Up to 30% of those infected, including 80% of the children under 5 years of age, died from the disease and one third of the survivors went blind.
- The Influenza Pandemic of 1918 (or the Spanish Flu) killed 25-50 million people (about 2% of world population of 1.7 billion). Today Influenza kills about 250,000 to 500,000 worldwide each year.

Emerging Diseases and Pandemics

In most cases, microorganisms live in harmony with their hosts. Such is the case for many tropical viruses and the insects, monkeys, or other animals in which they have lived and reproduced. Because the microbes and their hosts have co-evolved, the hosts gradually become resistant to the microorganisms. When a microbe jumps from a long-time animal host to a human being, it may cease to be a harmless parasite and become pathogenic.

With most new infectious diseases, some human action is involved, changing the environment so that an existing microbe can take up residence in a new niche. When that happens, a pathogen that had been confined to a remote habitat appears in a new or wider region, or a microbe that had infected only animals suddenly begins to cause human disease.

Several human activities have led to the emergence and spread of new diseases:

- Encroachment on wildlife habitats. The construction of new villages and housing developments in rural areas brings people into contact with animals—and the microbes they harbor.
- Changes in agriculture. The introduction of new crops attracts new crop pests and the microbes they carry to farming communities, exposing people to unfamiliar diseases.
- The destruction of rain forests. As countries make use of their rain forests, by building roads through forests and clearing areas for settlement or commercial ventures, people encounter insects and other animals harboring previously unknown microorganisms.
- Uncontrolled urbanization. The rapid growth of cities in many developing countries tends to concentrate large numbers of people into crowded areas with poor sanitation. These conditions foster transmission of contagious diseases.
- Modern transport. Ships and other cargo carriers often harbor unintended "passengers", that can spread diseases to faraway destinations. While with international jet-airplane travel, people infected with a disease can carry it to distant lands, or home to their families, before their first symptoms appear.

TRANSMISSION OF INFECTIOUS DISEASES DURING COMMERCIAL AIR TRAVEL

Over 1 billion passengers travel by air annually; 50 million of these travel to the developing world. Although infrequently reported and very difficult to assess accurately, there is a risk of disease transmission during commercial air travel and this risk has become the focus of heightened attention. The growing mobility of people and popularity of airline transportation has amplified the potential for disease to be transmitted to passengers not only during but also before and after flights. Here, we review knowledge about transmission of infectious diseases associated with commercial air travel, with particular emphasis on transmission within the aircraft passenger cabin.

The Aircraft Cabin Environment

During flight, the aircraft cabin is a ventilated, enclosed environment that exposes passengers to hypobaric hypoxia, dry humidity and close proximity to fellow passengers. This space is regulated by an environmental system that controls pressurisation, temperature, ventilation and air filtration on the aircraft. Although this system is wholly automated, the number of airconditioning packs in operation, zone temperatures and the mixture of fresh and re-circulated air delivered to the cabin can be manipulated by the flight deck.1 When parked at the terminal, fresh air is supplied to the aircraft by auxiliary power units.

During flight, fresh air is supplied into the cabin from the engines where the air is heated, compressed, cooled and passed into the cabin to be circulated by the ventilation system. The outside air is assumed to be sterile at typical cruising altitudes. Air circulation patterns aboard standard commercial aircraft are side-to-side (laminar) with air entering the cabin from overhead, circulating across the aircraft and exiting the cabin near the floor.

Little front-toback (longitudinal) airflow takes place. This air circulation pattern divides the air flow into sections within the cabin, thereby limiting the spread of airborne particles throughout the passenger cabin.

Most commercial aircraft in service recirculate 50% of the air delivered to the passenger cabin for improved control of cabin circulation, humidity and fuel efficiency. This recirculated air usually passes through high efficiency particulate air filters (HEPA) before delivery into the cabin. Normal airline cabin air exchange rates range from 15 to 20 air changes per hour compared with 12 air changes per hour for a typical office building. Ventilation capacity varies substantially, dependent on the aircraft type but typically averages 10 (4·7 L/s) cubic feet per minute.

Ventilation rates can also vary within the different cabin sections, such as first and economy class. In general, HEPA filters used on commercial airlines have a particle-removing efficiency of 99·97% at 0·3 microns. These filters remove dust, vapours, bacteria and fungi. HEPA filters also effectively capture viral particles because viruses usually spread by droplet nuclei.1 No ventilation operational standards for commercial aircraft are available.

Although a survey showed that most air carriers equip their large aircraft with HEPA filters, neither the Civil Aviation Authority nor the Federal Aviation Administration require their use. cabin air quality has been the focus of many media investigations and criticism from special interest groups. Most of this concern is associated with the perception that airborne particles are distributed throughout the entire cabin by the ventilation system. However, no peer-reviewed scientific work links cabin air quality and aircraft ventilation rates to heightened health risks compared with other modes of transport or with office buildings,so this work is limited.

These shortcomings motivated the UK government to form the Aviation Health Working Group in 2000 to further investigate health aspects of air travel, including cabin air quality and safety. In a detailed study into cabin air quality released in 2004, the group concluded that temperature, humidity, air speed and concentrations of carbon monoxide, carbon dioxide and microbiological flora aboard 14 commercial flights using British Aerospace and Boeing 300 aircraft were similar to other reported studies.

The European Cabin Air study coordinated by the Building Research and Consultancy continues to investigate environmental aspects within the passenger cabin. These efforts will probably lead to improved international regulations for the certification, inspection and maintenance of aircraft environmental control systems.

MODES OF DISEASE TRANSMISSION

Four routes for the spread of microorganisms exist: contact, airborne, common vehicle and vector-borne. Contact transmission involves direct contact in which body-to-body contact takes place, or indirect in which the susceptible person comes into contact with a contaminated intermediate host (fomite). Large droplet transmission is judged a form of contact transmission in which large droplets contaminated with microorganisms are generated when an infected person sneezes, coughs, or talks.

These droplets are propelled short distances and deposited on a susceptible host's conjunctiva or mucosa. Airborne transmission happens by aerolisation of an infectious agent through droplet nuclei.

These residual droplets become aerosolised and disperse widely, dependent on environmental conditions and remain suspended in air for indefinite periods. Common vehicle transmission involves one inanimate vehicle, which transmits infection to many hosts and typically applies to microorganisms spread by food and water.

Vector-borne transmission results from the spread of disease by insects and vermin. The infectious diseases that have been transmitted on commercial airlines. All types of disease transmission are relevant to commercial air travel. Large droplet and airborne mechanisms probably represent the greatest risk for passengers within the aircraft because of the high density and close proximity of passengers. In addition to proximity, successful spread of contagion to other hosts is dependent on many factors, including infectiousness of the source; pathogenicity of the microorganism; duration of exposure; environmental conditions (ventilation, humidity, temperature) and host-specific factors such as general health and immune status. How these factors affect risk of disease transmission within the aircraft cabin is unclear.

Risk of Transmission

The risk of disease transmission within the confined space of the aircraft cabin is difficult to determine. Insufficient data prohibits meta-analysis, which would allow an idea of the probability of disease transmission for each respective contagion. Many of the available epidemiological studies are compromised by reporting bias caused by incomplete passenger manifests, thereby complicating risk assessment. Despite these limitations, data suggest that risk of disease transmission to other symptom-free passengers within the aircraft cabin is associated with sitting within two rows of a contagious passenger for a flight time of more than 8 h.

This association is mainly derived from investigations of inflight transmission of tuberculosis, but is believed to be relevant to other airborne infectious diseases. Some variation in this association has been reported, with one outbreak of severe acute respiratory syndrome (SARS) in which passengers seated as far as seven rows from the source passenger were affected. Risk of disease transmission within the aircraft cabin also seems to be affected by cabin ventilation. In general, proper ventilation within any confined space reduces the concentration of airborne organisms in a logarithmic fashion and one air exchange removes 63% of airborne organisms suspended in that particular space.

The main laminar flow pattern within the aircraft cabin with the practice of frequent cabin air exchanges and use of HEPA filtration for recirculated air clearly limits transmission of contagion. Transmission becomes widespread within all sections of the passenger cabin when the ventilation system is nonoperational, as shown by an influenza outbreak when passengers were kept aboard a grounded aircraft with an inoperative ventilation system. Risk assessment incorporating epidemiological data into mathematical models may show how proximity and ventilation affects disease transmission aboard commercial airlines. Deterministic modelling with data from an in-flight tuberculosis investigation revealed that doubling ventilation rate within the cabin reduced infection risk by half. Risk also reduced exponentially to almost zero in passengers seated 15 seats from the infectious source. Clearly ventilation provides a crucial determinant of risk and efforts to increase ventilation will reduce risk.

AIRBORNE AND LARGE DROPLET-TRANSMITTED DISEASES

Tuberculosis

Tuberculosis is a serious global threat and estimates suggest that about a third of the world's population has the disease. The transmission of Mycobacterium tuberculosis is the most studied model of the spread of airborne pathogens aboard aircraft. Several studies about in-flight transmission of tuberculosis have been reported, with most being done in the mid-1990s. Two of the seven investigations revealed a probable link of onboard transmission. In the first occurrence, a flight attendant was the index case and two documented tuberculin skin test conversions occurred during 5 months in 1992 in 212 fellow crew members and 59 frequent flyer passengers. The second and largest, incident was of a passenger with pulmonary tuberculosis travelling from Baltimore to Chicago and then on to Honolulu.

Four of 15 fellow passengers seated within two rows of the index passenger had positive tuberculin skin test conversion. Although there is a risk of tuberculosis within the aircraft cabin, no cases of active disease have been reported as a result of air travel. Transmission within the aircraft cabin seems to be more likely with close proximity to a contagious passenger (within two rows) over a long time (greater than 8 h) and not as a result of the practice of recirculating 50% of the cabin air. An overall probability of infection in the order of one in 1000 when a symptomatic source is present has been suggested and this probability of risk is similar to, if not less than, those in other confined spaces.

SARS

SARS is a non-typical pneumonia caused by a coronavirus. The global spread by air travellers and in-flight spread of SARS has been documented. The disease is believed to usually be spread by large aerosolised droplets or by direct and indirect contact, but airborne or small droplet transmission better explains the distribution of SARS cases that has occurred on commercial airlines. Svidence suggests that transmission of SARS during the Amoy Gardens outbreak in Hong Kong was a result of airborne spread via a viral plume.

A total of 40 flights have been investigated for carrying SARS-infected passengers. ive of these flights have been associated with probable on-board transmission of SARS in 37 passengers. Most of those passengers were seated within five rows of the index case. One 3-hour flight carrying 120 passengers travelling from Hong Kong to Beijing on March 15, 2003, began a superspreading event accounting for 22 of the 37 people who contracted SARS after air travel.

Laboratory-confirmed SARS coronavirus infection occurred in 16 people, two passengers had a diagnosis of probable SARS and four were reported to have SARS but could not be interviewed. The number of secondary cases from that flight remains under investigation, but more than 300 people might have been affected. This pattern could be important because it did not follow the typical example of in-flight transmission of airborne pathogens—ie, risk of disease transmission is associated with a flight time of more than 8 h and sitting within two rows of the index passenger.

The duration of the Hong Kong to Beijing flight was 3 h and affected passengers were seated seven rows in front and five rows behind the index passenger. Possible explanations for this outbreak distribution include: airborne transmission rather than direct contact spread; a malfunctioning cabin filtration system; and passengers infected before or after the flight. No on-board transmissions have occurred since late March, 2003, when the WHO issued specific guidelines for in-flight containment of SARS. As the first severe contagious disease of the 21st century, SARS exemplifies the everpresent threat of new infectious diseases and the real potential for rapid spread made possible by the volume and speed of air travel. Finally, the distribution pattern of SARS transmission aboard the flight emphasises the need to study airborne transmission patterns aboard commercial aircraft.

Common Cold

Common cold outbreaks as a result of air travel have not been reported, which could be attributable to the difficulties of investigating such outbreaks in view of the ubiquitous nature of the common cold. One study compared the risk of developing an upper respiratory tract infection during air travel in passengers flying on aircraft that recirculated 50% cabin air versus aircraft using 100% fresh air in the passenger cabin.36 Recirculation of aircraft cabin air was not a risk factor for contracting upper respiratory tract infection symptoms.

Influenza

The aircraft as a vector for global spread of influenza strains is a greater concern than is in-flight transmission. The fact that influenza outbreaks worldwide have been affected by influenza strains imported by air travel is well established; however, only three studies of in-flight transmission of influenza have been reported. The first was in an outbreak of influenza A/Texas strain aboard a commercial carrier in 1979 that resulted in 72% of all passengers aboard the airline contracting influenza within 72 h. The secondary attack rate in their families was estimated to be 20% within 2 weeks. The high transmission rate in this particular case was believed attributable to passengers being kept aboard the aircraft for 3 h with an inoperative ventilation system while repair work was being done.

The second study described an outbreak of influenza A/ Taiwan/1/86 at a naval air station in 1989 in military personnel who were returning from temporary duty.55 Transmission of influenza occurred both on the ground and aboard two DC-9 aircraft that transported the squadron from Puerto Rico to a Florida naval station. The third outbreak happened in 1999 in mine workers travelling on a 75-seat aircraft. 15 passengers travelling with the index case developed symptoms within 4 days. Nine of the 15 were seated within two rows and all were seated within five rows, of the index case. No further influenza outbreaks aboard commercial aircraft have been reported since 1999.

Other Airborne Diseases

Meningococcal disease occurs after direct contact with respiratory secretions and is associated with high morbidity and mortality rates. A case of meningococcal disease associated with air travel is defined as the development of the illness within 14 days of travel on a flight lasting at least 8 h, including ground delay, take off and landing. The US Centers for Disease Control received 21 reports of suspected air travel-associated meningococcal disease from 1999 to 2001.

In all cases the index person was contagious while aboard a commercial flight, but no secondary cases of the disease were reported. People seated next to an ill passenger should be quickly contacted and given chemoprophylaxis within 24 h of identifying the index case. Chemoprophylaxis given more than 14 days after onset of illness in the index case is probably of little or no value.

Measles is an airborne and highly contagious viral infection with an attack rate of about 80%. Transmission can occur during the prodromal illness and passengers might be unaware of their diagnosis at the time of travel. Measles is no longer endemic in the USA and importations from developing countries account for most outbreaks. Imported measles and associated cases accounted for at least 17% of all reported cases within the USA during 1982. From 1996 to 2000 30% of all imported measles cases were estimated to be in people who flew while symptomatic with the disease.

Three case studies have described measles transmission during commercial air travel. A case report in 1982 identified seven secondary cases of measles epidemiologically linked to the index case as a result of in-transit exposure; one was a passenger flying on the same aircraft as the index case and five others were people who had visited at least one common departure gate with the index case. In the same year, another study reported an index passenger who infected two fellow passengers on a flight from Venezuela to Miami. A third report recorded eight cases of in-flight transmitted measles during a 10-h flight from New York to Tel Aviv in 1994. The source case was not identified but was speculated to be a crew member. In 2004, a passenger with measles travelled from Japan to Hawaii, but this did not result in any transmission to other air travellers. International adoptions have a significant role in the number of imported measles cases.

COMMON VEHICLE DISEASES

The most commonly reported diseases transmitted on aircraft have been spread by the fecal-oral route via contaminated food. A total of 41 in-flight food-borne outbreaks resulting in 11 deaths were documented between 1947 and 1999. Salmonella is the most usually reported food-borne pathogen spread via a commercial airline, with fifteen documented outbreaks between 1947 and 1999 infecting nearly 4000 passengers and resulting in seven deaths. Eight food-borne outbreaks caused by Staphylococcus and one associated death were reported between 1947 and 1999.

One of the largest cases involved 57% of the passengers served a ham omelette on an international flight in 1975. Surprisingly, only one viral-induced enteritis outbreak has been described. In this incident, contaminated orange juice was the vehicle of transmission and a Norwalk-like agent was isolated from fecal samples of 30 ill passengers.

There have been a few reported cases of ill passengers as a result of food or water contaminated with Vibrio cholerae consumed during international air travel. The first documented in-flight outbreak was in 1972 on a flight from London to Sydney via Singapore. Of the 47 people who developed cholera, which was attributed to a cold appetiser served during the flight, one died. The largest outbreak of airline-associated cholera occurred in 1992, during a cholera epidemic in Latin America. During a flight from Buenos Aires via Lima to Los Angeles, 75 passengers developed cholera, resulting in ten passengers being admitted and one death. A cold seafood dish, prepared in one of the cholera-affected countries, was implicated as the source of transmission. No food-borne or water-borne outbreaks have been reported over the past 5 years, which is probably attributable to greater use of prepackaged frozen meals, improved food handling and inspection, but might represent under-reporting by passengers or reporting bias.

VECTOR-BORNE AND ZOONOTIC DISEASES

Vector-borne diseases are very frequent causes of morbidity and mortality in many parts of the world and the potential of their importation via commercial aircraft remains a risk. Many cases of malaria occurring in and around airports all over the world in people who had not travelled to endemic areas, known as airport malaria, is evidence that malaria-carrying mosquitoes can be imported on aircraft. A total of 87 cases of airport malaria have been reported, 75 of which happened in Europe.

Dengue and yellow fever are both transmitted by mosquitoes of the genus Aedes. The Aedes mosquito has been introduced into countries where it had not previously been present and many of these mosquitoes were likely to have spread by aircraft.

A case of dengue fever was reported in Germany in a couple returning from a trip to Hawaii.71 Airport transmission of dengue fever was suspected in this particular case. The disinsection of aircraft—spraying aircraft before landing to kill insects and vector control around airports, as well as immunisation requirements, seem to have been effective in preventing outbreaks in non-endemic areas.

Although International Health Regulations recommend disinsecting aircraft travelling from countries with malaria and other vector-borne diseases, only five countries do so. Common and exotic animals are regularly transported on aircraft and may carry disease. Many zoonotic pathogens cause emerging and reemerging diseases. Up to now, no zoonotic outbreaks associated with air travel have been reported; however, continued monitoring of air transport of animals, especially from developing countries, is needed.

Bioterrorism Agents

The potential for spread of bioterrorism agents via air travel exists. A well written, comprehensive review of bioterrorism agents and the implications of air transport is available and a new interest in previously eradicated smallpox has emerged. In-flight spread of this disease has been documented and so the potential for in-flight spread of smallpox is of public health concern. In 1963 an epidemic was reported in Sweden, in which the index case was attributed to in-transit exposure, either at the air terminal or on the aeroplane and caused 24 secondary cases and four deaths.

Viruses causing viral haemorrhagic fever, such as ebola and lassa, have also been the focus of media attention and have been investigated for potential transmission while aboard an aircraft. Although aerosol spread of ebola has not been documented in man, this mode of transmission occurs in non-human primates. Lassa, known to be transmitted via large droplets, is thought to have an incubation period of up to 3 weeks, making infectious passengers potentially symptom-free and unaware of their status at the time of travel.

A study of passengers exposed to an index case of lassa fever in-flight noted no evidence of transmission, even in the 19 passengers seated within two rows of the index passenger. Because both lassa and ebola viruses have frequent fatal outcomes and no vaccine is available, appropriate infection control procedures should be followed to prevent the transmission of these diseases. Lassa fever is treatable with ribavirin if the drug is given within the first 6 days of illness, making early diagnosis especially important.

Management of Infectious Disease Exposures Aboard Commercial Aircraft

Early recognition and appropriate infection control measures are needed when passengers become exposed to an infectious or potentially infectious passenger. Government and international laws provide legal authority to control the movement of passengers with communicable diseases. This authority ranges from issuing travel alerts to quarantine of passengers arriving at airports.

Although air carriers have the right to refuse to take passengers who are ill with a communicable disease or medically unfit for air travel, systematic screening of passengers for contagious diseases and excluding passengers with infectious symptoms is impractical. Health care professionals are expected to identify individuals who are unfit for air travel, or advise the flying public of how to safely travel by air. Prevention of a disease outbreak is the most important means of control and travellers should therefore be advised to postpone any air travel when they are ill. Good hand hygiene has been proven to reduce the risk of disease transmission and air travellers should make it part of their normal travel routine. Although masks play a crucial part in infection control in health care settings, their use is unproven in disease control within the aircraft cabin.

However, a mask should be placed on a passenger suspected of having SARS and the passenger should be isolated. US CDC and WHO guidelines exist, on when and how to notify passengers and flight crew after they have been exposed to infectious diseases aboard commercial aircraft. Briefly, the airline is consulted whenever a health department determines that a passenger is infectious at time of flight. The airline then notifies passengers and flight crews in writing. Notification is typically limited to flights longer than 8 h and, in some cases, dependent upon the design of the aircraft, to passengers seated only in the same cabin area. Health officials have access to passenger manifests, but these lists are frequently incomplete or unavailable, making it difficult to locate potentially exposed passengers. Contact information in one large investigation into inflight tuberculosis transmission was inaccurate for 15% of passengers. Although air carriers are under no obligation to archive passenger manifests, most have internal policies to do so for up to 3 months.

RESPIRATORY DISEASES AND DISORDERS

Common Respiratory Ommon Respiratory Symptoms and Complaints

There is a common misperception that older persons tend to overestimate or exaggerate respiratory symptoms; however, the opposite is more often true. For example, many older persons and their physicians tend to underestimate the importance of dyspnea, which may go undiagnosed until advanced disease is evident. This is partly due to the fact that dyspnea is blamed on deconditioning and age.

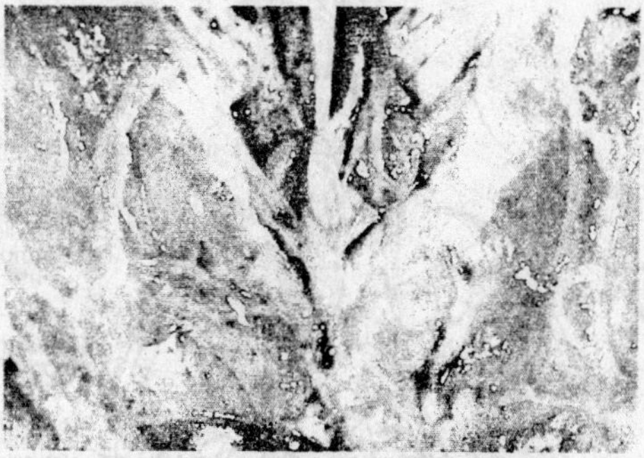

Fig. Respiratory Diseases

Older persons will often adjust their activity level to compensate for insidiously shrinking lung function and disabling dyspnea. Such changes in life style often go unnoticed by family, the patient's physician and even the patient. Pulmonary or cardiac disorders, or both, may underlie such modifications in life style and testing (eg, pulmonary function tests or chest radiography) may reveal major abnormalities such as asthma, emphysema, or pulmonary fibrosis.

Another complicating feature of symptom recognition in older persons is that older persons often have more than one explanation for their problems.

A patient may have overlapping symptoms of dyspnea, cough and wheezing because of a combination of diseases such as asthma or emphysema, obstructive sleep apnea, heart failure and gastroesophageal reflux.

DYSPNEA

Dyspnea becomes prominent in end-stage lung diseases such as chronic obstructive pulmonary disease (COPD) and idiopathic pulmonary fibrosis. Importantly, the level of dyspnea is the best predictor of quality of life, yet it does not correlate with either oxygenation or pulmonary function tests. A thorough history and physical examination can help tailor both testing and empirical treatment choices.

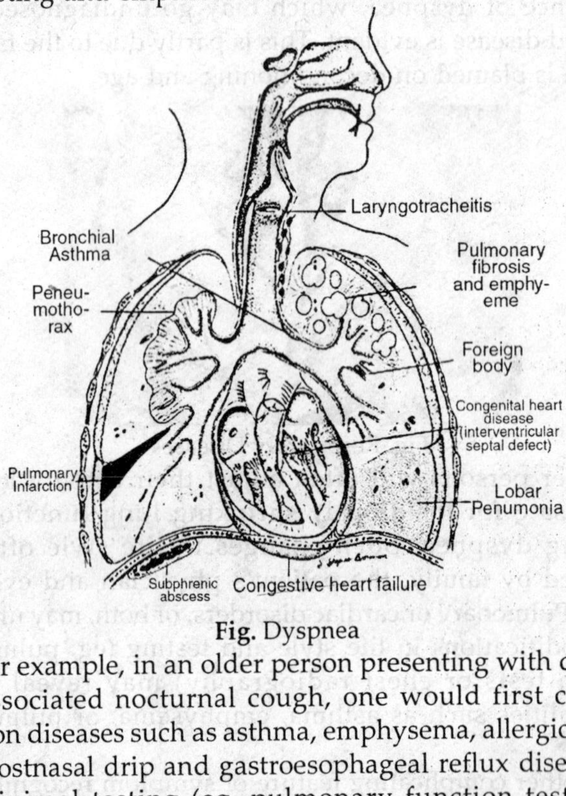

Fig. Dyspnea

For example, in an older person presenting with dyspnea and associated nocturnal cough, one would first consider common diseases such as asthma, emphysema, allergic rhinitis with postnasal drip and gastroesophageal reflux disease.

Minimal testing (eg, pulmonary function tests only) followed by an empiric trial directed toward the most likely cause would be a reasonable approach.

In the same patient, the presence of significant weight loss or constitutional symptoms (fever, night sweats) could suggest other diseases, such as malignancy or tuberculosis.

At times, the particular language the patient choses to

describe the dyspnea can be revealing, such as "heavy" for cardiac dysfunction or deconditioning or "Tight" for angina or asthma.

The common causes of dyspnea in older persons to consider include COPD, cardiac disease, asthma, interstitial lung disease and deconditioning.

CHRONIC COUGH

Fortunately, most patients can be reassured that chronic cough, though particularly annoying, usually has a benign cause.

By far, the most common causes of chronic cough are postnasal drip, asthma and gastroesophageal reflux. These three diagnoses account for over 90% of the causes identified in most series and a reasonable approach to the treatment of chronic cough, then, is empiric treatment for these conditions. Not infrequently, a combination of these conditions may contribute, so treatment for multiple causes may be warranted when single therapies are ineffective.

Less common yet important differential diagnostic considerations of cough in older persons would include drug effects (eg, angiotensin-converting enzyme inhibitors), heart failure, laryngeal dysfunction, Bordetella pertussis infection, chronic cough after viral upper respiratory tract infection or secondary bacterial infections, recurrent aspiration, or respiratory tract abnormalities such as bronchiectasis or airway tumors.

WHEEZING

Though asthma is a common cause of wheezing in all age groups, it is not the most common cause, particularly if the wheezing is not associated with cough or dyspnea. Postnasal drip is another common cause to consider; also, the rates of heart failure rise in the older age groups and associated pulmonary edema may present as "cardiac asthma." Finally, airway hyperresponsiveness from chronic bronchitis is not uncommon in older patients with a history of wheezing and sputum and tobacco use.

MAJOR PULMONARY DISEASES IN OLDER PERSONS

ASTHMA

After childhood, there is a second peak in the prevalence of asthma beyond the age of 65; 5% to 10% of older persons meet criteria for obstruction and bronchial hyperreactivity. The rate of death from asthma has increased most significantly in those aged 65 and over, accounting for up to 45% of all asthma deaths. This is likely due to reduced awareness of bronchial constriction on the part of the patient (with delays in seeking medical attention), as well as under-recognition and undertreatment on the part of clinicians.

Therapy of asthma in older and younger persons differs in several ways. Paramount to the care of the elderly asthmatic patient is adequate instruction in the proper use of peak expiratory flow monitoring (because of the older person's decreased perception of bronchoconstriction) and in the correct activation of the metered-dose inhaler.

Neurologic, muscular and arthritic diseases in older persons can lead to suboptimal timing and discoordination in the actuation of the inhaler device. The clinician should observe the patient actually using the inhaler. Inhaled corticosteroids (or other controller drugs such as leukotriene receptor antagonists) represent the mainstay of therapy in both older and younger persons.

Use the lowest effective dose and counsel regarding rinsing of the oropharynx to avoid thrush. In older people, theophylline is fraught with adverse effects and drug interactions and should be considered a third-line drug to be used only as a once-daily medication in the evening for severe asthma or COPD, targeting a serum level of 10 to 12 mg/dL if tolerated. Although controversy persists as to whether the response to relievers such as â-agonists varies with age, these drugs remain a mainstay as an as-needed reliever medication.

The potential for adverse effects of â-agonists—for example, hypokalemia or possible QT prolongation in cardiac patients on digoxin or other medications—warrants adequate controller use by asthmatic patients to minimize their overreliance on the â-agonist. The use of long-acting â-agonists is controversial and should be considered with caution and only in those patients who are reliably able to manage medications.

CHRONIC OBSTRUCTIVE PULMONARY DISEASE

COPD affects approximately 15 million people in the United States and is the fourth most common cause of death after heart disease, cancer and stroke. The prevalence and mortality rate from COPD is increasing, especially in older persons. Episodes of acute respiratory failure that require mechanical ventilation are associated with mortality rates ranging from 11% to 46%. The National Lung Health Education Programme Executive Committee has noted that the morbidity and mortality from COPD accounts for more than $15 billion per year in U.S. medical care expenditures. Hospitalization continues to represent the largest component of cost for COPD patients. The diagnosis of airflow limitation is challenging in that no single item or combination of items from the history and clinical examination excludes airflow limitation. The finding most strongly associated with a decreased likelihood of airflow limitation is a history of never having smoked cigarettes (especially in patients without a history of wheezing and without wheezing on examination).

Wheezing noted on physical examination is the most potent predictor of airflow limitation and patients with obstructive airflow limitation are 36 times more likely to have wheezing than are patients without this problem. Other findings associated with an increased likelihood of airflow limitation include a barrel-shaped chest, hyperresonance on percussion and a forced expiratory time of greater than 9 seconds measured during the clinical bedside examination.

Smoking cessation at any age has been shown to slow the decline in lung function and aggressive cessation efforts are appropriate even in the oldest-old patient. The basic elements of the approach are the "Five As" from the Agency for Health Care Policy and Research:

- Ask patients about use of tobacco at every office visit.
- Assess readiness to quit.
- Advise patients to quit.
- Assist patients in the quit attempt with aids such as a local cessation programme and pharmacologic agents such as bupropion or nicotine replacement.
- Arrange both a quit date and a follow-up visit

The chief components of daily drug therapy in emphysema consist of a â-agonist, ipratropium bromide or tiotropium, or both drugs in combination. For more severe disease, the use of long-acting â-agonists such as salmeterol along with a combined albuterol and ipratropium bromide metered-dose inhaler can achieve improved adherence and long-term control by reducing the number of inhalers by one (ie, the patient will have only the long-acting inhaler and the combination short-acting inhaler rather than three inhalers).

Inhaled corticosteroids have now been tested in several multicenter randomized controlled trials in COPD with negative results overall; however, there does appears to be a benefit in subgroup analyses of patients with "asthmatic COPD" as defined by spirometry testing with documented bronchodilator responsiveness.

A landmark investigation documented that the use of systemic corticosteroids (intravenous followed by oral) reduces the duration and recurrence of acute exacerbations of COPD for up to 6 months. Importantly, there is no benefit to a course of steroids longer than 14 days. For the few patients (5% to 10%) who do benefit or who require prolonged use of corticosteroids, the risks should be considered and documented in the patient's chart. These risks include peptic ulcer disease, hypertension, cataracts, diabetes mellitus, osteoporosis, psychosis, seizures, poor wound healing, infections and aseptic necrosis of the hip.

Appropriate preventive measures should also be taken in these circumstances of prolonged use, such as using the lowest possible dose of corticosteroids and using supplemental vitamin D, calcium and perhaps a bisphosphonate for those at risk for osteoporosis. Other possible beneficial interventions in older emphysema patients include pulmonary rehabilitation via exercise training and respiratory therapy and education. Both major depression and anxiety have been shown to be present in up to 40% of COPD patients and their treatment must be considered.

In fact, unprovoked anxiety attacks often result in patients' seeking help in emergency departments, being admitted to the hospital and being treated with potentially avoidable courses of oral corticosteroids when the anxiety is not diagnosed and treated.

OBSTRUCTIVE SLEEP APNEA

Sleep-related breathing disorders are very common in older persons and obstructive sleep apnea is the most common type of sleep-related breathing disorder. Obstructive sleep apnea has been associated with cerebrovascular accidents, myocardial infarctions and a threefold increase in mortality. Most patients with obstructive sleep apnea remain undiagnosed and therefore without treatment of this life-threatening, yet potentially correctable disease.

Treatment options include addressing upper-airway obstruction via weight loss, avoidance of alcohol and sedatives, sleeping on one's side or upright, correction of metabolic disorders such as hypothyroidism and continuous positive airway pressure (CPAP) via a nasal mask. To increase adherence with the use of CPAP, one might order the treatment with "nasal pillows" to increase comfort and "ramping technique" to give a delayed rise in the applied pressure after the patient has fallen asleep. Treatment issues are generally the same for the young and the old and the major consideration for the clinician is a high index of suspicion and clinical recognition of this disease.

IDIOPATHIC PULMONARY FIBROSIS

There are more than 100 causes of restrictive lung diseases; however, the history, examination, serologic testing and biopsy often leave the patient with the diagnosis of idiopathic pulmonary fibrosis. This disease is increasing in prevalence with the aging of our population. Rarely is it an inherited disorder. Pulmonary fibrosis is extremely frustrating for all involved because of its relentless progression. The median survival is 3 to 5 years.

The presentation is normally one of insidious dyspnea (often unrecognized because of a decrease in the activity level on the part of the patient) and cough. Clubbing is often a prominent finding on physical examination in pulmonary fibrosis, as opposed to emphysema, which rarely causes clubbing (prompting a search for another disease such as occult lung cancer).

Oral corticosteroids (0.5 mg/kg/day) for 3 to 6 months is the initial therapeutic maneuver most commonly taken, yet only 10% to 20% of patients respond and adverse effects are often prominent. Early referral to a subspecialist is warranted if the patient wishes to consider further therapeutic attempts so that steroid-sparing agents such as azathioprine or enrollment in a randomized controlled trial of newer pharmacologic agents (eg, interleukin-10 or interferon gamma) can be considered.

PULMONARY THROMBOEMBOLISM

The incidence of pulmonary thromboembolism triples between the ages of 65 and 90 years and has a reported 10% recurrence rate within 1 year. Age above 70 years has been independently associated with missed antemortem diagnosis. Importantly, 10% to 20% of patients with documented pulmonary embolism have an entirely normal blood gas (ie, normal Pao and normal A-a gradient for age).

Age-specific risk factors for pulmonary thromboembolism include hypercoagulability due to increases in fibrinogen, activated protein-C resistance due to factor-V Leiden gene mutation, malignancy, stasis (decreased mobility due to stroke, heart failure, or arthritis), or vessel injury (due to trauma or varicosities). The diagnostic work-up is not different for young and older patients.

Anticoagulants are central to therapy and generally guided by the same principles in younger and older patients.

Because of lessened cardiopulmonary reserve, it may be even more important to achieve therapeutic levels of heparinization quickly to avoid major adverse hemodynamic or oxygenation defects in older patients. The trend toward increased utilization of outpatient low-molecular-weight heparin preparations, while achieving anticoagulation with warfarin, is supported by large and appropriately designed randomized controlled trials.

There should be an overlap of approximately 1 to 3 days between heparinization and adequate warfarin therapy with an INR target of 2 to 3. Warfarin interacts with many drugs that are commonly used in the older age group. Studies have inconsistently shown that age itself is a risk factor for bleeding risk with use of warfarin.

Duration of therapy for at least 6 months has been shown to be superior to 3 months and the shorter duration should be used only for those patients with either a specific risk factor that is now removed or for those in whom the risk of prolonged anticoagulant therapy clearly outweighs that of completing 6 months of therapy. Indeed, patients with multiple ongoing risk factors for pulmonary thromboembolic disease are to be considered for anticoagulation therapy for up to 2 years or longer. Recurrent pulmonary thromboembolism is usually treated with lifelong anticoagulation.

INTENSIVE CARE OF THE CRITICALLY ILL

There is widespread belief that many patients do not want aggressive care at the end of life. Moreover, eight of ten health care professionals believe that the treatments they offer to patients are often overly burdensome. More consistency in our attempts to obtain patients' preferences for do-not-resuscitate orders might reduce the extent and burden of aggressive medical care for the elderly age group.

Clinicians receive little training regarding this topic. Data from the Study to Understand Prognoses and Preferences for Outcomes and Risks of Treatments (SUPPORT) showed that physician error rates in approximating patients' preferences for mechanical ventilation increased from 36% for patients younger than 50 up to 79% for patients older than 80. Furthermore, SUPPORT data also showed that age was a predictor of less aggressive care in the intensive care unit (ICU) even after adjusting for severity of illness, gender, race and diagnosis. A report from a cohort of mechanically ventilated patients in a medical ICU, on the other hand, showed that persons older than 75 years spent the same amount of time on the ventilator as did younger patients but had a lower cost of ICU and in-hospital care.

These outcomes were not explained by differences in mortality, since both groups had similar survival rates. Accordingly, the decision to use mechanical ventilation should not be based on age alone and the appropriate use of ventilatory support in the elderly patient requires further prospective evaluation.

When ICU care and mechanical ventilation are chosen to treat an older person with respiratory failure, special issues related to liberation from the ventilator may be important. In contrast to the general medical ICU population, acute lung injury and acute respiratory distress syndrome patients do appear to have a higher mortality with advancing age. Interestingly, three separate studies of more than 1500 patients have shown that older persons recover from their pulmonary physiologic abnormalities such as hypoxemia and ventilatory disturbances at a rate equal to their younger counterparts after acute lung injury.

These patients do, however, have a particularly difficult time during the latter stages of ventilatory support and require nearly twice as long to be successfully liberated from the ventilator and discharged from the ICU. Often comorbid conditions contribute to these patients' deaths despite the apparent correction of their acute lung injury-induced physiologic disturbances.

Sedative and analgesic medications must be used judiciously and delirium, heart failure, hypothyroidism, electrolyte disturbances, oropharyngeal dysfunction and aspiration should be considered in patients with difficulty liberating from the ventilator. For many older people in good physical condition who succumb to an acute illness, cognitive decline is the main threat to their ability to recover their former functional abilities. For those whose physical activities were already limited, cognitive decline may dramatically worsen after an ICU stay and become the major additional threat to quality of life.

Establishing end-of-life care in critically ill patients is treacherous. Numerous studies have documented the tremendous inaccuracy in formulating the prognosis for patients with serious illnesses. At times, in caring for critically ill patients, it may become apparent to the patient, family and clinician that further intervention would not likely be of substantial benefit, depending on the individual patient's goals, values and hopes.

Thus, a strict definition of futility is impossible. The American Medical Association recommends a standardized "fair process" rather than a strict definition of futility. As much as possible, physicians should base futility decisions on factors such as clinical efficacy of treatment, likelihood of mortality and subsequent quality-of-life considerations rather than on chronologic age alone.

A recent analysis of 6303 ICU-related deaths found that 26% of these deaths occurred in patients receiving full ICU care, including failed cardiopulmonary resuscitation; 24% received full ICU care without cardiopulmonary resuscitation; 14% had life support withheld; and 36% (the single largest group) had life support actively withdrawn. Although there was a wide variation in the practice pattern of different ICUs, the authors concluded that the limitation of life support prior to death is a common practice in teaching ICUs across the country.

In another study of 851 patients receiving mechanical ventilation, 63% were successfully liberated, 17% died while receiving mechanical ventilation and 20% had mechanical ventilation withdrawn. The top four reasons given for withdrawing the ventilator were

- Physicians' perception that the patient did not want life support,

- Physicians' prediction that ICU survival was lower than 10%,
- Physicians' prediction that future cognitive function would be severely impaired, the ongoing need for an inotrope or vasopressors.

The inability for physicians to accurately prognosticate beyond hours of impending death must again be acknowledged as challenging such decisions and should temper concerns about the application of technology in critically ill older patients.

SEXUALLY TRANSMITTED DISEASE (STD)

The Venereal Diseases Act of 1917 defined 3 such diseases. They were syphilis, gonorrhoea and chancroid. In the UK chancroid is unimportant and often forgotten although it is still troublesome in some parts of Africa. Syphilis ebbed but has been resurgent and the prevalence of gonorrhoea is often taken as an index of the degree of promiscuity in the community.

The generic term for the venereal diseases (VD) was changed to sexually transmitted diseases (STD) and the VD clinics became special clinics as new euphemisms were devised to hide the embarrassment of society. More recently the speciality was called genito-urinary medicine (GUM). They are also referred to as sexually transmitted infections (STI). The range of diseases that would meet that classification is now much wider than the 3 that were named in the early part of the last century.

Epidemiology

After a surge in sexually transmitted diseases during the First World War there was a lull between wars and another surge during the Second World War. When young men are about to go off to war and there is a realistic chance that they may never return, there is a great urge to "take comfort" before embarkation and this may not be under judicious conditions.

The 1960s saw an unprecedented ease in contraception with the ready availability of the pill and a soaring rate of promiscuity, without barrier contraception. The slogan of the young was "make love, not war", judgment was often impaired by chemicals and perhaps there was an unrealistic feeling that antibiotics could cure such diseases. The advent of AIDS was a chilling reminder of the limitation of antibiotics and whilst promiscuity fell, as indicated by the incidence of gonorrhoea, this lull was only temporary.

Modes of Transmission

To make a meaningful analysis of the facts and figures, it is important to look at the various ways by which the diseases may be spread. There may be heterosexual activity.There may be homosexual activity. Not all such diseases are always spread by sexual activity. There may be drug taking activity that leads to transmission of such diseases.

The geographical variation in the importance of various forms of transmission with regard to acquired immune deficiency syndrome. Such means of dissemination are not discrete as people may be bisexual, indulging in sexual activity with both sexes and intravenous drug abusers may turn to prostitution to feed their habit. A number of diseases can be vertically transmitted from mother to child.

Another very important group is those who harbour sexually transmitted diseases but they are unaware that they have the disease or have refused to present for diagnosis and treatment. Such people can account for a very large discrepancy between the number known to have the disease and the number who are actually infected. With regard to HIV infection, they may represent more than 30% of cases. They are a particular problem because, if they are unaware of the infection, they are more liable to spread it.

In many cases, infection remains undiagnosed as the individual is asymptomatic. The disease may be more likely to produce symptoms in one sex than the other, but it remains contagious. A man may readily see a primary chancre of syphilis on his glans whilst a woman is unaware of one on her cervix. Candida may cause vaginal discharge and pruritis vulvae but often in men it causes no symptoms.

Recent Figures

The Health Protection Agency's report for 2005, based on figures from 2004, gives much cause for concern. To summarize much of it:

- There is:
  - An undiminished and high level of transmission of HIV and other STIs among homosexuals
  - A steady increase in the number of HIV-infected black Africans in the UK
  - Limited but compelling evidence that heterosexual transmission of HIV within the UK is slowly rising
  - Continuing high transmission of other STIs, especially chlamydia among young people.
- By the end of 2004 there were an estimated 58,300 people living with HIV in the UK, of whom 34% were unaware of their infection.
- The incidence of gonorrhoea also remained high amongst homosexuals in 2004, with 3,977 infections diagnosed. More than a quarter were resistant to ciprofloxacin. The epidemic of syphilis in this group continued to grow and there has been a significant increase in the number of cases of lymphogranuloma venereum (LGV). The last was previously thought of as a rare disease.
- Young people in the UK have a high prevalence of chlamydia, gonorrhoea and genital warts. Rates of diagnoses continued to increase among young people in 2004, with the highest rates of gonorrhoea diagnoses seen among men aged 20-24 and women aged 16-19. Three-quarters of chlamydia diagnoses in women were in young women and 56% of diagnoses in men were in young men. Some of the increases in gonorrhoea and chlamydia diagnoses among young people may reflect increased attendance of young people at GUM clinics and for chlamydia, increased and more sensitive testing. Diagnoses however, will underestimate the true level of infection in the UK, as many infections are asymptomatic.

- HIV, gonorrhoea and syphilis have marked variation in geographical distribution within the British Isles and within England. For chlamydia, genital warts and genital herpes simplex virus (HSV), there was much less variation across the UK.

Risk Factors

The risk of acquiring sexually transmitted diseases is greatest in those who are promiscuous, especially if they do not use barrier contraceptives. Those who are celibate or who have a faithful, monogamous, long-standing relationship are not at risk. This applies to both homosexuals and heterosexuals. However, it is possible to acquire such diseases by other means.

The following may be seen as relative risk factors, although no group should be seen as immune:

- Being young
- Homosexuals
- Substance abusers
- People from sub-Sahara Africa
- Social deprivation
- Failure to use barrier contraceptives

Nowadays most prostitutes will insist that their clients wear a condom but if their motive is to feed a drug habit, they may be less discriminating. Hence they are at great risk of both getting and giving sexually transmitted diseases and they are at risk of having acquired a sexually transmissible disease from their intravenous drug abuse.

Sexually Transmissible Diseases

The number of diseases that are usually transmitted by sexual intercourse or which may be transmitted by that route, is much greater than the 3 defined by the Venereal Disease Act. Some on this list may seem a little surprising to be classified as sexually transmitted diseases:

- Syphilis
- Gonorrhoea
- Chancroid

- HIV/AIDS
- Chlamydia
- Nonspecific urethritis (NSU) including Reiter's disease
- Lymphogranuloma venereum
- Genital herpes simplex
- Genital warts
- Carcinoma of cervix
- Vaginal candidiasis
- Hepatitis B and, to a rather lesser extent, hepatitis C
- Pelvic inflammatory disease
- Trichomonas Vaginalis
- Epididymo-orchitis

Many of these diseases have their own articles that will give a much fuller account. Most STIs are best treated in a GUM clinic as they have expertise in diagnosis and treatment as well as the ability to perform contact tracing.

Anyone who is being treated for a STI should abstain from sexual activity until treatment is complete. It is commonly believed by the lay public, that anyone who is receiving a course of antibiotics should avoid alcohol although metronidazole is the only antibiotic with an adverse reaction with alcohol. The group of doctors who most commonly give this advice is GUM specialists.

This may be because they believe that those who indulge in alcohol are more likely to transmit the disease before it is fully treated. There is a great tendency to treat such disease with a single dose of antibiotics where possible, to improve compliance.

SYPHILIS

The origins of syphilis are uncertain, as described in the final section. It has long been known to be a sexually transmitted disease that can also be passed from mother to child. It can also be transmitted by blood transfusion but blood is routinely screened for the disease. Pregnant women are also routinely screened for syphilis.

If not stopped by adequate treatment, it can produce a variety of conditions and over the years the disease became known as the great mimicker. The introduction of penicillin transformed the management of the disease and it declined. In the 1960s around 70% of syphilis was in homosexuals.

There would be no reliable figures about the number who caught in from homosexual practices before that time as such a confession would make the patient liable to imprisonment. In recent decades, syphilis has become resurgent, including in the heterosexual population.

It is not uncommon for it to be acquired at the same time as other sexually transmitted diseases and as it has an incubation period of around 21 days, it may not have presented when other diseases are diagnosed.

It is a disease that is very easy to overlook these days, whereas 50 years ago the great mimicker would have been in almost every differential diagnosis. If a patient has an unexpected reaction to an antibiotic, it is worth bearing in mind that this could be a Jarisch-Herxheimer reaction due to latent syphilis. Another disease that can also give that reaction and can also be difficult to diagnose is Lyme disease.

Gonorrhoea

Gonorrhoea has almost certainly been around for many centuries and possibly back to biblical times although it is impossible to be certain. The incubation period is 3 to 5 days. Men tend to suffer urethral discharge, frequency of micturition and dysuria. Women tend to get purulent vaginal discharge, frequency of micturition and anorectal discomfort.

There may be bartholinitis. Men are usually symptomatic but only around 60% of infected women show the characteristic features. However, they may develop salpingitis and even the Fitz-Hugh-Curtis syndrome. Because of variety of sexual practice, swabs should be taken from the endocervix, the urethra in both men and women, the throat and anus. Anal infection can cause proctitis. A suitable transport medium must be used. Co-infection with Chlamydia is common and should be sought.

A number of complications have been mentioned above, including pelvic inflammatory disease. Arthritis may also occur. It is generally polyarticular and migratory, usually settling in one or two joints. The knees are most often involved, followed by the ankles, wrists, tarsi, MTP joints and dorsal tendon sheaths of the fingers. Infection from the mother's genital tract can cause ophthalmia neonatorum. Treatment is usually a single, high dose of antibiotic. Amoxicillin is no longer favoured and resistance patterns may make ciprofloxacin obsolete in favour of a new generation cephalosporin such as cefixime.

Chancroid

This is the 3rd of the classical "venereal diseases". Chancroid is caused by the Gram-negative bacillus Haemophilus ducreyi. It may be difficult to isolate in culture, but can be seen microscopically as short Gram-negative rods. Chancroid presents as a soft, painful chancre that resembles the lesions of genital herpes. It is endemic in Africa, Asia and South America and is more common in men, particularly uncircumcised men.

HIV may also be present in as many as 60% in Africa. It may facilitate infection with HIV. Other common co-existing infections are syphilis and HSV-2. Painful lymphadenopathy in the groin occurs in 50%. There are a number of treatment regimes including azithromycin 1gram as a single dose and ciprofloxacin 500mg, also as a single dose.

HIV/AIDS

HIV/AIDS has replaced syphilis as the STD to be most feared in that it is both a killer disease and a great mimic. Highly active antiretroviral therapy (HAART) has transformed the prognosis but it remains a disease that has annihilated many people in Africa and, as explained in the article on the subject, unless it is taken very much more seriously in the UK, it will kill many more here too.

Chlamydia

Genitourinary chlamydia infection is a disease whose importance is being recognised much more recently. It can cause PID and infertility in women and non-gonococcal urethritis in men. It is symptomatic in only around 80% or women and 50% of men. There is evidence that a screening programme would be useful and cost-effective. At present there are a series of pilots with the intention to establish a national programme based upon their results by 2007. The recommended treatment is doxycycline 100mg twice daily for 7 days or azithromycin 1g as a single dose.

Non-specific Urethritis

Non-specific urethritis (NSU) is also called non-gonococcal urethritis. It affects males with symptoms of dysuria and possibly urethral discharge. It does not affect females although the implicated organisms affect females in other ways. Usually chlamydia is involved although the term NSU is often reserved for when there is male urethritis but neither the gonococcus nor chlamydia can be isolated. It is usually successfully treated with tetracycline or erythromycin.

Reiter's syndrome is a form of reactive arthritis. It can result from a number of infections, one of which is chlamydia. There is symmetrical oligoarthritis, urethritis and conjunctivitis. It can arise after food poisoning or after a sexually transmitted infection. It is one of the seronegative arthropathies associated with the HLA-B27 antigen.

Lymphogranuloma Venereum

This is a disease caused by a type of chlamydia. It is rare in the UK but the HPA has reported outbreaks amongst homosexual men. It derives its name from painful, swollen inguinal lymph nodes.

HERPES GENITALIS

The herpes simplex virus has two types, HSV1 and HSV2. As a general rule, HSV1 causes cold sores on lips whilst HSV2 causes genital herpes but occasionally their role can be reversed, especially after oro-genital sex. The HSV1 virus is very common in children but the HSV2 is more indicative of sexual transmission.

Genital Warts

They are caused by the human papilloma virus. They may affect the genitals or anus and are also known as condylomata acuminata. Human papilloma virus and genital.

Carcinoma of Cervix

For many years it has been known that carcinoma of the cervix is associated with an early age of onset of sexual intercourse and multiple partners. It is also more common in smokers. About 95% of cases show evidence of infection with the HPV virus types 16 and 18. Types 31 and 33 may also be implicated. This has led to the proposition that the disease is sexually transmitted. A vaccine against HPV16 has been developed and it could be in part of the immunisation schedule for all girls at around the age of 12. Needless to say, this has brought outrage from the usual quarters, suggesting that girls will see it as a permit for sexual promiscuity. It has been suggested that it may eventually make cervical smears unnecessary. There is no plan to give it to boys although boys spread the virus to girls. This is reminiscent of the early days when rubella vaccine was given only to girls.

Candidiasis

Candida albicans is a common commensal and may become dominant and cause symptoms when the ecology is upset as after a course of antibiotics. The reason for including it in this list is as a reminder that if a woman who is sexually active gets the infection, it is essential to treat her partner too or they will reinfect each other. It should not normally be seen as one of the STIs. For example, if a child gets a disease that is normally sexually transmitted, there is a very strong inference of sexual abuse but if a girl gets vaginal candida after a course of antibiotics, this is no cause for alarm. Candida is far more likely to cause symptoms in the female than the male.

Viral Hepatitis

Hepatitis A is usually spread by the faeco-oral route whilst hepatitis B and C are most often spread through body fluids, especially amongst intravenous drug abusers. However, the sexual transmission of hepatitis B is well established (as explained in Prevention of Hepatitis B). In England and Wales from 1995 to 2000 laboratory data suggested that the annual incidence of HBV infection was 7.4 per 100,000, with injecting drug use as the commonest cause of transmission.

The number of cases attributed to heterosexual contact was fairly stable, whereas the number of cases in homosexual men decreased. Sexual transmission of the hepatitis C virus is possible but uncommon. Less than 5% of the regular sexual partners of people with HCV infection will become infected. This means that infectivity by the sexual route is fairly low. It does not mean that it does not occur.There are specify guidelines for the management of hepatitis B with HIV and hepatitis C with HIV.

Pelvic Inflammatory Disease

Most cases of pelvic inflammatory disease are acquired from ascending infection from the genital tract. The gonococcus and chlamydia are most commonly involved. There is often mixed infection. It can cause chronic pelvic pain and infertility.

Trichomonas Vaginalis

Trichomonas vaginalis is a flagellated protozoan that may be found in the urethra and genital tract of both men and women. It is asymptomatic in up to 50% of both sexes but may cause vaginal discharge, vulval itching, dysuria, or an offensive odour in women.

Sometimes the presenting complaint is of low abdominal discomfort. In men it may cause urethral discharge and dysuria but discharge is rarely profuse. Vaginal discharge occurs in up to 70% of women, but the classical discharge of frothy yellow occurs in only 10 to 30%. Vulvitis and vaginitis are reported. Around 10% have no abnormalities on examination. Men may have urethral discharge. Treatment is metronidazole, 400mg twice daily for 5 to 7 days or 2g as a single oral dose.

EPIDIDYMO-ORCHITIS

Epididymo-orchitis is usually sexually transmitted in men of less than 35 years old. Chlamydia trachomatis and Neisseria gonorrhoeae are the common organisms. There is unilateral testicular pain and swelling and sometimes urethral discharge if it was sexually transmitted. A scrotal support and NSAIDs give symptomatic relief but antibiotics should be ceftriaxone 250mg intramuscularly as a single dose or ciprofloxacin 500mg as a single oral dose plus doxycycline 100mg twice daily, orally for 10 to 14 days.

Prevention

Sexually transmitted diseases occur as a result of sexual promiscuity. They do not occur within faithful monogamous relationships or in celibacy. Hence, when such diseases are diagnosed, society's response can be both sanctimonious and punitive. Fortunately, a great deal of pragmatism has prevailed. In the armed forces, contracting a sexually transmitted disease used to be an offence punishable at court marshall. This did not stop servicemen from taking injudicious rest and recreation but it did mean that they would hide the disease if it occurred rather than presenting to the medical officer for treatment.

The social stigma of having a sexually transmitted disease is enormous and the GUM clinics, since their origin, have been discrete and kept anonymity for their patients. Patients are often given a number so that they do not have to present with a name. They tend to have a discrete entrance in a distant part of the hospital and, to ensure compliance, the drugs they dispense are exempt from prescription charges. It is the only part of the NHS that will not routinely inform the patient's GP that he has attended.

Despite this cloak of secrecy and anonymity, they will attempt to trace contacts of those with sexually transmissible diseases. This is a very important role that must require the utmost tact and diplomacy.

Prevention of sexual transmission of diseases must depend upon education and attitude. Getting the content right for sex education is not easy. A number of factors have been identified as contributing to a successful programme. Adverse attitudes to the wearing of condoms must also be overcome. As explained in the article on AIDS, the Roman Catholic Church has much to answer for in terms of spreading disinformation about condoms. They even advocate that if one of a married couple have HIV that it is better for the other partner to catch the disease than for them to use condoms.

HISTORY OF SEXUALLY TRANSMITTED DISEASES

There is uncertainty about the origins of syphilis. Some people claim that it was brought back from America by Columbus' sailors but others argue that it originated in the Old World. An archeological find of the bones of an Essex woman suggested that she had advanced syphilis, between 1300 and 1450. This would predate Columbus' voyage of 1492.

It is often claimed that Henry VIII died of syphilis but the evidence for or against is poor. In those days any ulceration was called pox and syphilis was the great pox. The ulcer on his leg that failed to heal may well have been from type 2 diabetes.

The legendary Giacomo Casanova (1725-1798) was said to have invented the condom, fashioned from a sheep's intestine, not as a contraceptive but as protection against syphilis. He may have used them but he was not the first. Condoms appear in some of the pictures of A Harlot's Progress by William Hogarth (1697-1764). This series of paintings was completed in 1731, when Casanova would have been 6 years old. The term syphilis was not part of the English language until 1717, following the translation of an Italian work. The disease had been the pox and in London rhyming slang, the acquisition of any GUM infection is still referred to as "a dose of the Surrey Docks".

The clap refers to gonorrhoea, from the old French word clapier, meaning a brothel. It was long thought that the two diseases were one. In 1767, John Hunter inoculated a subject with matter of gonorrhoea on the prepuce and glans but the inoculum was from a patient suffering from both diseases and syphilis developed. In 1793, Benjamin Bell, an Edinburgh surgeon confirmed that the diseases were distinct as a result of experimentation on medical students. This was in the days before Local Research Ethics Committees. The eminent Guy's surgeon Sir Asley Cooper recognised the two diseases as distinct and in 1824 he wrote in The Lancet "a man who gives mercury in gonorrhoea deserves to be flogged out of the profession because he must be quite ignorant of the principle in which the disease is cured."

The etymology of gonorrhoea is probably from the Greek for seed flow in that it was assumed that the urethral discharge was semen. Folk etymology suggests an association with the biblical town of Gomorrah. Its equally evil neighbour, Sodom, is the origin for the term for anal intercourse.

In 1864 Parliament passed the Contagious Diseases Act. This legislation allowed policeman to arrest prostitutes in ports and army towns and bring them in to have compulsory checks for venereal disease. If the women were suffering from sexually transmitted diseases they were placed in a locked hospital until cured. It was claimed that this was the best way to protect men from infected women.

Many of the women arrested were not prostitutes but they still were forced to go to the police station to undergo a humiliating medical examination.The traditional treatment for syphilis was heavy metals such as bismuth and mercury. They are of dubious efficacy but highly toxic. Hence the adage "one night with venus and a lifetime with mercury." The discovery that penicillin can treat syphilis has revolutionized its management.

AIDS

Acquired immune deficiency syndrome or acquired immunodeficiency syndrome (AIDS or Aids) is a set of symptoms and infections resulting from the damage to the human immune system caused by the human immunodeficiency virus (HIV). This condition progressively reduces the effectiveness of the immune system and leaves individuals susceptible to opportunistic infections and tumors.

HIV is transmitted through direct contact of a mucous membrane or the bloodstream with a bodily fluid containing HIV, such as blood, semen, vaginal fluid, preseminal fluid and breast milk. This transmission can involve anal, vaginal or oral sex, blood transfusion, contaminated hypodermic needles, exchange between mother and baby during pregnancy, childbirth, or breastfeeding, or other exposure to one of the above bodily fluids.

AIDS is now a pandemic. In 2007, an estimated 33.2 million people lived with the disease worldwide and it killed an estimated 2.1 million people, including 330,000 children. Over three-quarters of these deaths occurred in sub-Saharan Africa, retarding economic growth and destroying human capital. Most researchers believe that HIV originated in sub-Saharan Africa during the twentieth century.

The disease was first identified by the U.S. Centers for Disease Control and Prevention in 1981 and its cause identified by American and French scientists in the late 1980s.

Although treatments for AIDS and HIV can slow the course of the disease, there is currently no vaccine or cure. Antiretroviral treatment reduces both the mortality and the morbidity of HIV infection, but these drugs are expensive and routine access to antiretroviral medication is not available in all countries. Due to the difficulty in treating HIV infection, preventing infection is a key aim in controlling the AIDS epidemic, with health organizations promoting safe sex and needle-exchange programmes in attempts to slow the spread of the virus.

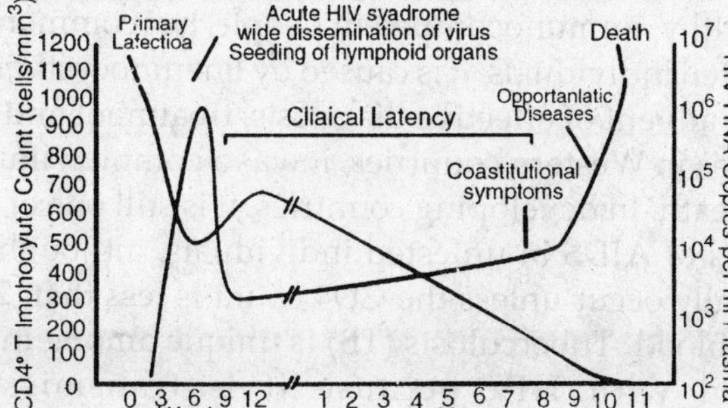

Symptoms

A generalized graph of the relationship between HIV copies (viral load) and CD4 counts over the average course of untreated HIV infection; any particular individual's disease course may vary considerably. $CD4^+$ T Lymphocyte count (cells/mm^3) HIV RNA copies per mL of plasma

The symptoms of AIDS are primarily the result of conditions that do not normally develop in individuals with healthy immune systems. Most of these conditions are infections caused by bacteria, viruses, fungi and parasites that are normally controlled by the elements of the immune system that HIV damages. Opportunistic infections are common in people with AIDS.

HIV affects nearly every organ system. People with AIDS also have an increased risk of developing various cancers such as Kaposi's sarcoma, cervical cancer and cancers of the immune system known as lymphomas. Additionally, people with AIDS often have spromoting safe sex and needle-exchange programmes in attempts to ystemic symptoms of infection like fevers, sweats (particularly at night), swollen glands, chills, weakness and weight loss. The specific opportunistic infections that AIDS patients develop depend in part on the prevalence of these infections in the geographic area in which the patient lives.

Pulmonary Infections

Pneumocystis pneumonia (originally known as Pneumocystis carinii pneumonia and still abbreviated as PCP, which now stands for Pneumocystis pneumonia) is relatively rare in healthy, immunocompetent people, but common among HIV-infected individuals. It is caused by Pneumocystis jirovecii. Before the advent of effective diagnosis, treatment and routine prophylaxis in Western countries, it was a common immediate cause of death. In developing countries, it is still one of the first indications of AIDS in untested individuals, although it does not generally occur unless the CD4 count is less than 200 cells per μL of blood. Tuberculosis (TB) is unique among infections associated with HIV because it is transmissible to immunocompetent people via the respiratory route, is easily treatable once identified, may occur in early-stage HIV disease and is preventable with drug therapy.

However, multidrug resistance is a potentially serious problem. Even though its incidence has declined because of the use of directly observed therapy and other improved practices in Western countries, this is not the case in developing countries where HIV is most prevalent. In early-stage HIV infection (CD4 count >300 cells per μL), TB typically presents as a pulmonary disease. In advanced HIV infection, TB often presents atypically with extrapulmonary (systemic) disease a common feature. Symptoms are usually constitutional and are not localized to one particular site, often affecting bone marrow, bone, urinary and gastrointestinal tracts, liver, regional lymph nodes and the central nervous system.

Gastrointestinal Infections

Esophagitis is an inflammation of the lining of the lower end of the esophagus (gullet or swallowing tube leading to the stomach). In HIV infected individuals, this is normally due to fungal (candidiasis) or viral (herpes simplex-1 or cytomegalovirus) infections. In rare cases, it could be due to mycobacteria. Unexplained chronic diarrhea in HIV infection is due to many possible causes, including common bacterial (Salmonella, Shigella, Listeria or Campylobacter) and parasitic infections; and uncommon opportunistic infections such as cryptosporidiosis, microsporidiosis, Mycobacterium avium complex (MAC) and viruses, astrovirus, adenovirus, rotavirus and cytomegalovirus, (the latter as a course of colitis). In some cases, diarrhea may be a side effect of several drugs used to treat HIV, or it may simply accompany HIV infection, particularly during primary HIV infection.

It may also be a side effect of antibiotics used to treat bacterial causes of diarrhea (common for Clostridium difficile). In the later stages of HIV infection, diarrhea is thought to be a reflection of changes in the way the intestinal tract absorbs nutrients and may be an important component of HIV-related wasting.

Neurological Diseases

Toxoplasmosis is a disease caused by the single-celled parasite called Toxoplasma gondii; it usually infects the brain causing toxoplasma encephalitis but it can infect and cause disease in the eyes and lungs.

Progressive multifocal leukoencephalopathy (PML) is a demyelinating disease, in which the gradual destruction of the myelin sheath covering the axons of nerve cells impairs the transmission of nerve impulses.

It is caused by a virus called JC virus which occurs in 70% of the population in latent form, causing disease only when the immune system has been severely weakened, as is the case for AIDS patients.

It progresses rapidly, usually causing death within months of diagnosis. AIDS dementia complex (ADC) is a metabolic encephalopathy induced by HIV infection and fueled by immune activation of HIV infected brain macrophages and microglia which secrete neurotoxins of both host and viral origin. Specific neurological impairments are manifested by cognitive, behavioral and motor abnormalities that occur after years of HIV infection and is associated with low $CD4^{+}$ T cell levels and high plasma viral loads.

Prevalence is 10–20% in Western countries but only 1–2% of HIV infections in India. This difference is possibly due to the HIV subtype in India. Cryptococcal meningitis is an infection of the meninx (the membrane covering the brain and spinal cord) by the fungus Cryptococcus neoformans. It can cause fevers, headache, fatigue, nausea and vomiting. Patients may also develop seizures and confusion; left untreated, it can be lethal.

Tumors and Malignancies

Patients with HIV infection have substantially increased incidence of several malignant cancers. This is primarily due to co-infection with an oncogenic DNA virus, especially Epstein-Barr virus (EBV), Kaposi's sarcoma-associated herpesvirus (KSHV) and human papillomavirus (HPV).

Kaposi's sarcoma (KS) is the most common tumor in HIV-infected patients. The appearance of this tumor in young homosexual men in 1981 was one of the first signals of the AIDS epidemic.

Caused by a gammaherpes virus called Kaposi's sarcoma-associated herpes virus (KSHV), it often appears as purplish nodules on the skin, but can affect other organs, especially the mouth, gastrointestinal tract and lungs. High-grade B cell lymphomas such as Burkitt's lymphoma, Burkitt's-like lymphoma, diffuse large B-cell lymphoma (DLBCL) and primary central nervous system lymphoma present more often in HIV-infected patients. These particular cancers often foreshadow a poor prognosis. In some cases these lymphomas are AIDS-defining. Epstein-Barr virus (EBV) or KSHV cause many of these lymphomas.

Cervical cancer in HIV-infected women is considered AIDS-defining. It is caused by human papillomavirus (HPV).

In addition to the AIDS-defining tumors listed above, HIV-infected patients are at increased risk of certain other tumors, such as Hodgkin's disease and anal and rectal carcinomas. However, the incidence of many common tumors, such as breast cancer or colon cancer, does not increase in HIV-infected patients. In areas where HAART is extensively used to treat AIDS, the incidence of many AIDS-related malignancies has decreased, but at the same time malignant cancers overall have become the most common cause of death of HIV-infected patients.

Other Opportunistic Infections

AIDS patients often develop opportunistic infections that present with non-specific symptoms, especially low-grade fevers and weight loss.

These include infection with Mycobacterium avium-intracellulare and cytomegalovirus (CMV). CMV can cause colitis, as described above and CMV retinitis can cause blindness. Penicilliosis due to Penicillium marneffei is now the third most common opportunistic infection (after extrapulmonary tuberculosis and cryptococcosis) in HIV-positive individuals within the endemic area of Southeast Asia.

CAUSE

AIDS is the most severe acceleration of infection with HIV. HIV is a retrovirus that primarily infects vital organs of the human immune system such as $CD4^+$ T cells (a subset of T cells), macrophages and dendritic cells. It directly and indirectly destroys $CD4^+$ T cells. Once HIV has killed so many $CD4^+$ T cells that there are fewer than 200 of these cells per microliter (μL) of blood, cellular immunity is lost. Acute HIV infection progresses over time to clinical latent HIV infection and then to early symptomatic HIV infection and later to AIDS, which is identified either on the basis oı the amount of $CD4^+$ T cells remaining in the blood and/or the presence of certain infections, as noted above.

In the absence of antiretroviral therapy, the median time of progression from HIV infection to AIDS is nine to ten years and the median survival time after developing AIDS is only 9.2 months. However, the rate of clinical disease progression varies widely between individuals, from two weeks up to 20 years. Many factors affect the rate of progression.

These include factors that influence the body's ability to defend against HIV such as the infected person's general immune function. Older people have weaker immune systems and therefore have a greater risk of rapid disease progression than younger people. Poor access to health care and the existence of coexisting infections such as tuberculosis also may predispose people to faster disease progression.

The infected person's genetic inheritance plays an important role and some people are resistant to certain strains of HIV. An example of this is people with the homozygous CCR5-Ä32 variation are resistant to infection with certain strains of HIV. HIV is genetically variable and exists as different strains, which cause different rates of clinical disease progression.

Sexual Transmission

Sexual transmission occurs with the contact between sexual secretions of one person with the rectal, genital or oral mucous membranes of another. Unprotected receptive sexual acts are riskier than unprotected insertive sexual acts and the risk for transmitting HIV through unprotected anal intercourse is greater than the risk from vaginal intercourse or oral sex. However, oral sex is not entirely safe, as HIV can be transmitted through both insertive and receptive oral sex.

The risk of HIV transmission from exposure to saliva is considerably smaller than the risk from exposure to semen, one would have to swallow liters of saliva from a carrier to run a significant risk of becoming infected. Sexual assault greatly increases the risk of HIV transmission as protection is rarely employed and physical trauma to the vagina frequently occurs, facilitating the transmission of HIV.

Other sexually transmitted infections (STI) increase the risk of HIV transmission and infection, because they cause the disruption of the normal epithelial barrier by genital ulceration and/or microulceration; and by accumulation of pools of HIV-susceptible or HIV-infected cells (lymphocytes and macrophages) in semen and vaginal secretions. Epidemiological studies from sub-Saharan Africa, Europe and North America suggest that genital ulcers, such as those caused by syphilis and/or chancroid, increase the risk of becoming infected with HIV by about four-fold. There is also a significant although lesser increase in risk from STIs such as gonorrhea, Chlamydial infection and trichomoniasis, which all cause local accumulations of lymphocytes and macrophages.

Transmission of HIV depends on the infectiousness of the index case and the susceptibility of the uninfected partner. Infectivity seems to vary during the course of illness and is not constant between individuals. An undetectable plasma viral load does not necessarily indicate a low viral load in the seminal liquid or genital secretions. However, each 10-fold increase in the level of HIV in the blood is associated with an 81% increased rate of HIV transmission. Women are more susceptible to HIV-1 infection due to hormonal changes, vaginal microbial ecology and physiology and a higher prevalence of sexually transmitted diseases. People who have been infected with one strain of HIV can still be infected later on in their lives by other, more virulent strains.

Exposure to Blood-borne Pathogens

This transmission route is particularly relevant to intravenous drug users, hemophiliacs and recipients of blood transfusions and blood products. Sharing and reusing syringes contaminated with HIV-infected blood represents a major risk for infection with HIV.

Needle sharing is the cause of one third of all new HIV-infections in North America, China and Eastern Europe. The risk of being infected with HIV from a single prick with a needle that has been used on an HIV-infected person is thought to be about 1 in 150. Post-exposure prophylaxis with anti-HIV drugs can further reduce this risk.

This route can also affect people who give and receive tattoos and piercings. Universal precautions are frequently not followed in both sub-Saharan Africa and much of Asia because of both a shortage of supplies and inadequate training. The WHO estimates that approximately 2.5% of all HIV infections in sub-Saharan Africa are transmitted through unsafe healthcare injections. Because of this, the United Nations General Assembly has urged the nations of the world to implement precautions to prevent HIV transmission by health workers.

The risk of transmitting HIV to blood transfusion recipients is extremely low in developed countries where improved donor selection and HIV screening is performed. However, according to the WHO, the overwhelming majority of the world's population does not have access to safe blood and between 5% and 10% of the world's HIV infections come from transfusion of infected blood and blood products.

Perinatal Transmission

The transmission of the virus from the mother to the child can occur in utero during the last weeks of pregnancy and at childbirth. In the absence of treatment, the transmission rate between a mother and her child during pregnancy, labour and delivery is 25%. However, when the mother takes antiretroviral therapy and gives birth by caesarean section, the rate of transmission is just 1%.

The risk of infection is influenced by the viral load of the mother at birth, with the higher the viral load, the higher the risk. Breastfeeding also increases the risk of transmission by about 4 %.

AIDS Denialism

A small group of activists, including several scientists who do not study HIV/AIDS, question the connection between HIV and AIDS, the existence of HIV itself, or the validity of current testing and treatment methods. Though these claims have been examined and thoroughly rejected by the scientific community, they continue to be promulgated through the Internet and have had a significant political impact, particularly in South Africa, where President Thabo Mbeki's embrace of AIDS denialism has been blamed for an ineffective response to that country's AIDS epidemic.

Misconceptions

A number of misconceptions have arisen surrounding HIV/AIDS. Three of the most common are that AIDS can spread through casual contact, that sexual intercourse with a virgin will cure AIDS and that HIV can infect only homosexual men and drug users. Other misconceptions are that any act of anal intercourse between gay men can lead to AIDS infection.

PATHOPHYSIOLOGY

The pathophysiology of AIDS is complex, as is the case with all syndromes. Ultimately, HIV causes AIDS by depleting $CD4^+$ T helper lymphocytes. This weakens the immune system and allows opportunistic infections. T lymphocytes are essential to the immune response and without them, the body cannot fight infections or kill cancerous cells.

The mechanism of $CD4^+$ T cell depletion differs in the acute and chronic phases. During the acute phase, HIV-induced cell lysis and killing of infected cells by cytotoxic T cells accounts for $CD4^+$ T cell depletion, although apoptosis may also be a factor. During the chronic phase, the consequences of generalized immune activation coupled with the gradual loss of the ability of the immune system to generate new T cells appear to account for the slow decline in $CD4^+$ T cell numbers.

Although the symptoms of immune deficiency characteristic of AIDS do not appear for years after a person is infected, the bulk of $CD4^+$ T cell loss occurs during the first weeks of infection, especially in the intestinal mucosa, which harbors the majority of the lymphocytes found in the body. The reason for the preferential loss of mucosal $CD4^+$ T cells is that a majority of mucosal $CD4^+$ T cells express the CCR5 coreceptor, whereas a small fraction of $CD4^+$ T cells in the bloodstream do so. HIV seeks out and destroys CCR5 expressing $CD4^+$ cells during acute infection. A vigorous immune response eventually controls the infection and initiates the clinically latent phase. However, $CD4^+$ T cells in mucosal tissues remain depleted throughout the infection, although enough remain to initially ward off life-threatening infections.

Continuous HIV replication results in a state of generalized immune activation persisting throughout the chronic phase. Immune activation, which is reflected by the increased activation state of immune cells and release of proinflammatory cytokines, results from the activity of several HIV gene products and the immune response to ongoing HIV replication. Another cause is the breakdown of the immune surveillance system of the mucosal barrier caused by the depletion of mucosal $CD4^+$ T cells during the acute phase of disease. This results in the systemic exposure of the immune system to microbial components of the gut's normal flora, which in a healthy person is kept in check by the mucosal immune system.

The activation and proliferation of T cells that results from immune activation provides fresh targets for HIV infection. However, direct killing by HIV alone cannot account for the observed depletion of $CD4^+$ T cells since only 0.01-0.10% of $CD4^+$ T cells in the blood are infected.

A major cause of $CD4^+$ T cell loss appears to result from their heightened susceptibility to apoptosis when the immune system remains activated. Although new T cells are continuously produced by the thymus to replace the ones lost, the regenerative capacity of the thymus is slowly destroyed by direct infection of its thymocytes by HIV. Eventually, the minimal number of $CD4^+$ T cells necessary to maintain a sufficient immune response is lost, leading to AIDS

Cells affected

The virus, entering through which ever route, acts primarily on the following cells:

- Lymphoreticular system:
  - CD_4+ T-Helper cells
  - CD_4+ Macrophages
  - CD_4+ Monocytes
  - B-lymphocytes
- Certain endothelial cells
- Central nervous system:
  - Microglia of the nervous system
  - Astrocytes
  - Oligodendrocytes
  - Neurones - indirectly by the action of cytokines and the gp-120

The Effect

The virus has cytopathic effects but how it does it is still not quite clear. It can remain inactive in these cells for long periods, though. This effect is hypothesized to be due to the CD_4-gp120 interaction.

- The most prominent effect of the HIV virus is its T-helper cell suppression and lysis. The cell is simply killed off or deranged to the point of being functionless (they do not respond to foreign antigens). The infected B-cells can not produce enough antibodies either. Thus the immune system collapses leading to the familiar AIDS complications, like infections and neoplasms (vide supra).
- Infection of the cells of the CNS cause acute aseptic meningitis, subacute encephalitis, vacuolar

myelopathy and peripheral neuropathy. Later it leads to even AIDS dementia complex.

- The CD_4-gp120 interaction (vide supra) is also permissive to other viruses like Cytomegalovirus, Hepatitis virus, Herpes simplex virus, etc. These viruses lead to further cell damage i.e. cytopathy.

DIAGNOSIS

The diagnosis of AIDS in a person infected with HIV is based on the presence of certain signs or symptoms. Since June 5, 1981, many definitions have been developed for epidemiological surveillance such as the Bangui definition and the 1994 expanded World Health Organization AIDS case definition. However, clinical staging of patients was not an intended use for these systems as they are neither sensitive, nor specific. In developing countries, the World Health Organization staging system for HIV infection and disease, using clinical and laboratory data, is used and in developed countries, the Centers for Disease Control (CDC) Classification System is used.

WHO Disease Staging System

In 1990, the World Health Organization (WHO) grouped these infections and conditions together by introducing a staging system for patients infected with HIV-1. An update took place in September 2005. Most of these conditions are opportunistic infections that are easily treatable in healthy people.

- *Stage I:* HIV infection is asymptomatic and not categorized as AIDS
- *Stage II:* Includes minor mucocutaneous manifestations and recurrent upper respiratory tract infections
- *Stage III:* Includes unexplained chronic diarrhea for longer than a month, severe bacterial infections and pulmonary tuberculosis
- *Stage IV:* Includes toxoplasmosis of the brain, candidiasis of the esophagus, trachea, bronchi or lungs and Kaposi's sarcoma; these diseases are indicators of AIDS.

CDC Classification System

There are two main definitions for AIDS, both produced by the Centers for Disease Control and Prevention (CDC). The older definition is to referring to AIDS using the diseases that were associated with it, for example, lymphadenopathy, the disease after which the discoverers of HIV originally named the virus. In 1993, the CDC expanded their definition of AIDS to include all HIV positive people with a $CD4^+$ T cell count below 200 per μL of blood or 14% of all lymphocytes. The majority of new AIDS cases in developed countries use either this definition or the pre-1993 CDC definition. The AIDS diagnosis still stands even if, after treatment, the $CD4^+$ T cell count rises to above 200 per μL of blood or other AIDS-defining illnesses are cured.

HIV test

Many people are unaware that they are infected with HIV. Less than 1% of the sexually active urban population in Africa has been tested and this proportion is even lower in rural populations. Furthermore, only 0.5% of pregnant women attending urban health facilities are counseled, tested or receive their test results. Again, this proportion is even lower in rural health facilities. Therefore, donor blood and blood products used in medicine and medical research are screened for HIV.

HIV tests are usually performed on venous blood. Many laboratories use fourth generation screening tests which detect anti-HIV antibody (IgG and IgM) and the HIV p24 antigen. The detection of HIV antibody or antigen in a patient previously known to be negative is evidence of HIV infection. Individuals whose first specimen indicates evidence of HIV infection will have a repeat test on a second blood sample to confirm the results.

The window period (the time between initial infection and the development of detectable antibodies against the infection) can vary since it can take 3–6 months to seroconvert and to test positive. Detection of the virus using polymerase chain reaction (PCR) during the window period is possible and evidence suggests that an infection may often be detected earlier than when using a fourth generation EIA screening test. Positive results obtained by PCR are confirmed by antibody tests. Routinely used HIV tests for infection in neonates, born to HIV-positive mothers, have no value because of the presence of maternal antibody to HIV in the child's blood. HIV infection can only be diagnosed by PCR, testing for HIV pro-viral DNA in the children's lymphocytes.

Prevention

Estimated per act risk for acquisitionof HIV by exposure route

Exposure Route	*Estimated infectionsper 10,000 exposuresto an infected source*
Blood Transfusion	9,000
Childbirth	2,500
Needle-sharing injection drug use	67
Percutaneous needle stick	30
Receptive anal intercourse	50
Insertive anal intercourse	6.5
Receptive penile-vaginal intercourse	10
Insertive penile-vaginal intercourse	5
Receptive oral intercourse	1
Insertive oral intercourse	0.5

assuming no condom use source refers to oral intercourseperformed

on a man

The three main transmission routes of HIV are sexual contact, exposure to infected body fluids or tissues and from mother to fetus or child during perinatal period. It is possible to find HIV in the saliva, tears and urine of infected individuals, but there are no recorded cases of infection by these secretions and the risk of infection is negligible.

Sexual Contact

The majority of HIV infections are acquired through unprotected sexual relations between partners, one of whom has HIV. The primary mode of HIV infection worldwide is through sexual contact between members of the opposite sex. During a sexual act, only male or female condoms can reduce the chances of infection with HIV and other STDs and the chances of becoming pregnant. The best evidence to date indicates that typical condom use reduces the risk of heterosexual HIV transmission by approximately 80% over the long-term, though the benefit is likely to be higher if condoms are used correctly on every occasion. The male latex condom, if used correctly without oil-based lubricants, is the single most effective available technology to reduce the sexual transmission of HIV and other sexually transmitted infections. Manufacturers recommend that oil-based lubricants such as petroleum jelly, butter and lard not be used with latex condoms, because they dissolve the latex, making the condoms porous. If necessary, manufacturers recommend using water-based lubricants. Oil-based lubricants can however be used with polyurethane condoms.

The female condom is an alternative to the male condom and is made from polyurethane, which allows it to be used in the presence of oil-based lubricants. They are larger than male condoms and have a stiffened ring-shaped opening and are designed to be inserted into the vagina. The female condom contains an inner ring, which keeps the condom in place inside the vagina – inserting the female condom requires squeezing this ring. However, at present availability of female condoms is very low and the price remains prohibitive for many women. Preliminary studies suggest that, where female condoms are available, overall protected sexual acts increase relative to unprotected sexual acts, making them an important HIV prevention strategy. Studies on couples where one partner is infected show that with consistent condom use, HIV infection rates for the uninfected partner are below 1% per year. Prevention strategies are well-known in developed countries, however, recent epidemiological and behavioral studies in Europe and North America have suggested that a substantial minority of young people continue to engage in high-risk practices and that despite HIV/AIDS knowledge, young people underestimate their own risk of becoming infected with HIV.

Randomized controlled trials have shown that male circumcision lowers the risk of HIV infection among heterosexual men by up to 60%. It is expected that this procedure will be actively promoted in many of the countries affected by HIV, although doing so will involve confronting a number of practical, cultural and attitudinal issues. Some experts fear that a lower perception of vulnerability among circumcised men may result in more sexual risk-taking behaviour, thus negating its preventive effects.

Exposure to Infected Body Fluids

Health care workers can reduce exposure to HIV by employing precautions to reduce the risk of exposure to contaminated blood. These precautions include barriers such as gloves, masks, protective eyeware or shields and gowns or aprons which prevent exposure of the skin or mucous membranes to blood borne pathogens. Frequent and thorough washing of the skin immediately after being contaminated with blood or other bodily fluids can reduce the chance of infection. Finally, sharp objects like needles, scalpels and glass, are carefully disposed of to prevent needlestick injuries with contaminated items. Since intravenous drug use is an important factor in HIV transmission in developed countries, harm reduction strategies such as needle-exchange programmes are used in attempts to reduce the infections caused by drug abuse.

Mother-to-child Transmission (MTCT)

Current recommendations state that when replacement feeding is acceptable, feasible, affordable, sustainable and safe, HIV-infected mothers should avoid breast-feeding their infant. However, if this is not the case, exclusive breast-feeding is recommended during the first months of life and discontinued as soon as possible.

Treatment

There is currently no vaccine or cure for HIV or AIDS. The only known methods of prevention are based on avoiding exposure to the virus or, failing that, an antiretroviral treatment directly after a highly significant exposure, called post-exposure prophylaxis (PEP). PEP has a very demanding four week schedule of dosage. It also has very unpleasant side effects including diarrhea, malaise, nausea and fatigue.

Antiviral Therapy

Current treatment for HIV infection consists of highly active antiretroviral therapy, or HAART. This has been highly beneficial to many HIV-infected individuals since its introduction in 1996 when the protease inhibitor-based HAART initially became available. Current optimal HAART options consist of combinations (or "cocktails") consisting of at least three drugs belonging to at least two types, or "classes," of antiretroviral agents. Typical regimens consist of two nucleoside analogue reverse transcriptase inhibitors (NARTIs or NRTIs) plus either a protease inhibitor or a non-nucleoside reverse transcriptase inhibitor (NNRTI). Because HIV disease progression in children is more rapid than in adults and laboratory parameters are less predictive of risk for disease progression, particularly for young infants, treatment recommendations are more aggressive for children than for adults. In developed countries where HAART is available, doctors assess the viral load, rapidity in CD4 decline and patient readiness while deciding when to recommend initiating treatment.

HAART allows the stabilization of the patient's symptoms and viremia, but it neither cures the patient of HIV, nor alleviates the symptoms and high levels of HIV-1, often HAART resistant, return once treatment is stopped. Moreover, it would take more than the lifetime of an individual to be cleared of HIV infection using HAART. Despite this, many HIV-infected individuals have experienced remarkable improvements in their general health and quality of life, which has led to the plummeting of HIV-associated morbidity and mortality. In the absence of HAART, progression from HIV infection to AIDS occurs at a median of between nine to ten years and the median survival time after developing AIDS is only 9.2 months. HAART is thought to increase survival time by between 4 and 12 years.

For some patients, which can be more than fifty percent of patients, HAART achieves far less than optimal results, due to medication intolerance/side effects, prior ineffective antiretroviral therapy and infection with a drug-resistant strain of HIV. Non-adherence and non-persistence with therapy are the major reasons why some people do not benefit from HAART. The reasons for non-adherence and non-persistence are varied. Major psychosocial issues include poor access to medical care, inadequate social supports, psychiatric disease and drug abuse. HAART regimens can also be complex and thus hard to follow, with large numbers of pills taken frequently. Side effects can also deter people from persisting with HAART, these include lipodystrophy, dyslipidaemia, diarrhoea, insulin resistance, an increase in cardiovascular risks and birth defects. Anti-retroviral drugs are expensive and the majority of the world's infected individuals do not have access to medications and treatments for HIV and AIDS.

Future Research

It has been postulated that only a vaccine can halt the pandemic because a vaccine would possibly cost less, thus being affordable for developing countries and would not require daily treatments. However, even after almost 30 years of research, HIV-1 remains a difficult target for a vaccine.

Research to improve current treatments includes decreasing side effects of current drugs, further simplifying drug regimens to improve adherence and determining the best sequence of regimens to manage drug resistance. A number of studies have shown that measures to prevent opportunistic infections can be beneficial when treating patients with HIV infection or AIDS. Vaccination against hepatitis A and B is advised for patients who are not infected with these viruses and are at risk of becoming infected. Patients with substantial immunosuppression are also advised to receive prophylactic therapy for Pneumocystis jiroveci pneumonia (PCP) and many patients may benefit from prophylactic therapy for toxoplasmosis and Cryptococcus meningitis as well.

Alternative Medicine

Various forms of alternative medicine have been used to treat symptoms or alter the course of the disease. Acupuncture has been used to alleviate some symptoms, such peripheral neuropathy, but cannot cure the HIV infection. Several randomized clinical trials testing the effect of herbal medicines have shown that there is no evidence that these herbs have any effect on the progression of the disease, but may instead produce serious side-effects.

Some data suggest that multivitamin and mineral supplements might reduce HIV disease progression in adults, although there is no conclusive evidence on if they reduce mortality among people with good nutritional status. Vitamin A supplementation in children probably has some benefit. Daily doses of selenium can suppress HIV viral burden with an associated improvement of the CD4 count. Selenium can be used as an adjunct therapy to standard antiviral treatments, but cannot itself reduce mortality and morbidity.

Current studies indicate that that alternative medicine therapies have little effect on the mortality or morbidity of the disease, but may improve the quality of life of individuals afflicted with AIDS. The psychological benefits of these therapies are the most important use.

EPIDEMIOLOGY

The AIDS pandemic can also be seen as several epidemics of separate subtypes; the major factors in its spread are sexual transmission and vertical transmission from mother to child at birth and through breast milk. Despite recent, improved access to antiretroviral treatment and care in many regions of the world, the AIDS pandemic claimed an estimated 2.1 million (range 1.9–2.4 million) lives in 2007 of which an estimated 330,000 were children under 15 years. Globally, an estimated 33.2 million people lived with HIV in 2007, including 2.5 million children. An estimated 2.5 million (range 1.8–4.1 million) people were newly infected in 2007, including 420,000 children.

Sub-Saharan Africa remains by far the worst affected region. In 2007 it contained an estimated 68% of all people living with AIDS and 76% of all AIDS deaths, with 1.7 million new infections bringing the number of people living with HIV to 22.5 million and with 11.4 million AIDS orphans living in the region. Unlike other regions, most people living with HIV in sub-Saharan Africa in 2007 (61%) were women. Adult prevalence in 2007 was an estimated 5.0% and AIDS continued to be the single largest cause of mortality in this region. South Africa has the largest population of HIV patients in the world, followed by Nigeria and India. South & South East Asia are second worst affected; in 2007 this region contained an estimated 18% of all people living with AIDS and an estimated 300,000 deaths from AIDS. India has an estimated 2.5 million infections and an estimated adult prevalence of 0.36%. Life expectancy has fallen dramatically in the worst-affected countries; for example, in 2006 it was estimated that it had dropped from 65 to 35 years in Botswana.

PROGNOSIS

Without treatment, the net median survival time after infection with HIV is estimated to be 9 to 11 years, depending on the HIV subtype and the median survival rate after diagnosis of AIDS in resource-limited settings where treatment is not available ranges between 6 and 19 months, depending on the study. In areas where it is widely available, the development of HAART as effective therapy for HIV infection and AIDS reduced the death rate from this disease by 80% and raised the life expectancy for a newly-diagnosed HIV-infected person to about 20 years.

As new treatments continue to be developed and because HIV continues to evolve resistance to treatments, estimates of survival time are likely to continue to change. Without antiretroviral therapy, death normally occurs within a year.

Most patients die from opportunistic infections or malignancies associated with the progressive failure of the immune system. The rate of clinical disease progression varies widely between individuals and has been shown to be affected by many factors such as host susceptibility and immune function health care and co-infections, as well as which particular strain of the virus is involved.

AIDS was first reported June 5, 1981, when the U.S. Centers for Disease Control and Prevention recorded a cluster of Pneumocystis carinii pneumonia (now still classified as PCP but known to be caused by Pneumocystis jirovecii) in five homosexual men in Los Angeles. In the beginning, the Centers for Disease Control and Prevention (CDC) did not have an official name for the disease, often referring to it by way of the diseases that were associated with it, for example, lymphadenopathy, the disease after which the discoverers of HIV originally named the virus.

They also used Kaposi's Sarcoma and Opportunistic Infections, the name by which a task force had been set up in 1981. In the general press, the term GRID, which stood for Gay-related immune deficiency, had been coined. However, after determining that AIDS was not isolated to the homosexual community, the term GRID became misleading and AIDS was introduced at a meeting in July 1982. By September 1982 the CDC started using the name AIDS and properly defined the illness. A more controversial theory known as the OPV AIDS hypothesis suggests that the AIDS epidemic was inadvertently started in the late 1950s in the Belgian Congo by Hilary Koprowski's research into a poliomyelitis vaccine. According to scientific consensus, this scenario is not supported by the available evidence.

SOCIETY AND CULTURE

Stigma

AIDS stigma exists around the world in a variety of ways, including ostracism, rejection, discrimination and avoidance of HIV infected people; compulsory HIV testing without prior consent or protection of confidentiality; violence against HIV infected individuals or people who are perceived to be infected with HIV; and the quarantine of HIV infected individuals. Stigma-related violence or the fear of violence prevents many people from seeking HIV testing, returning for their results, or securing treatment, possibly turning what could be a manageable chronic illness into a death sentence and perpetuating the spread of HIV.

AIDS stigma has been further divided into the following three categories:

- *Instrumental AIDS stigma:* A reflection of the fear and apprehension that are likely to be associated with any deadly and transmissible illness.
- *Symbolic AIDS stigma:* The use of HIV/AIDS to express attitudes toward the social groups or lifestyles perceived to be associated with the disease.
- *Courtesy AIDS stigma:* Stigmatization of people connected to the issue of HIV/AIDS or HIV- positive people.

Often, AIDS stigma is expressed in conjunction with one or more other stigmas, particularly those associated with homosexuality, bisexuality, promiscuity and intravenous drug use. In many developed countries, there is an association between AIDS and homosexuality or bisexuality and this association is correlated with higher levels of sexual prejudice such as anti-homosexual attitudes. There is also a perceived association between AIDS and all male-male sexual behaviour, including sex between uninfected men.

Economic Impact

HIV and AIDS affects economic growth by reducing the availability of human capital. Without proper nutrition, health care and medicine that is available in developed countries, large numbers of people are falling victim to AIDS. They will not only be unable to work, but will also require significant medical care. The forecast is that this will likely cause a collapse of economies and societies in countries with a significant AIDS populationi. In some heavily infected areas, the epidemic has left behind many orphans cared for by elderly grandparents. The increased mortality in this region will result in a smaller skilled population and labour force. This smaller labour force will be predominantly young people, with reduced knowledge and work experience leading to reduced productivity.

An increase in workers' time off to look after sick family members or for sick leave will also lower productivity. Increased mortality will also weaken the mechanisms that generate human capital and investment in people, through loss of income and the death of parents. By killing off mainly young adults, AIDS seriously weakens the taxable population, reducing the resources available for public expenditures such as education and health services not related to AIDS resulting in increasing pressure for the state's finances and slower growth of the economy.

This results in a slower growth of the tax base, an effect that will be reinforced if there are growing expenditures on treating the sick, training (to replace sick workers), sick pay and caring for AIDS orphans. This is especially true if the sharp increase in adult mortality shifts the responsibility and blame from the family to the government in caring for these orphans.

On the level of the household, AIDS results in both the loss of income and increased spending on healthcare by the household. The income effects of this lead to spending reduction as well as a substitution effect away from education and towards healthcare and funeral spending. A study in Côte d'Ivoire showed that households with an HIV/AIDS patient spent twice as much on medical expenses as other households.

DISEASES TRANSMITTED BY ANIMALS

Animal diseases are of great concern to humans for several reasons. Diseases can reduce the productivity of animals used to produce food, such as hens and dairy cows. Animals that are raised as food, such as pigs and beef cattle, that become ill may affect the economic well-being of many industries. Some animal diseases can be transmitted to humans, and control of these types of diseases, known as zoonoses, is vital to public health. In the wild, animal populations reduced by disease can upset the ecological balance of an area. And, in the case of pets, prevention and treatment of animal diseases helps pets live long and healthy lives, enhancing the companionship shared by a pet and its human owner.

Animal diseases are characterized as infectious and noninfectious. Infectious diseases are caused by an agent, such as bacteria or a virus, that penetrates the body's natural defense mechanisms, while noninfectious diseases are caused by factors such as diet, environment, injury, and heredity. Sometimes the cause of a disease is unknown. An animal may also experience one disease or a combination of diseases at any one time.

To identify a disease, a veterinarian (a doctor who treats animals) first determines the animal's signalment—its species, breed, age, and sex. This information helps to identify a disease because some diseases are more prevalent in certain species, or a disease may preferentially affect one sex or age group. The veterinarian then gathers a complete history of the animal and its problem. This history includes the symptoms the animal is displaying and when they first appeared, as well as whether the animal has been exposed to something new in its surroundings or to other animals. The veterinarian gives the animal a thorough physical examination, which may include measuring its body temperature, listening to its heart, checking its pulse, and feeling its abdomen and lymph nodes. The veterinarian then creates a list of possible diseases that may be making the animal sick. The list may be narrowed by running diagnostic tests such as X rays, electrocardiograms, blood analyses, and bacterial or fungal cultures. Once the disease is identified, the doctor develops a treatment plan for the animal.

INFECTIOUS DISEASES

Many microscopic organisms naturally and peacefully exist in enormous quantities within animal bodies. For example, the multichambered stomach of a cow contains bacteria that help the animal digest its food. But many other microscopic organisms, known as pathogens, cause diseases in animals. Pathogens include bacteria, viruses, fungi, prions—newly identified mutated proteins—and parasites. Pathogens are easily spread: an animal may consume food or drink something that has been contaminated with infected fecal material, for example. If the ground is contaminated by *Salmonella* bacteria, for instance, infection can travel from barn to barn on the soles of a farmer's boots. Or an animal may be exposed while walking across contaminated ground. Some diseases are transmitted by biting insects; others are spread by sexual contact.

In addition to reducing the productivity of livestock, some infectious diseases pose a danger to humans. More than 100 zoonoses are recognized. Most cases are transmitted from animals that have close contact with humans, such as pets, farm animals, or rats. Examples of zoonoses include toxocariasis, a disease caused by a parasitic worm transmitted by infective eggs within canine feces; psittacosis, a respiratory disease caused by the bacteria-like *Chlamydia psittaci* and transmitted from infected birds; hantavirus pulmonary syndrome, spread by contact with rodent feces and urine; and rabies, a viral infection transmitted in the saliva of infected animals, typically foxes, bats, and raccoons, that causes damage to the brain and spinal cord.

As the human population grows and expands into wilderness territories, humans are coming into closer contact with other animals that carry pathogens dangerous to humans. Some of these pathogens are carried by insects, as in the case of yellow fever, spread from monkeys to humans via mosquito bites. Some hemorrhagic fevers, such as that caused by the Ebola virus, are recognized as zoonoses, but the exact transmission route from animal to human is still unknown.

BACTERIAL DISEASES

Salmonellosis is any disease caused by the *Salmonella* bacteria, characterized by septicemia and severe diarrhea. In its many forms, it is one of the major diseases of wild and domestic mammals, birds, and reptiles, as well as humans. *Salmonella* bacteria usually enter the body through the mouth, most commonly along with food or water contaminated by infected feces. Transmission also may occur through direct contact with an infected animal. In addition, *salmonella* bacteria can be spread by contact with objects, such as bowls and cutting boards, that have been contaminated by infected animal products, such as eggs or meat.

Anthrax is one of the oldest and most destructive diseases recorded in history. Caused by the bacterium *Bacillus anthracis*, anthrax can affect virtually all warm-blooded animals and humans.

The onset of anthrax may be sudden and death may occur before symptoms are observed. In other cases, typical symptoms include restlessness, lethargy, appetite loss, fever, rapid breathing, and unsteady gait. The disease is contracted from contaminated soil, feed, or water. It can also spread when the skin is penetrated by insect bites or by objects contaminated with anthrax spores.

Leptospirosis, caused by spiral *Leptospira* bacteria, affects cattle, dogs, pigs, sheep, goats, and humans. Ponds, lakes, and other bodies of water are common sources of leptospirosis, and rodents may carry the infection. This infection causes kidney disease and destruction of red blood cells with potential anemia; it may also cause abortion. Brucellosis also causes abortion, as well as swelling of the reproductive organs in males. Caused by the *Brucella* bacterium, it occurs primarily in cattle, pigs, sheep, dogs, and goats, and may be transmitted to humans.

Tuberculosis (TB) is a chronic disease of animals and humans, caused by bacteria of the genus *Mycobacterium* and transmitted by inhalation of droplets from an infected animal's cough or sneeze, or by wound infection. TB infection causes lesions called tubercles to develop in certain tissues, such as the lung or liver. Symptoms include fever, emaciation, and progressive loss of strength.

Kennel cough is a respiratory disease of dogs that is caused by the bacterium *Bordetella bronchiseptica*, with or without the aid of various viruses. Symptoms include a harsh, dry cough, appetite loss, discharge from the nose or eyes, and lethargy. It typically spreads when dogs are grouped together, such as at dog shows or boarding kennels.

Viral Diseases

Viruses are unable to grow and reproduce outside of the living cells from other hosts. Viruses attach and invade a cell and replicate, and then the newly created viruses destroy the host cell and seek out other cells to continue replication.

Feline leukemia is caused by the feline leukemia virus. Often fatal, it can seriously impair the immune system and, in some cases, cause the growth of life-threatening tumors. Spread from direct contact with an infected cat, symptoms of the disease include lethargy, weight loss, anemia, and fever. A cat may not appear ill until years after exposure.

Foot-and-mouth disease is caused by a virus found in the saliva of cattle, pigs, and other hoofed animals. Highly contagious, it is spread from direct contact with an infected animal. It may also spread in milk or in garbage that contains contaminated meat. Typical symptoms include blisters that appear on the mouth and feet; animals may become lame when their hooves degenerate.

Canine distemper is a highly contagious disease caused by the paramyxovirus, which is transmitted in discharges from the nose and eyes. Symptoms begin with fever, malaise, and nasal and ocular discharges and may progress to convulsions and other nervous system disorders. Parvoviruses affect dogs and in some cases cattle, pigs, and humans. Usually fatal if left untreated, canine parvovirus causes inflammation of the intestines, producing diarrhea, vomiting, fever, and loss of appetite.

Fungal Diseases

A fungal infection typically develops slowly and recurs more frequently than a bacterial infection. Histoplasmosis, characterized by a chronic cough and diarrhea, is contracted by inhaling the *Histoplasma capsulatum* fungus, which grows in soil. In the Central United States histoplasmosis is the most widespread fungal disease diagnosed in dogs, although it also affects other animals. Ringworm, a common skin disease of many species, causes circular patches of hair loss and scaly, reddened skin. It readily spreads by direct contact with an infected animal.

Yeast, another type of fungus, grows in warm and moist places, such as the ear canals of dogs. It may cause otitis externa, an infection of the outer ear. The yeast *Candida albicans* is commonly found in the intestinal tract of birds and other animals. It may be the primary cause of disease, or it may be a secondary invader in an animal already sick with another infection.

Parasitic Infections

Diseases caused by parasites are widespread in domestic animals and wildlife. Parasites may be internal or external. Internal parasites include Coccidia, a microscopic protozoal (single-celled) organism that causes diarrhea and extreme weight loss in many young animals.

Other internal parasites include the roundworm, tapeworm, and fluke. Larval roundworms can cause considerable damage to lungs and other organs in some animals. For instance, *Capillaria* worms may attack the lining of the digestive tract of chickens and turkeys; they parasitize the respiratory and urinary tracts of dogs.

Adults of the heartworm *Dirofilaria immitis*, another roundworm, live in the heart of dogs and produce microscopic larval stages, which swim in the blood. Symptoms of heartworm disease include coughing, fatigue, and weight loss. If left untreated, an animal may experience heart failure. Tapeworms may have very damaging larval stages. In echinococcosis, the larval tapeworms may form large cysts in liver, lungs, and other organs of humans and animals. Flukes may directly damage the liver, lungs, or intestines, or they may act as carriers of other disease agents, as in the case of salmon poisoning of dogs in which the fluke, encysted in the body of a salmon, carries a virulent rickettsial agent.

External parasites live or feed on the surface of the animal's body. This group includes bloodsucking insects, such as mosquitoes, gnats, some flies, fleas, and some lice. Some insects are bloodsuckers in larval stages, such as ear maggots of hawk nestlings. Others, including some larval flies and some lice, eat tissue. Great damage to the meat and hides of cattle is caused by larval flies such as the ox warble, which migrates through the tissues and, after boring breathing holes through the skin, leaves the body to reproduce. Bloodsucking flies can transmit parasitic blood protozoans and some viruses.

Lice are of two types, those with chewing mouthparts and those with sucking mouthparts. Lice cause irritation, carry disease agents, and may cause anemia. Fleas are all bloodsuckers, and may transmit larval tapeworms, roundworms, and other disease agents. The sticktight flea may kill young birds by excessive bloodsucking. Mites may be external bloodsuckers, such as the red mite of birds, or they may be internal parasites, such as the *Sternostoma* mites of the lungs and air passages of canaries and other birds. Ticks, larger than mites, feed on blood and can carry serious infectious agents such as the bacteria that cause Q Fever and Lyme disease, which can be transmitted to humans.

Prion Diseases]

Newly identified protein particles called prions have been found in the brains of animals that have died from diseases such as scrapie and bovine spongiform encephalopathy, more commonly known as mad cow disease.

How prions act is unclear, but scientists theorize that prions attach to normal proteins in the brain. Once attached, the prions cause the normal proteins to change into an abnormal shape, leading to progressive destruction of brain cells and death.

Prion diseases are thought to spread by means of feed supplements derived from infected animals. In recent years, public health officials have been concerned about the possibility that prion diseases may be transmitted to humans. This happens when humans eat contaminated beef or organs, causing them to contract such rare neurological diseases as Creutzfeldt-Jakob disease.

Prevention and Treatment

Controlling the spread of infectious animal diseases begins with isolating, or quarantining, animals with threatening infections, such as salmonella, to prevent further transmission. Many bacterial diseases can be treated with various antibiotics, such as penicillin and streptomycin.

But as with all disease, prevention is more important than treatment, and a major activity for veterinarians is immunization of animals. Immunization commonly involves an injection of a weakened or killed pathogen for a specific disease that encourages the immune system to fight off infection.

Many infectious diseases, including rabies, canine distemper, feline leukemia, anthrax, and brucellosis, can be prevented by immunization. In the case of severe outbreaks of infectious disease, public health officials may work with animal owners to destroy large groups of animals. This was the case in the early 1990s, when an outbreak of bovine spongiform encephalopathy triggered the slaughter of many beef cattle in Britain.

Transmission of animal diseases to humans is a constant concern of public health officials. To protect people from disease, veterinarians inspect food animals for wholesomeness; quarantine and examine animals brought into the United States from other countries; test animals for the presence of disease; and actively work to prevent and eradicate diseases that threaten human health.

NONINFECTIOUS DISEASES

Even if it were possible, a world without pathogens would not be disease-free. Many animal diseases are caused by noninfectious factors such as an animal's environment, genetics, and nutrition. Heatstroke, for example, occurs when an animal is forced to endure high temperatures without access to water, adequate ventilation, or suitable shade. A common scenario involves an animal that has been locked inside a car without air-conditioning during hot weather. Conversely, extreme cold can lead to hypothermia or frostbite. Other environmental hazards include the vast array of products humans use to eliminate pests and weeds from homes, farms, and gardens.

For example, rodenticide, poison used to kill rats and mice, can cause fatal internal hemorrhaging in any animal that ingests this toxic substance. Improper use of flea powders, sprays, dips, and collars can also cause illness. Automobile antifreeze is another well-known poison. Its sweet taste appeals to some animals, such as cats and dogs, but consuming only a small amount can result in death. Many plant species are also toxic to animals. Some, such as pokeweed and yew, commonly grow in pastures and yards.

Poor feeding practices can lead to diseases such as nutritional secondary hyperparathyroidism, a condition involving the muscles and bones of dogs that is associated with an all-meat diet. Large, rapidly growing puppies that consume too many calories and too much calcium can develop hypertrophic osteodystrophy, a disease resulting in lameness. Cats need sufficient amounts of an essential amino acid called taurine in their diets. Without it, they may develop eye problems. Not enough iodine intake can cause a goiter, or enlargement of the thyroid gland, in cows, horses, and other animals.

Trauma is a leading cause of injury and premature death in animals, especially pets that are allowed to roam free outdoors. Many animals are hit by cars or bitten by other animals. Farm animals may be attacked by predators, or they may harm themselves on sharp fencing or discarded nails. Untreated wounds can become infected and cause permanent damage. Hip dysplasia, a painful and debilitating skeletal condition, is a noninfectious disease caused in part by heredity. Certain defects of the heart or palate, the roof of the mouth, may also be inherited. Some animals are genetically predisposed to diseases such as generalized demodectic mange, a skin disease caused by mites and characterized by hair loss and scaling around the eyelids, mouth, and front legs.

An animal's immune system is designed to detect and

eliminate invading organisms. Occasionally, however, it behaves as though the animal's own body were the attacker, and it destroys healthy tissue. Diseases caused by this response, known as autoimmune diseases, include pemphigus foliaceous, a skin disease of dogs, cats, and horses; and rheumatoid arthritis, a severe type of arthritis that involves inflammation of the joints. In the autoimmune disease hemolytic anemia, the animal's own red blood cells are destroyed by its immune system.

Cancer exists in all animals. It is classified as either benign—that is, relatively noninvasive and unlikely to return after treatment; or as malignant—that is, aggressive and likely to spread. Any organ or system can be affected, either directly or through metastasis—when cancer cells from one part of the body spread to other areas of the body. Some forms of cancer are more widespread in animals of a particular breed, age, or sex, and even individuals of a specific color. For example, cancer of the mammary gland occurs more often in female animals, while melanoma, or skin cancer, is the most frequent tumor of elderly gray horses, and lymphosarcomas, tumors of the lymph nodes, are the most common type of specific tumor in cats.

RABIES VIRUS

Rabies is a viral zoonotic neuroinvasive disease that causes acute encephalitis (inflammation of the brain) in mammals. It is most commonly caused by a bite from an infected animal, but occasionally by other forms of contact. If left untreated in humans it is almost invariably fatal. In some countries it is a significant killer of livestock.

The rabies virus makes its way to the brain by following the peripheral nerves. The incubation period of the disease depends on how far the virus must travel to reach the central nervous system, usually taking a few months. Once the infection reaches the central nervous system and symptoms begin to show, the untreated infection is almost inevitably fatal within days.

In the beginning stages of rabies, the symptoms are

malaise, headache, and fever, while in later stages it includes acute pain, violent movements, uncontrolled excitements, depressions, and the inability to swallow water. In the final stages, the patient begins to have periods of mania and lethargy, and coma. Death generally occurs due to respiratory insufficiency.

In non-vaccinated humans, rabies is almost invariably fatal after neurological symptoms have developed, but prompt post-exposure vaccination may prevent the virus from progressing. Rabies kills around 55,000 people a year, mostly in Asia and Africa. There are only six known cases of a person surviving symptomatic rabies, and only two known cases of survival in which the patient received no rabies-specific treatment either before or after illness onset.

RHABDOVIRUSES

The family Rhabdoviridae consists of more than 100 single-stranded, negative-sense, nonsegmented viruses that infect a wide variety of hosts, including vertebrates, invertebrates, and plants. Common to all members of the family is a distinctive rod- or bullet-shaped morphology. Human pathogens of medical importance are found in the genera *Lyssavirus* and *Vesiculovirus*. Only rabies virus, medically the most significant member of the genus *Lyssavirus*, is reviewed in this chapter.

Clinical Manifestations

Five general stages of rabies are recognized in humans: incubation, prodrome, acute neurologic period, coma, and death. No specific antirabies agents are useful once clinical signs or symptoms develop. The incubation period in rabies, usually 30 to 90 days but ranging from as few as 5 days to longer than 2 years after initial exposure, is more variable than in any other acute infection.

Incubation periods may be somewhat shorter in children and in individuals bitten close to the central nervous system (e.g., the head). Clinical symptoms are first noted during the prodromal period, which usually lasts from 2 to 10 days. These symptoms are often nonspecific (general malaise, fever, and fatigue) or suggest involvement of the respiratory system (sore throat, cough, and dyspnea), gastrointestinal system (anorexia, dysphagia, nausea, vomiting, abdominal pain, and diarrhea), or central nervous systems (headache, vertigo, anxiety, apprehension, irritability, and nervousness).

More remarkable abnormalities (agitation, photophobia, priapism, increased libido, insomnia, nightmares, and depression) may also occur, suggesting encephalitis, psychiatric disturbances, or brain conditions. Pain or paresthesia at the site of virus inoculation, combined with a history of recent animal bite, should suggest a consideration of rabies.

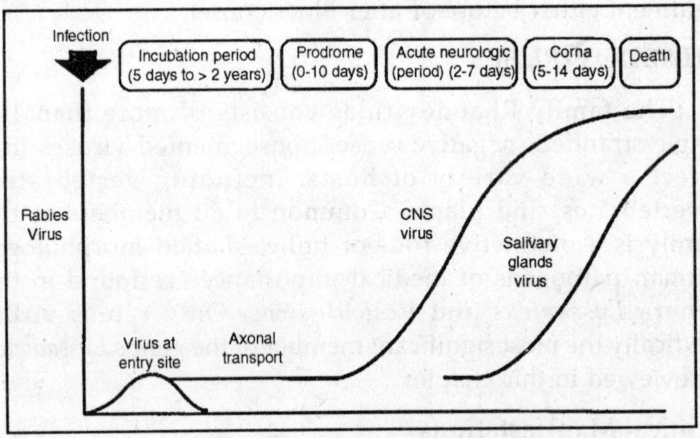

Fig. Pathogenesis of Rabies.

The acute neurologic period begins with objective signs of central nervous system dysfunction. The disease may be classified as furious rabies if hyperactivity (i.e., hydrophobia) predominates and as dumb rabies if paralysis dominates the clinical picture. Fever, paresthesia, nuchal rigidity, muscle fasciculations, focal and generalized convulsions, hyperventilation, and hypersalivation may occur in both forms of the disease.

At the end of the acute neurologic phase, periods of rapid, irregular breathing may begin; paralysis and coma soon follow. Respiratory arrest may occur thereafter, unless the patient is receiving ventilatory assistance, which may prolong survival for days, weeks, or longer, with death due to other complications.

Although life support measures can prolong the clinical course of rabies, rarely will they affect the outcome of disease. The possibility of recovery, however, must be recognized, and when resources permit, every effort should be made to support the patient. At least seven cases of human "recovery" have been documented.

Structure

The rabies virus is a negative-sense, non-segmented, single-stranded RNA virus measuring approximately 60 nm × 180 nm. It is composed of an internal protein core or nucleocapsid, containing the nucleic acid, and an outer envelope, a lipid-containing bilayer covered with transmembrane glycoprotein spikes.

The virus genome encodes five proteins associated with either the ribonucleoprotein (RNP) complex or the viral envelope. The L (transcriptase), N (nucleoprotein), and NS (transcriptase-associated) proteins comprise the RNP complex, together with the viral RNA. These aggregate in the cytoplasm of virus-infected neurons and compose Negri bodies, the characteristic histopathologic finding of rabies virus infection. The M (matrix) and G (glycoprotein) proteins are associated with the lipid envelope. The G protein forms the protrusions that cover the outer surface of the virion envelope and is the only rabies virus protein known to induce virus-neutralizing antibody.

Classification and Antigenic Types

The genus *Lyssavirus* includes rabies virus and the antigenically- and genetically-related rabies- like viruses: Lagos bat, Mokola, and Duvenhage viruses, and two suggested subtypes of European bat lyssaviruses. Cross-protection studies suggest that animals immunized with traditional rabies vaccines may not be fully protected if challenged with other lyssaviruses.

Rabies viruses may be categorized as either fixed (adapted by passage in animals or cell culture) or street (wild type). The use of monoclonal antibodies and genetic sequencing to differentiate street rabies viruses has been helpful in identifying viral variants originating in major host reservoirs throughout the world and suggesting the likely sources of human exposure when a history of definitive animal bite was otherwise missing from a patient's case history.

Multiplication

The replication of rabies virus is believed to be similar to that of other negative-stranded RNA viruses. The virus attaches to the host cell membranes via the G protein, penetrates the cytoplasm by fusion or pinocytosis, and is uncoated to RNP.

The core initiates primary transcription of the five complementary monocistronic messenger RNAs by using the virion-associated RNA-dependent RNA polymerase. Each RNA is then translated into an individual viral protein. After viral proteins have been synthesized, replication of the genomic RNA continues with the synthesis of full length, positive-stranded RNA, which acts as a template for the production of progeny negative-stranded RNA.

Pathogenesis

Rabies virus is most commonly transmitted through the bite of an infected mammal, all of which may be susceptible, but to greatly varying degrees. The virus may enter the peripheral nervous system directly, or may replicate in muscle tissue after entering the host, remaining at or near the site of introduction for most of the incubation period. However, the precise sites of viral sequestration remain unknown, since neither antigen nor virus can usually be found in any organ during this phase.

Virus may enter the peripheral nervous system via the neuromuscular junctions, and moves rapidly centripetally to the central nervous system for replication; symptoms may develop shortly thereafter. The virus then begins to pass centrifugally to many tissues and organs, such as the salivary glands.

In general, gross examination of the brain shows mild congestion of the meningeal vessels; microscopic examination usually demonstrates slight perivascular cuffing, limited tissue necrosis, acidophilic intracytoplasmic neuronal inclusions, and rarely, neuronophagia.

Host Defenses

The host animal species, viral variant, inoculum concentration, body location and severity of exposure, and host immune status have been associated with overt susceptibility to infection and with different incubation periods. The association of virus-neutralizing antibody, principally IgG, and protective immunity is well known. Production of cytokine, such as interferon, induced during rabies virus infection or vaccination, has been reported to abort the disease if it occurs shortly after viral infection. In one clinical trial, however, all subjects died despite experimental treatment with high doses of alpha interferon.

Recently it has been demonstrated that animals immunized with purified RNP complexes or recombinant nucleoprotein vaccines resisted lethal challenge with rabies virus, although the role of N protein in protection, illness, or recovery is unclear.

Epidemiology

Rabies has been recognized for over 4,000 years. Today it is found in most countries, with the exception of those regions from which it has not been naturally reported, including many Australian islands, or areas achieving secondary elimination, such as the United Kingdom. Almost all human rabies is caused by the bite of a rabid animal. In the United States and Europe, domestic animal rabies was largely controlled during the 1940-50s and now represents less than 10% of all animal rabies recorded. Wildlife rabies in the United States occurs primarily among wild terrestrial carnivores, such as raccoons, skunks, foxes, and coyotes, and in insectivorous bats.

Human rabies is almost always attributable to a bite. Nonbite exposures (contamination of an open wound or a mucous membrane via scratches, licks, and inhalation of aerosol) rarely cause rabies in humans. In the United States, nonbite exposures were reported as the source of infection for only 5 (3 %) of the 154 cases reported from 1950 through 1980. Of these five human cases, four were apparently attributable via exposure to aerosols containing highly concentrated live rabies virus: two in spelunkers and two in rabies research laboratory workers.

The fifth case occurred in the recipient of a cornea transplanted from a patient dying of unsuspected rabies encephalitis. An increasing proportion of current human rabies patients in the United States have had no known exposure to the virus; since 1980, 19 of 25 (90%) of the rabies patients have had no definitive rabid animal exposure. This may be attributable either to an inability of the patient to recognize actual rabies exposure at the time, or a failure to properly question the patient concerning potential animal contact. Although it is now a rare human disease in the United States, its actual incidence may be higher than generally believed.

DIAGNOSIS

Differential Diagnosis

The diagnosis of human rabies is usually suggested by epidemiologic and clinical findings and confirmed in the laboratory. The diagnosis is not difficult if there is a history of animal bite exposure and if a full spectrum of symptoms and signs has appeared. Otherwise, careful but rapid assessment of the epidemiologic and clinical features of less typical cases is essential before special laboratory tests are performed. Every patient with neurologic signs or symptoms or unexplained encephalitis should be questioned about the possibility of animal exposure in a rabies-endemic area inside or outside the country of residence.

The failure to suspect rabies in several of the recent human deaths in the United States may have occurred because no thorough exposure history had been sought. Early in the course of illness, rabies can mimic numerous infectious and noninfectious diseases. Many other encephalitides, such as those caused by herpesviruses and arboviruses, resemble rabies. Other infectious diseases also may resemble rabies, such as tetanus, cerebral malaria, rickettsial diseases, and typhoid. Paralytic infectious illnesses that may be confused with rabies include poliomyelitis, botulism, and simian herpes type B encephalitis.

Noninfectious diseases that may be confused with rabies encompass a number of neurologic syndromes, especially acute inflammatory polyneuropathy as well as allergic postvaccinal encephalomyelitis secondary to vaccination with nervous-tissue rabies vaccines, intoxication with poisons or drugs, withdrawal from alcohol, acute porphyria, and rabies hysteria. Guillain-Barre syndrome may be mistaken for the paralytic form of rabies, and vice versa.

Laboratory Diagnosis

The detection of rabies antigen, antibody, viral RNA, or the isolation of virus establishes a diagnosis of rabies. Because any individual test may not be positive in a patient with rabies, serial serum specimens for detection of rabies antibodies, saliva specimens for culture of virus, and skin biopsies for direct immunofluorescence testing for virus antigen are sometimes necessary, especially when rabies is strongly suspected.

One of the most rapid methods to diagnose rabies antemortem in humans is to perform a direct immunofluorescence test on a skin biopsy from the nape of the neck for evidence of rabies antigen. The direct immunofluorescence test is the most sensitive and specific method of detecting rabies antigen in skin and other fresh tissue although the results may occasionally be negative in early stages of the disease. If fresh tissue is unavailable, enzyme digestion of fixed tissues may enhance the reactivity of the immunofluorescence test; however, sensitivity may be unacceptably low.

The diagnosis can also be established if virus is isolated from saliva after inoculation of neuroblastoma cells or laboratory rodents; this is generally most successful during the first 2 to 3 weeks of illness. The detection of rabies virus-neutralizing antibody, as typically performed by the rapid fluorescent focus inhibition test in the serum of unvaccinated individuals is also diagnostic.

The presence of antibody in the cerebrospinal fluid confirms the diagnosis, but it may appear 2 to 3 days later than serum antibody and may there fore be less useful early in the disease. Whereas the serological response after vaccination cannot be generally differentiated from that due to disease, vaccination does not typically produce cerebrospinal fluid antibody.

Only seven "recoveries" from rabies, all in the past 25 years, have been well documented. Although rabies virus was not isolated in any of the patients, the high rabies-neutralizing antibody titer in serum samples and the presence of neutralizing antibodies in cerebrospinal fluid strongly supported the diagnoses.

TREATMENTS

Post-exposure Prophylaxis

Treatment after exposure, known as post-exposure prophylaxis or "P.E.P.", is highly successful in preventing the disease if administered promptly, generally within six days of infection. Thoroughly washing the wound as soon as possible with soap and water for approximately five minutes is very effective at reducing the number of viral particles. "If available, a virucidal antiseptic such as povidone-iodine, iodine tincture, aqueous iodine solution or alcohol should be applied after washing." Exposed mucous membranes such as eyes, nose or mouth should be flushed well with water. In the United States, patients receive one dose of immunoglobulin and five doses of rabies vaccine over a twenty-eight day period.

One-half the dose of immunoglobulin is injected in the region of the bite, if possible, with the remainder injected intramuscularly away from the bite. This is much less painful compared with administering immunoglobulin through the abdominal wall with a large needle, as was done in the past. The first dose of rabies vaccine is given as soon as possible after exposure, with additional doses on days three, seven, fourteen, and twenty-eight after the first. Patients that have previously received pre-exposure vaccination do not receive the immunoglobulin, only the post-exposure vaccinations on day 0 and 3.

Since the widespread vaccination of domestic dogs and cats and the development of effective human vaccines and immunoglobulin treatments, the number of recorded deaths in the U.S. from rabies has dropped from one hundred or more annually in the early twentieth century, to 1–2 per year, mostly caused by bat bites, which may go unnoticed by the victim and hence untreated.

In instances when post-exposure prophylaxis is administered as a precaution, it is now mainly given in the gluteal region and deltoid. The number of shots delivered to the gluteal area on the first day is determined by weight, and it is not uncommon to require three of these shots. Subsequent shots of the immunoglobulin (to build longer term immunity to rabies) are given to the arm. Recipients of the vaccine have reported that these shots are no more painful than normal shots.

Most official documentation on rabies on the internet and otherwise warn that treatment becomes futile with the onset of prodrome. These texts are written to convince the layman not to delay seeking treatment. However, this may also lead them to falsely conclude that their situation is not an urgency and that treatment is possible up until the very end of the incubation period, as it may last 1 to 3 months on average; or it may at least convince them that it is safe to delay treatment by a few days.

While the virus is treatable only during the incubation period, it is important to note that it is not treatable during its entirety. The disease becomes untreatable long before the onset of the prodromal stage, or the first signs of symptoms. Rabies is treatable while the virus is present in tissues composed of cells other than neurons, such as skin and muscle. However, once the infection spreads to a neuron, the virus is sequestered from the immune system and will eventually make its way to the spinal cord and then to the brain.

Treatment at this point is futile, even though symptoms may begin to appear weeks or even months later. Therefore, it is essential that P.E.P. be administered as soon as possible. Begun without delay, or very little delay, P.E.P. is highly effective against rabies. In the case where there has been a significant delay in administering P.E.P., the treatment should be administered regardless of that delay.

Blood-brain Barrier

Some recent works have shown that during lethal rabies infection the blood-brain barrier (BBB) does not allow anti-viral immune cells to enter the brain, the primary site of rabies virus replication. This aspect contributes to the pathogenicity of the virus and artificially increasing BBB permeability promotes viral clearance. Opening the BBB during rabies infection has been suggested as a possible novel approach to treat the disease.

PLAGUE

Plague is a deadly infectious disease caused by the enterobacteria *Yersinia pestis (Pasteurella pestis)*. Plague is a zoonotic, primarily carried by rodents (most notably rats) and spread to humans via fleas. Plague is notorious throughout history, due to the unprecedented scale of death and devastation it brought. Plague is still endemic in some parts of the world. The epidemiological use of the term *plague* is currently applied to bacterial infections that cause *buboes*, although historically the medical use of the term plague has been applied to pandemic infections in general. Plague is often synonymous with "bubonic plague" but this only describes one of its manifestations.

Infection and Transmission

Bubonic plague is mainly a disease in rodents and fleas

(*Xenopsylla cheopis*). Infection in a human occurs when a person is bitten by a flea that has been infected by biting a rodent that itself has been infected by the bite of a flea carrying the disease.

The bacteria multiply inside the flea, sticking together to form a plug that blocks its stomach and causes it to begin to starve. The flea then voraciously bites a host and continues to feed, even though it cannot quell its hunger, and consequently the flea vomits blood tainted with the bacteria back into the bite wound. The bubonic plague bacterium then infects a new victim, and the flea eventually dies from starvation. Serious outbreaks of plague are usually started by other disease outbreaks in rodents, or a rise in the rodent population.

In 1894, two bacteriologists, Alexandre Yersin of France and Shibasaburo Kitasato of Japan, independently isolated the bacterium in Hong Kong responsible for the Third Pandemic. Though both investigators reported their findings, a series of confusing and contradictory statements by Kitasato eventually led to the acceptance of Yersin as the primary discoverer of the organism.

Yersin named it *Pasteurella pestis* in honor of the Pasteur Institute, where he worked, but in 1967 it was moved to a new genus, renamed *Yersinia pestis* in honor of Yersin. Yersin also noted that rats were affected by plague not only during plague epidemics but also often preceding such epidemics in humans, and that plague was regarded by many locals as a disease of rats: villagers in China and India asserted that, when large numbers of rats were found dead, plague outbreaks in people soon followed.

PATHOLOGY

Bubonic Plague

When a flea bites a human and contaminates the wound with regurgitated blood, the plague carrying bacteria are passed into the tissue. *Y. pestis* can reproduce inside cells, so even if phagocytosed, they can still survive. Once in the body, the bacteria can enter the lymphatic system, which drains interstitial fluid. Plague bacteria secrete several toxins, one of which is known to cause dangerous beta-adrenergic blockade. *Y. pestis* spreads through the lymphatics of the infected human until it reaches a lymph node, where it stimulates severe haemorrhagic inflammation that causes the lymph nodes to expand. The expansion of lymph nodes is the cause of the characteristic "bubo" associated with the disease.

Septicemic Plague

Lymphatics ultimately drain into the bloodstream, so the plague bacteria may enter the blood and travel to almost any part of the body. In septicemic plague, bacterial endotoxins cause disseminated intravascular coagulation (DIC), causing tiny clots throughout the body and possibly ischaemic necrosis (tissue death due to lack of circulation/perfusion to that tissue) from the clots.

DIC results in depletion of the body's clotting resources, so that it can no longer control bleeding. Consequently, there is bleeding into the skin and other organs, which can cause red and/or black patchy rash and hemoptysis/haemoptysis (coughing up or vomiting of blood).

There are bumps on the skin that look somewhat like insect bites; these are usually red, and sometimes white in the center. Untreated, septicemic plague is usually fatal. Early treatment with antibiotics reduces the mortality rate to between 4 and 15 percent. People who die from this form of plague often die on the same day symptoms first appear.

Pneumonic Plague

The pneumonic plague infects the lungs, and with that infection comes the possibility of person-to-person transmission through respiratory droplets. The incubation period for pneumonic plague is usually between two and four days, but can be as little as a few hours. The initial symptoms, of headache, weakness, and coughing with hemoptysis, vomiting blood, are indistinguishable from other respiratory illnesses.

Without diagnosis and treatment, the infection can be fatal

in one to six days; mortality in untreated cases is 50–90%.There are a few other rare manifestations of plague, including asymptomatic plague and abortive plague. *Cellulocutaneous plague* sometimes results in infection of the skin and soft tissue, often around the bite site of a flea. *Plague meningitis* can occur in very rare cases of septicemic plague.

RICKETTSIA

Rickettsia is a genus of motile, Gram-negative, non-sporeforming, highly pleomorphic bacteria that can present as cocci (0.1 ìm in diameter), rods (1-4 ìm long) or thread-like (10 ìm long). Obligate intracellular parasites, the *Rickettsia* depend on entry, growth, and replication within the cytoplasm of eukaryotic host cells (typically endothelial cells).

Because of this, *Rickettsia* cannot live in artificial nutrient environments and are grown either in tissue or embryo cultures (typically, chicken embryos are used). In the past they were regarded as microorganisms positioned somewhere between viruses and true bacteria. The majority of *Rickettsia* bacteria are susceptible to antibiotics of the tetracycline group.

Rickettsia species are carried as parasites by many ticks, fleas, and lice, and cause diseases such as typhus, rickettsialpox, Boutonneuse fever, African Tick Bite Fever, Rocky Mountain spotted fever, Australian Tick Typhus, Flinders Island Spotted Fever and Queensland Tick Typhus in human beings. They have also been associated with a range of plant diseases. Like viruses, they only grow inside living cells. The name rickettsia is often used for any member of the Rickettsiales. They are thought to be the closest living relatives to bacteria that were the origin of the mitochondria organelle that exists inside most eukaryotic cells.

Genomics

Certain segments of Rickettsial genomes resemble that of mitochondria. The deciphered genome of *R. prowazekii* is 1,111,523 bp long and contains 834 protein-coding genes. Unlike free-living bacteria, it contains no genes for anaerobic glycolysis or genes involved in the biosynthesis and regulation of amino acids and nucleosides. In this regard it is similar to mitochondrial genomes; in both cases, nuclear (host) resources are used. ATP production in *Rickettsia* is the same as that in mitochondria.

In fact, of all the microbes known, the *Rickettsia* is probably the closest "relative" (in phylogenetic sense) to the mitochondria. Unlike the latter, the genome of *R. prowazekii*, however, contains a complete set of genes encoding for the tricarboxylic acid cycle and the respiratory chain complex. Still, the genomes of the Rickettsia as well as the mitochondria are frequently said to be "small, highly derived products of several types of reductive evolution".

The recent discovery of another parallel between *Rickettsia* and viruses may become a basis for fighting HIV infection. Human immune response to the scrub typhus pathogen, *Orientia tsutsugamushi* rickettsia, appears to provide a beneficial effect against HIV infection progress, negatively influencing the virus replication process.

A probable reason for this actively studied phenomenon is a certain degree of homology between the rickettsia and the virus - namely, common epitope(s) due to common genome fragment(s) in both pathogens. Surprisingly, the other infection reported to be likely to provide the same effect (decrease in viral load) is the virus-caused illness dengue fever.

MALARIA

Malaria is a vector-borne infectious disease caused by protozoan parasites. It is widespread in tropical and subtropical regions, including parts of the Americas, Asia, and Africa. Each year, there are approximately 515 million cases of malaria, killing between one and three million people, the majority of whom are young children in Sub-Saharan Africa. Ninety percent of malaria-related deaths occur in Sub-Saharan Africa. Malaria is commonly associated with poverty, but is also a cause of poverty and a major hindrance to economic development.

Malaria is one of the most common infectious diseases and an enormous public health problem. The disease is caused by protozoan parasites of the genus *Plasmodium*. Only four types of the plasmodium parasite can infect humans; the most serious forms of the disease are caused by *Plasmodium falciparum* and *Plasmodium vivax*, but other related species (*Plasmodium ovale, Plasmodium malariae*) can also affect humans. This group of human-pathogenic *Plasmodium* species is usually referred to as *malaria parasites*.

Usually, people get malaria by being bitten by an infective female *Anopheles* mosquito. Only Anopheles mosquitoes can transmit malaria, and they must have been infected through a previous blood meal taken on an infected person. When a mosquito bites an infected person, a small amount of blood is taken, which contains microscopic malaria parasites.

About one week later, when the mosquito takes its next blood meal, these parasites mix with the mosquito's saliva and are injected into the person being bitten. The parasites multiply within red blood cells, causing symptoms that include symptoms of anemia (light-headedness, shortness of breath, tachycardia, etc.), as well as other general symptoms such as fever, chills, nausea, flu-like illness, and, in severe cases, coma, and death. Malaria transmission can be reduced by preventing mosquito bites with mosquito nets and insect repellents, or by mosquito control measures such as spraying insecticides inside houses and draining standing water where mosquitoes lay their eggs.

Although some are under development, no vaccine is currently available for malaria; preventive drugs must be taken continuously to reduce the risk of infection. These prophylactic drug treatments are often too expensive for most people living in endemic areas. Most adults from endemic areas have a degree of long-term infection, which tends to recur, and also possess partial immunity (resistance); the resistance reduces with time, and such adults may become susceptible to severe malaria if they have spent a significant amount of time in non-endemic areas.

They are strongly recommended to take full precautions if they return to an endemic area. Malaria infections are treated through the use of antimalarial drugs, such as quinine or artemisinin derivatives. However, parasites have evolved to be resistant to many of these drugs. Therefore, in some areas of the world, only a few drugs remain as effective treatments for malaria.

Symptoms

Symptoms of malaria include fever, shivering, arthralgia, vomiting, anemia (caused by hemolysis), hemoglobinuria, retinal damage, and convulsions. The classic symptom of malaria is cyclical occurrence of sudden coldness followed by rigor and then fever and sweating lasting four to six hours, occurring every two days in *P. vivax* and *P. ovale* infections, while every three for *P. malariae*. *P. falciparum* can have recurrent fever every 36–48 hours or a less pronounced and almost continuous fever.

For reasons that are poorly understood, but that may be related to high intracranial pressure, children with malaria frequently exhibit abnormal posturing, a sign indicating severe brain damage. Malaria has been found to cause cognitive impairments, especially in children. It causes widespread anemia during a period of rapid brain development and also direct brain damage. This neurologic damage results from cerebral malaria to which children are more vulnerable.

Severe malaria is almost exclusively caused by *P. falciparum* infection and usually arises 6–14 days after infection. Consequences of severe malaria include coma and death if untreated—young children and pregnant women are especially vulnerable.

Splenomegaly (enlarged spleen), severe headache, cerebral ischemia, hepatomegaly (enlarged liver), hypoglycemia, and hemoglobinuria with renal failure may occur. Renal failure may cause blackwater fever, where hemoglobin from lysed red blood cells leaks into the urine. Severe malaria can progress extremely rapidly and cause death within hours or days.

In the most severe cases of the disease fatality rates can exceed 20%, even with intensive care and treatment. In endemic areas, treatment is often less satisfactory and the overall fatality rate for all cases of malaria can be as high as one in ten. Over the longer term, developmental impairments have been documented in children who have suffered episodes of severe malaria.

Chronic malaria is seen in both *P. vivax* and *P. ovale,* but not in *P. falciparum.* Here, the disease can relapse months or years after exposure, due to the presence of latent parasites in the liver. Describing a case of malaria as cured by observing the disappearance of parasites from the bloodstream can, therefore, be deceptive. The longest incubation period reported for a *P. vivax* infection is 30 years. Approximately one in five of *P. vivax* malaria cases in temperate areas involve overwintering by hypnozoites (i.e., relapses begin the year after the mosquito bite).

CAUSES

Malaria Parasites]

Malaria is caused by protozoan parasites of the genus *Plasmodium* (phylum Apicomplexa). In humans malaria is caused by *P. falciparum, P. malariae, P. ovale, P. vivax* and *P. knowlesi. P. falciparum* is the most common cause of infection and is responsible for about 80% of all malaria cases, and is also responsible for about 90% of the deaths from malaria. Parasitic *Plasmodium* species also infect birds, reptiles, monkeys, chimpanzees and rodents.

There have been documented human infections with several simian species of malaria, namely *P. knowlesi, P. inui, P. cynomolgi, P. simiovale, P. brazilianum, P. schwetzi* and *P. simium;* however, with the exception of *P. knowlesi,* these are mostly of limited public health importance. Although avian malaria can kill chickens and turkeys, this disease does not cause serious economic losses to poultry farmers.

MOSQUITO VECTORS AND THE *PLASMODIUM* LIFE CYCLE

The parasite's primary (definitive) hosts and transmission vectors are female mosquitoes of the *Anopheles* genus. Young mosquitoes first ingest the malaria parasite by feeding on an infected human carrier and the infected *Anopheles* mosquitoes carry *Plasmodium* sporozoites in their salivary glands.

A mosquito becomes infected when it takes a blood meal from an infected human. Once ingested, the parasite gametocytes taken up in the blood will further differentiate into male or female gametes and then fuse in the mosquito gut.

This produces an ookinete that penetrates the gut lining and produces an oocyst in the gut wall. When the oocyst ruptures, it releases sporozoites that migrate through the mosquito's body to the salivary glands, where they are then ready to infect a new human host. This type of transmission is occasionally referred to as anterior station transfer. The sporozoites are injected into the skin, alongside saliva, when the mosquito takes a subsequent blood meal.

Only female mosquitoes feed on blood, thus males do not transmit the disease. The females of the *Anopheles* genus of mosquito prefer to feed at night. They usually start searching for a meal at dusk, and will continue throughout the night until taking a meal. Malaria parasites can also be transmitted by blood transfusions, although this is rare.

PATHOGENESIS

Malaria in humans develops via two phases: an exoerythrocytic and an erythrocytic phase. The exoerythrocytic phase involves infection of the hepatic system, or liver, whereas the erythrocytic phase involves infection of the erythrocytes, or red blood cells.

When an infected mosquito pierces a person's skin to take a blood meal, sporozoites in the mosquito's saliva enter the bloodstream and migrate to the liver. Within 30 minutes of being introduced into the human host, the sporozoites infect hepatocytes, multiplying asexually and asymptomatically for a period of 6–15 days. Once in the live, these organisms differentiate to yield thousands of merozoites, which, following rupture of their host cells, escape into the blood and infect red blood cells, thus beginning the erythrocytic stage of the life cycle. The parasite escapes from the liver undetected by wrapping itself in the cell membrane of the infected host liver cell.

Within the red blood cells, the parasites multiply further, again asexually, periodically breaking out of their hosts to invade fresh red blood cells. Several such amplification cycles occur. Thus, classical descriptions of waves of fever arise from simultaneous waves of merozoites escaping and infecting red blood cells. Some *P. vivax* and *P. ovale* sporozoites do not immediately develop into exoerythrocytic-phase merozoites, but instead produce hypnozoites that remain dormant for periods ranging from several months (6–12 months is typical) to as long as three years.

After a period of dormancy, they reactivate and produce merozoites. Hypnozoites are responsible for long incubation and late relapses in these two species of malaria.

The parasite is relatively protected from attack by the body's immune system because for most of its human life cycle it resides within the liver and blood cells and is relatively invisible to immune surveillance. However, circulating infected blood cells are destroyed in the spleen. To avoid this fate, the *P. falciparum* parasite displays adhesive proteins on the surface of the infected blood cells, causing the blood cells to stick to the walls of small blood vessels, thereby sequestering the parasite from passage through the general circulation and the spleen.

This "stickiness" is the main factor giving rise to hemorrhagic complications of malaria. High endothelial venules (the smallest branches of the circulatory system) can be blocked by the attachment of masses of these infected red blood cells. The blockage of these vessels causes symptoms such as in placental and cerebral malaria. In cerebral malaria the sequestrated red blood cells can breach the blood brain barrier possibly leading to coma.

Although the red blood cell surface adhesive proteins are exposed to the immune system, they do not serve as good immune targets because of their extreme diversity: there are at least 60 variations of the protein within a single parasite and perhaps limitless versions within parasite populations. Like a thief changing disguises or a spy with multiple passports, the parasite switches between a broad repertoire of PfEMP1 surface proteins, thus staying one step ahead of the pursuing immune system.

Some merozoites turn into male and female gametocytes. If a mosquito pierces the skin of an infected person, it potentially picks up gametocytes within the blood. Fertilization and sexual recombination of the parasite occurs in the mosquito's gut, thereby defining the mosquito as the definitive host of the disease. New sporozoites develop and travel to the mosquito's salivary gland, completing the cycle.

LYME DISEASE

Lyme disease, or borreliosis, is an emerging infectious disease caused by at least three species of bacteria belonging to the genus *Borrelia*. *Borrelia burgdorferi* is the predominant cause of Lyme disease in the United States, whereas *Borrelia afzelii* and *Borrelia garinii* are implicated in most European cases. Lyme disease is the most common tick-borne disease in the Northern Hemisphere. *Borrelia* is transmitted to humans by the bite of infected hard ticks belonging to several species of the genus *Ixodcs*.

Early manifestations of infection may include fever, headache, fatigue, depression, and a characteristic skin rash called erythema migrans. Left untreated, late manifestations involving the joints, heart, and nervous system can occur. In most cases, the infection and its symptoms are eliminated with antibiotics, especially if diagnosis and treatment occur early in the course of illness. Late, delayed, or inadequate treatment can lead to late manifestations of Lyme disease which can be disabling and difficult to treat.

Some Lyme disease patients who have completed a course of antibiotic treatment continue to have symptoms such as severe fatigue, sleep disturbance, and cognitive difficulties. Some groups have argued that "chronic" Lyme disease is responsible for a range of medically unexplained symptoms beyond the objectively recognized manifestations of late Lyme disease, and that additional, long-term antibiotic treatment is warranted in such cases. Of four randomized controlled trials of long-term antibiotic courses in patients with ongoing symptoms, two found no benefit, and two found inconsistent benefits and significant side effects and risks from further antibiotic treatment.

Most expert groups including the Infectious Diseases Society of America and the American Academy of Neurology have found that existing scientific evidence does not support a role for *Borrelia* nor ongoing antibiotic treatment in such cases.

Symptoms

Lyme disease can affect multiple body systems, producing a range of potential symptoms. Not all patients with Lyme disease will have all symptoms, and many of the symptoms are not specific to Lyme disease but can occur in other diseases as well. The incubation period from infection to the onset of symptoms is usually 1–2 weeks, but can be much shorter (days), or much longer (months to years).

Symptoms most often occur from May through September because the nymphal stage of the tick is responsible for most cases. Asymptomatic infection exists but is found in less than 7% of infected individuals in the United States. Asymptomatic infection may be much more common among those infected in Europe.

Stage 1 – Early Localized Infection

The classic sign of early local infection is a circular, outwardly expanding rash called erythema chronicum migrans (also *erythema migrans* or *EM*), which occurs at the site of the tick bite 3 to 32 days after being bitten. The rash is red, and may be warm, but is generally painless. Classically, the innermost portion remains dark red and becomes indurated; the outer edge remains red; and the portion in between clears – giving the appearance of a bullseye.

However, the partial clearing is uncommon, and thus a true bullseye occurs in as few as 9% of cases.Erythema migrans is thought to occur in about 80% of infected patients. Patients can also experience flu-like symptoms such as headache, muscle soreness, fever, and malaise.Lyme disease can progress to later stages even in patients who do not develop a rash.

Stage 2 – Early Disseminated Infection

Within days to weeks after the onset of local infection, the borrelia bacteria may begin to spread through the bloodstream. Erythema migrans may develop at sites across the body that bear no relation to the original tick bite. Another skin condition, which is apparently absent in North American patients, is borrelial lymphocytoma, a purplish lump that develops on the ear lobe, nipple, or scrotum. Other discrete symptoms include migrating pain in muscles, joint, and tendons, and heart palpitations and dizziness caused by changes in heartbeat.

Acute neurological problems, which appear in 15% of untreated patients, encompasses a spectrum of disorders. One is facial or Bell's palsy, which is the loss of muscle tone on one or both sides of the face. Another common neurologic manifestation is meningitis, characterized by severe headaches, neck stiffness, and sensitivity to light. Radiculoneuritis causes shooting pains that may interfere with sleep and abnormal skin sensations. Mild encephalitis may lead to memory loss, sleep disturbances, or changes in mood or affect. In addition, simple altered mental status as the sole presenting symptom has been reported in early neuroborreliosis.

Stage 3 – Late Persistent Infection

After several months, untreated or inadequately treated patients may go on to develop severe and chronic symptoms affecting many organs of the body including the brain, nerves, eyes, joints and heart. Myriad disabling symptoms can occur.

Chronic neurologic symptoms occur in up to 5% of untreated patients. A polyneuropathy manifested primarily as shooting pains, numbness, and tingling in the hands or feet may develop. A neurologic syndrome called Lyme encephalopathy is associated with subtle cognitive problems such as difficulties with concentration and short term memory. Such patients may also experience profound fatigue. Other problems such as depression and fibromyalgia are no more common in people who have been infected with Lyme than in the general population.

Chronic encephalomyelitis, which may be progressive, may involve cognitive impairment, weakness in the legs, awkward gait, facial palsy, bladder problems, vertigo, and back pain.

In rare cases, frank psychosis has been attributed to chronic Lyme disease effects, including mis-diagnoses of schizophrenia and bipolar disorder. Panic attack and anxiety can occur, also delusional behavior, including somatoform delusions, sometimes accompanied by a depersonalization or derealization syndrome similar to what was seen in the past in the prodromal or early stages of general paresis.

Lyme arthritis usually affects the knees. In a minority of patients arthritis can occur in other joints, including the ankles, elbows, wrist, hips, and shoulders. Pain is often mild or moderate, usually with swelling at the involved joint. Baker's cysts may form and rupture. In some cases joint erosion occurs.

Acrodermatitis chronica atrophicans (ACA) is a chronic skin disorder observed primarily in Europe. ACA begins as a reddish-blue patch of discolored skin, usually in sun-exposed regions of the upper or lower limbs. The lesion slowly atrophies, and the skin may become so thin that it resembles wrinkled cigarette paper.

Transmission

Hard-bodied ticks of the genus *Ixodes* are the primary vectors of Lyme disease. The majority of infections are caused by ticks in the nymph stage, since adult ticks are more easily detected and removed as a consequence of their relatively large size.

Transmission is relatively rare, with only about 1% of recognized tick bites resulting in Lyme disease: this may be due to the fact that an infected tick has to be attached for at least a day for transmission to occur.

TICK BORNE CO-INFECTIONS

Ticks that transmit *B. burgorferi* to humans can also carry and transmit several other parasites such as *Theileria microti* and *Anaplasma phagocytophilum*, which cause the diseases babesiosis and human granulocytic anaplasmosis (HGA), respectively.

Among early Lyme disease patients, depending on their location, 2-12% will also have HGA and 2-40% will have babesiosis. Cat scratch fever is another common co-infection, although there is debate among experts on this topic on tick-to-human transmission.

Co-infections complicate Lyme symptoms, especially diagnosis and treatment. It is possible for a tick to carry and transmit one of the co-infections and not Borrelia, making diagnosis difficult and often elusive. The Centers for Disease Control (CDC)'s emerging infections diseases department did a study in rural New Jersey of 100 ticks and found that 55% of the ticks were infected with at least one of the pathogens.

DIAGNOSIS

Lyme disease is diagnosed clinically based on symptoms, objective physical findings, a history of possible exposure to infected ticks, as well as serological tests. When making a diagnosis of Lyme disease, health care providers should consider other diseases that may cause similar illness. Most but not all patients with Lyme disease will develop the characteristic bulls-eye rash, and many may not recall a tick bite.

Laboratory testing is not recommended for persons who do not have symptoms of Lyme disease.

Because of the difficulty in culturing *Borrelia* bacteria in the laboratory, diagnosis of Lyme disease is typically based on the clinical exam findings and a history of exposure to endemic Lyme areas. The EM rash, which does not occur in all cases, is considered sufficient to establish a diagnosis of Lyme disease even when serologies are negative. Serological testing can be used to support a clinically suspected case but is not diagnostic.

Diagnosis of late-stage Lyme disease is often difficult because of the multi-faceted appearance which can mimic symptoms of many other diseases. For this reason, a reviewer called Lyme the new "great imitator." Lyme disease may be misdiagnosed as multiple sclerosis, rheumatoid arthritis, fibromyalgia, chronic fatigue syndrome (CFS), lupus, or other autoimmune and neurodegenerative diseases.

Laboratory testing

Several forms of laboratory testing for Lyme disease are available, some of which have not been adequately validated. Most recommended tests are blood tests that measure antibodies made in response to the infection. These tests may be falsely negative in patients with early disease, but they are quite reliable for diagnosing later stages of disease.

The serological laboratory tests most widely available and employed are the Western blot and ELISA. A two-tiered protocol is recommended by the CDC: the more sensitive ELISA is performed first, if it is positive or equivocal, the more specific Western blot is run. The reliability of testing in diagnosis remains controversial, however studies show the Western blot IgM has a specificity of 94–96% for patients with clinical symptoms of early Lyme disease.

Erroneous test results have been widely reported in both early and late stages of the disease. These errors can be caused by several factors, including antibody cross-reactions from other infections including Epstein-Barr virus and cytomegalovirus, as well as herpes simplex virus.

Polymerase chain reaction (PCR) tests for Lyme disease have also been developed to detect the genetic material (DNA) of the Lyme disease spirochete. PCR tests are susceptible to false-positive results from poor laboratory technique. Even when properly performed, PCR often shows false-negative results with blood and CSF specimens. Hence PCR is not widely performed for diagnosis of Lyme disease. However PCR may have a role in diagnosis of Lyme arthritis because it is highly sensitive in detecting *ospA* DNA in synovial fluid. With the exception of PCR, there is no currently practical means for detection of the presence of the organism, as serologic studies only test for antibodies of *Borrelia*. High titers of either immunoglobulin G (IgG) or immunoglobulin M (IgM) antibodies to *Borrelia* antigens indicate disease, but lower titers can be misleading. The IgM antibodies may remain after the initial infection, and IgG antibodies may remain for years.

Western blot, ELISA and PCR can be performed by either blood test via venipuncture or cerebrospinal fluid (CSF) via lumbar puncture.

Though lumbar puncture is more definitive of diagnosis, antigen capture in the CSF is much more elusive; reportedly CSF yields positive results in only 10–30% of patients cultured. The diagnosis of neurologic infection by *Borrelia* should not be excluded solely on the basis of normal routine CSF or negative CSF antibody analyses.

New techniques for clinical testing of *Borrelia* infection have been developed, such as LTT-MELISA, which is capable of identifying the active form of *Borrelia* infection (Lyme disease).

Others, such as *focus floating microscopy*, are under investigation. New research indicates chemokine CXCL13 may also be a possible marker for neuroborreliosis.

Some laboratories offer Lyme disease testing using assays whose accuracy and clinical usefulness have not been adequately established. These tests include urine antigen tests, immunofluorescent staining for cell wall-deficient forms of Borrelia burgdorferi, and lymphocyte transformation tests. In general, CDC does not recommend these tests.

NIACIN

Nicotinic acid was synthesized in 1873 but was left on the shelf as an organic compound unrelated to the severe pellagra afflictions occurring throughout the world at that time. Sixty years later the compound was shown to be present in coenzymes I and II and two years thereafter Elvehjem cured 'black tongue' in dogs with the vitamin. Niacin was postulated to be part of factor H for fish in 1937, but deficiency symptoms were not adequately described until reported in trout in 1947.

Nicotinic Acid

Niacin is the preferred nomenclature. Nicotinic acid amide or niacinamide is the common form in which the vitamin is physiologically active. Niacin is a white, crystalline solid, soluble in water and alcohol.

It is stable in the dry state and may be autoclaved for short periods without destruction. It is also stable to heat in mineral acids and alkali. Niacin is both a carboxylic acid and an amine and forms quaternary ammonium compounds because of its basic nature.

Acidic characteristics include salt formation with alkali. Niacin can be esterified easily, then converted to amides. Niacinamide is a crystalline powder soluble in water and ethanol and the dry material is stable up to about 60°C. In aqueous solutions it is stable for a short period when autoclaved. It is the form in which the vitamin is normally found in niacinamide adenine dinucleotide (NAD) and in niacinamide adenine dinucleotide phosphate (NADP).

POSITIVE FUNCTIONS

The major function of niacin in NAD and NADP is hydrogen transport in intermediary metabolism. Most of these enzyme systems function by alternating between the oxidized and reduced state of the coenzymes NAD-NADH and NADP-NADPH. Oxidation-reduction reactions may be anaerobic as when pyruvate acts as the hydrogen acceptor and lactate is formed, or the reactions may be coupled to electron transport systems with oxygen as the ultimate hydrogen acceptor; such aerobic reactions occur in respiration.

Both NAD and NADP are involved in the synthesis of high energy phosphate bonds which furnish energy for certain steps in glycolysis, in pyruvate metabolism, amino acid and protein metabolism and in photosynthesis. An interrelationship between thiamine and niacin exists since both vitamins are coenzymes in intermediary metabolism whereby food is oxidized to furnish energy for physiological functions, to maintain homeostasis and for body temperature maintenance in homeotherms.

DEFICIENCY SYNDROME

The deficiency signs in fish niacin are more slowly exhausted under experimental conditions than are some of the other vitamins resulting in less defined and more slowly developing symptoms.

Niacin deficiencies in fish were experimentally induced in the late forties and early fifties by using basal diets which had a low niacin content. Loss of appetite and poor food conversion were the first signs noted. Then the fish turned dark and went off feed, followed by the appearance of lesions in the colon, erratic motion, oedema of the stomach and colon and muscle spasms while fish were apparently resting. A predisposition to sunburn in fish confined in the open in shallow ponds or raceways was described. Common carp showed a congestion of the skin with subcutaneous hemorrhages. Common symptoms of niacin deficiency in most fish studied were muscular weakness and spasms, coupled with poor growth and poor food conversion.

Requirements

In homeotherms on a balanced test ration the niacin requirement is generally estimated to be about ten times that of the thiamine requirement. These rations generally contain considerable carbohydrate material to furnish energy to maintain body temperature. In fish the requirement appears to be twenty to thirty times that of the thiamine needs determined for the same test conditions and test rations. This difference may be due to low carbohydrate content of young fish diets and the higher protein content of these rations. Conversion of tryptophan to niacin occurs in the mammalian liver and possibly also in the liver of fish.

This conversion may account for the slow development of the niacin-deficiency syndrome in fish. However, after 10-14 weeks on diets devoid of niacin, deficiency symptoms did occur in several species of fish. The symptoms were reduced by replacement of niacin in the ration even when high protein diets containing an excess of tryptophan were fed. Too much niacin inhibits growth.

Sources and Protection

Niacin is found in most animal and plant tissues. Rich sources are yeast, liver, kidney, heart, legumes and green vegetables. Wheat contains more niacin than corn and the vitamin is also found in milk and egg products. The vitamin is very stable since it is generally found in coenzyme form in raw materials. Niacin added to the diet as a supplement remains relatively unaltered during diet manufacture, processing and storage.

Antimetabolites and Inactivation

Pyridine-3-sulphonic acid and 3-acetylpyridine are compounds structurally related to niacin and are antimetabolites for this vitamin in animals and in micro-organisms. Additional niacin can overcome the anti-metabolic effect. Deficiency symptoms in rats may be induced by 6-amino niacinamide. The symptoms are reversed by addition of ten times more niacinamide than the antimetabolite. Thioacetamide has been reported to be a niacin antagonist in fish.

Clinical Assessment

Salmon feeding actively in the oceans showed liver niacin content of 70-80ωg of niacin/g of fresh tissue. About half of this amount is present in fingerling salmon raised in fresh water at 12-15°C and fed test rations containing 40-50 percent protein and niacin supplements 500-750 mg/kg dry diet. Urinary metabolities of niacin have been measured in other animals on standard niacin load in test rations containing a standard tryptophan load. The technique is well developed to measure the N1-methyl derivative in mammalian urine. These data have not been reported for fish but metabolism chambers are available for collecting branchial and urinary wastes from large fish intubated with different diet material.

BIOTIN

Certain foods, notably liver and kidney, were found to be protective against a form of dermatitis resulting from the consumption of egg white by rats, in the early thirties. It was soon learned that a specific component of egg albumin, avidin, rendered dietary biotin unavailable, hence producing the symptoms. Biotin was variously called coenzyme R and 'vitamin H'. It was isolated by Du Vigneaud in 1941 and synthesized by workers of Merck and Company in 1943. Biotin was once thought to be part of factor H for fish. Blue slime patch disease due to biotin deficiency was reported in trout.

Biotin

Biotin is a monocarboxylic acid slightly soluble in water and alcohol and insoluble in organic solvents. Salts of the acid are soluble in water. Aqueous solutions or the dry material are stable at 100 C and to light. The vitamin is destroyed by acids and alkalis and by oxidizing agents such as peroxides or permanganate. Biocytin is a bound form of biotin isolated from yeast, plant and animal tissues.

Other bound forms of the vitamin can generally be liberated by peptic digestion. Oxybiotin has partial vitamin activity but oxybiotin sulphonic acid and other analogues are antimetabolites inhibiting the growth of bacteria.

Avidin, a protein found in raw egg white, binds biotin and makes it unavailable to fish and other animals. Heating to denature the protein makes the bound biotin available again to the fish. Biocytin or -biotinyl lysine (the epsilon amino group of lysine and the carboxyl of biotin being combined in a peptide bond) is hydrolyzed by the enzyme biotinase making the protein-bound biotin available.

POSITIVE FUNCTIONS

Biotin is required in several specific carboxylation and decarboxylation reactions, including the carboxylation of pyruvic acid to form oxaloacetic acid. It is part of the coenzyme of several carboxylating enzymes fixing CO_2 such as propionyl coenzyme A involved in the conversion of propionic acid to succinic acid in methylmalonyl coenzyme A. Biotin is also involved in the conversion of acetyl CO_2 to malonyl coenzyme A in the formation of long chain fatty acids. It has possible involvement in citrulline synthesis and may have effects on purine and pyrimidine synthesis. It is involved in the conversion of unsaturated fatty acids to the stable cis form in the synthesis of biologically active fatty acids.

DEFICIENCY SYNDROME

Some signs of biotin deficiency in salmonids are skin disorders, muscle atrophy, lesions in the colon, loss of appetite and spastic convulsions. Haematology discloses fragmentation of erythrocytes. Poor growth is a common symptom and has been reported for salmonids, common carp, goldfish (Carassius auratus) and eel. Blue slime patch disease in brook trout deficient in biotin appears typical for this species. Fish reared in 10 -15 C water exhaust biotin stores in 8-12 weeks and the first signs are anorexia, poor food conversion and general listlessness before the more acute deficiency symptoms become detectable.

Sources and Protection

Rich sources of biotin are liver, kidney, yeast, milk products and egg yolks. Nut meats contain good supplies of biotin. The diet should be protected from strong oxidizing agents or conditions which promote oxidation of ingredients. Raw egg white should not be incorporated into moist fish diets. Cooking will inactivate the avidin.

Antimetabolites and Inactivation

Raw egg white has already been discussed which irreversibly binds biotin and makes it unavailable to young fish. Many biotin homologues with different side chain lengths inhibit the growth of bacteria. Oxybiotin, in which sulphur is replaced by oxygen, has about the same biological activity as the natural biotin. On the other hand, oxybiotin sulphonic acid inhibits biotin activity.

Clinical Assessment

Measurement or urinary excretion of biotin in animals does not prove a good clinical method since biotin is synthesized by several organisms in the gut. In fish, this technique needs to be explored and may well be valuable as a clinical tool since the gastrointestinal tract of many freshwater fish contains only small quantities of bacteria. Biotin is one of the most expensive vitamins to add to fish rations. Salmon feeding actively in the oceans had liver biotin concentrations of 10-12 mg of wet liver tissue.

The concentrations in the liver of young salmon fingerlings in fresh water fed test diets containing an excess of biotin were between 6-8 mg of the vitamin/g of tissue. Fish with these levels of biotin in the liver should probably be in sound biotin nutritional status.

FOLIC ACID (FOLACIN)

A factor effective in curing metaloblastic anaemia in monkeys was found in yeast and liver concentrates and was designated vitamin M in 1935. In 1939 an anti-anaemic factor was found in liver and called vitamin BC. These were later shown to be the same substance with folic acid as the active ingredient. Folic acid was synthesized in 1946 and was soon used in fish diets for preventing purified diet anaemia.

Folic Acid

Folic acid crystallizes into yellow spear-shaped leaflets which are soluble in water and dilute alcohol. It can be precipitated with heavy metal salts.

It is stable to heat in neutral or alkaline solution, but unstable in acid solution. It deteriorates when exposed to sunlight, or during prolonged storage. Several analogues have biological activity including pteroic acid, rhizopterin, folinic acid, xanthopterin and several formyltetrahydropteroyl-glutamic acid derivatives.

These have closely allied ring structures and many have been isolated as derivatives in various animals or microbiological preparations. One simple form, xanthopterin, present in the pigments of insects, is shown below and is of special interest because of early work with this compound as the anti-anaemic factor H for fish.

Positive Functions

Folic acid is required for normal blood cell formation and is involved as a coenzyme in one-carbon transfer mechanisms.

In the presence of ascorbic acid, folic acid is transformed into the active (5-formyl-5, 6,7,8) tetrahydrofolic acid. Folic acid is involved in many one-carbon metabolism systems such as serine and glycine interconversion, methionine-homocysteine synthesis, histidine synthesis and pyrimidine synthesis.

Several coenzyme forms of the active vitamin have been isolated. Folic acid is involved in the conversion of megaloblastic bone marrow to normoblastic type. It has a role in blood glucose regulation and improves cell membrane function and hatchability of eggs.

Deficiency Syndrome

Macrocytic normochromic anaemia occurs in several experimental animals, including fish fed diets devoid of the folic acid. Increasing numbers of senile cells are observed as the deficiency progresses until only a few old and degenerating cells are found in the blood of deficient fish. Anterior kidney imprints disclose only adult cells and no preforms present. Other signs observed have been poor growth, anorexia, general anaemia, lethargy, fragile fins, dark skin pigmentation and infarction of spleen.

SOURCES AND PROTECTION

Yeast, green vegetables, liver, kidney, glandular tissue, fish tissue and fish viscera are good sources of folic acid. Insects contain xanthopterin which has folic acid activity. At one time the yellow pigment of xanthopterin was identified as the fish anti-anaemic factor H, but subsequent experiments showed only partial activity and that folic acid itself was a much more potent antimacrocytic anaemia factor.

Insects may contribute significantly to the folic acid requirements of wild fish, but in scientific fish husbandry artificial diets are more reliable sources. Activity is lost during extended storage and when material is exposed to sunlight. Therefore, dry feeds should be carefully protected during manufacture and moist diet rations should be carefully preserved. Both types of fish diets should be fed soon after manufacture to assure minimal loss of folic acid activity.

ANTIMETABOLITES AND INACTIVATION

One antagonist of folacin is 4-aminopteroylglutamic acid or aminopterin. This material, when incorporated in the diet of guinea pigs and rats, induces anaemia and leucopenia and has been used to treat leukemia in man. Amethopterin (4-amino-N10-methylpteroylglutamic acid) can also be used to induce deficiency by inhibiting purine, pyrimidine and nucleic acid protection; viz., inhibition of nucleic acid synthesis results in macroytic anaemia.

CLINICAL ASSESSMENT

Haematology is used as a simple clinical tool to assess haemopoiesis in fish. Anterior kidney imprints easily disclose normal distribution of immature cells and preforms undergoing reticulosis. Salmon actively feeding in the ocean and young salmon fingerlings raised in fresh water and fed diets rich in folacin show liver storage of 3-4ωg of folic acid/g of wet tissue. Microbiological assay is preferred for assessment of total folic acid in dietary raw materials because the total biological activity it measures includes all the various coenzyme forms and folic acid analogues.

Assessment of the dietary intake of folic acid is important for intensive cold water fish husbandry. In pond culture, aquatic and terrestrial insects, algae, etc., may also be available and folic acid in the supplementary feed may not be as critical. Since folic acid is labile in storage, excess amounts are generally added to manufactured feed in anticipation of storage losses.

However, prudent fish husbandry dictates rapid use of manufactured rations with minimum storage. Routine periodic haematology of fish assures proper nutritional status for maximum production and sound health. The author has noted in several series of experiments that when fish diseases occur through inadvertent contamination of the water supply, those groups of fish partially or completely deficient in folic acid were among the first to show acute disease symptoms. Therefore, folic acid must also play an important role in resistance to disease.

VITAMIN B12

The antipernicious anaemia factor found to be contained in liver was isolated and crystallized in the mid forties. This substance, named vitamin B12 by its discoverers, was later to be recognized as essential for growth of chicken fed diets entirely of plant origin and was designated animal protein factor (APF). When anaemic salmon were injected with crystalline B12 in combination with folic acid and xanthopterin positive haemopoiesis occurred within a few days and the salmon showed rapid recovery from the anaemia.

CHEMISTRY

Vitamin B12 or cynacobalamin has the following approximate chemical structure.

Cyanocobalamin

The molecule has a planar group and a nucleotide group lying nearly at right angles to one another. This cobalt-containing vitamin has a net charge of one at the central cobalt atom to which is attached a replaceable cyano group. Vitamin B12 is stable to mild heat in neutral solution, but is rapidly destroyed by heating in dilute acid or alkali. Crude concentrates are more unstable and rapidly lose activity.

The compound is similar to the porphyrins in its spatial configuration with a central cobalt atom linked to four reduced pyrrole rings in the haeme series. Replacing the cyanide ion with a variety of anions produce derivatives which have comparable biological activities; viz., hydroxocobalamin, nitritocobalamin, chlorocobalamin and sulphatocobalamin.

Positive Functions

Cyanocobalamin is involved with folic acid in haemopoiesis. It is required by many micro-organisms and is a growth factor for many animals. The animal protein factor present in fish and animal by-products was not recognized until crystalline vitamin B12 was injected into anaemic chinook salmon fingerlings in 1949 and positive haemopoiesis was observed. A coenzyme incorporating vitamin B12 is involved in the reversible isomerization of methyl-malonyl coenzyme A to succinyl coenzyme A and in the isomerization of methylaspartate to glutomate.

Cyanocobalamin is involved in the coenzyme for the methylation of homocystine to form methionine. It is also involved in several other one-carbon reactions and in the synthesis of labile methyl compounds.

One vitamin B12 containing coenzyme acts in methylation of the purine ring during thymine synthesis. Vitamin B12 is also involved in cholesterol metabolism, in purine and pyrimidine biosynthesis and in the metabolism of glycols.

Deficiency Syndrome

Deficiency signs in young pigs, chicks and rats show abnormal blood elements, poor growth and pernicious anaemia. An intrinsic factor is necessary for good absorption of the vitamin from the gut. This factor is a low molecular weight mucoprotein which normally occurs in gastric juice and especially in hog gut mucosa.

Pernicious anaemia in chinook and coho salmon is characterized by fragmented, erythrocytes with many aberrant forms present. Haemoglobin levels are inconsistent and erythrocyte counts have a range extending from frank anaemia to a near normal blood pattern.

Cyanocobalamin stores in fish tissues are slowly exhausted and only after 12-16 weeks on test do the symptoms appear in deficient salmon populations. Poor appetite, poor growth, poor food conversion and some dark pigmentation can be observed before frank anaemia is detected.

Sources and Protection

Rich sources of vitamin B12 are found in fish meal, fish viscera, liver, kidney, glandular tissues and slaughter house wastes.

Since vitamin B12 is labile on storage and in mild acid

solution is easily destroyed by heating, care must be exercised in diet preparation containing flesh or meat scraps.

Antimetabolites and Inactivation

The vitamin B12 coenzymes are very unstable under light, which rapidly decomposes the coenzymes. Photo-sensitivity is increased in dilute acid solutions.

Clinical Assessment

Generalized anaemia with fragmentation of erythrocytes and extremely variable haemoglobin levels and erythrocyte counts indicate possible B12 deficiency. Prompt response in individual fish is obtained by injecting B12 alone or in combination with folic acid in the ratio of 1 part vitamin B12 to 100 parts folic acid. Careful interpretation of haemotological data will enable one to distinguish one form of anaemia from the other.

ASCORBIC ACID

Experimental work to cure scurvy with fruit juice was described by Lind in 1753, but nearly 200 years elapsed before the exact chemical compound responsible for reducing the symptoms was defined. The isolation of ascorbic acid was due to Szent-Györgyi of Hungary and C.G. King of the U.S.A. Vitamin C synthesis was accomplished in 1933 after the chemical structure of ascorbic acid was established by British and Swiss workers.

McCay and Tunison reported scoliosis in brook trout fed formalin-preserved meat in 1934 and McLaren observe haemorrhages in trout fed rations low in ascorbic acid. It was not until the sixties that a critical need for L-ascorbic acid by trout and salmon was demonstrated. L-ascorbic acid is a white, odourless, crystalline compound, soluble in water but insoluble in fat solvents.

It is readily oxidized to dehydroascorbic acid, the less biologically potent form. Ascorbic acid is very stable in acid solution because of the preservation of the lactone ring, but in alkaline solution hydrolysis occurs rapidly and vitamin activity is lost. It is very heat labile and prone to atmospheric oxidation, especially in the presence of copper, iron, or several other metallic catalysts. The reduced form is the most biologically active but several derivatives or salts are obtainable which have varying degrees of ascorbate activity.

Positive Functions

L-ascorbic acid acts as a biological reducing agent in hydrogen transport. It is involved in many enzyme systems for hydroxylation; i.e., hydroxylation of tryptophan, tyrosine, or proline. It is involved in the detoxification of aromatic drugs and also acts in the production of adrenal steroids. Ascorbic acid is necessary for the formation of hydroxy proline which is a constituent of collagen, a component of intercellular material in bones and soft tissues.

Ascorbic acid plays a synergistic role with vitamin E as intracellular antioxidants and free radical traps. The conversion of folic acid to folinic acid requires vitamin C. Ascorbic acid is involved in the formation of chondroitin sulphate and intercellular ground substance.

Labelled ascorbic acid fed to fish previously deficient in the vitamin was shown to be rapidly mobilized and fixed in areas of rapid collagen synthesis and became concentrated in the thick collagen of the skin and in cartilagenous bones, as well as in the glands of the anterior kidney. Ascorbic acid is also involved in erythrocyte maturation.

Deficiency Syndrome

Scurvy with impaired collagen formation, perifollicular haemorrhages, loose teeth and poor bone formation, anaemia and oedema have been reported in other animals. Deficiency signs in fish are generally related to impaired collagen formation.

Fish soon show hyperplasia of jaw and snout. The same symptoms have been observed in trout, salmon, yellowtail, carp, guppies and char. Histologically, hypertrophy of the adrenal tissues and haemorrhage at the bases of fins have been observed in coho salmon.

Deficiency signs cease to develop and new growth

becomes normal upon replacement of ascorbic acid in the ration. Anaemia eventually develops in extremely deficient fish and scoliosis and lordosis do not repair but are walled off by new growth around the afflicted areas of the spine when ascorbic acid is once again added to the ration.

Requirements

In the rainbow trout, reasonable blood and anterior kidney storage levels were obtained with an intake of about 100 mg of vitamin C/kg of dry ration in 10°, 12°, or 15° C water systems. When wound repair experiments were initiated, however, or when fish were exposed to other stress then the requirements doubled or tripled.

When severe abdominal or intramuscular wounds were inflicted, young fish needed at least 500 mg of active ascorbate for tissue repair comparable with control fish receiving 1 g or more of ascorbate in the diet/kg of dry diet. Coho salmon appear to need about half of these requirements for adequate tissue levels and for maximum severe wound repair rates.

The requirement for ascorbic acid is related to stress, growth rate and size of the animal, as well as to the other nutrients present in the diet. A compromise value of about 200 mg of ascorbic acid/kg diet for trout and salmon raised in freshwater systems between 10-15°C would ensure reasonable tissue storage levels and furnish some excess for mild stress conditions and for ascorbic acid loss from the diet through oxidation. Large common carp can synthesize some ascorbate and the requirement for this species may be dependent on fish size and the environment in which they are reared.

Sources and Protection

Ascorbic acid is widely distributed in nature with citrus fruits, cabbage, liver and kidney tissue good sources for the vitamin. Fresh insects and fish tissues contain reasonable amounts of the vitamin. Synthetic ascorbic acid is also readily available. Fish food should be protected from oxidizing agents and kept sealed or frozen until used to prevent loss of the vitamin.

Antimetabolites and Inactivation

D-ascorbic acid, the optical isomer of the active form, has no vitamin activity and competes for sites of several enzyme reactions mediated by L-ascorbic acid. 6-deoxy-L-ascorbic acid, dehydroascorbic acid and L-glucoascorbic acid have very little activity.

Clinical Assessment

Ascorbic acid status for experimental animals is normally attempted by tissue ascorbate analysis. Most of the assays previously used measure total ascorbate and not biologically active L-ascorbic acid.

Consequently these are fraught with errors and misconception of true vitamin C status. In fish tissues, blood and liver do not adequately reflect the ascorbic acid intake and status, whereas assay of the anterior kidney which contains adrenal tissue provides a fairly representative picture of tissue storage of the vitamin.

Stress rapidly reduces the ascorbic acid content of this tissue with concurrent production of adrenal steroids. Conversely, dietary repletion is reflected by up to four or five-fold storage levels from the deficient state. Examination of fragile support cartilage in the gill filaments under low magnification will detect early hypovitaminosis before clinically acute symptoms become noticeable. However, the best tissue for routine clinical analysis to assess vitamin C status in trout and salmon appears to be the anterior kidney with samples selected from the junction of the two forward wings. Tissues are blotted free of blood with filter paper and then assayed for total ascorbate by one of the improved quick methods to determine total ascorbate.

INOSITOL

Muscle 'sugar' was discovered by Scherer in 1850 and was characterized as inositol in 1887. Inositol was shown to be an alopecia (a form of hairlessness) preventing factor for mice in 1940. Poor growth and poor food passage in inositol-deficient fish was observed in salmon and carp.

Inositol

Seven optically inactive and two optically active isomers of hexahydroxycyclohexane can exist. One of the optically active forms, myo-inositol, is a white crystalline powder soluble in water and insoluble in alcohol and ether. The material can be synthesized, but is easily isolated from biological material in free or combined forms. The mixed calcium-magnesium salt of the hexophosphate is phytin. Isomers have little biological activity but will compete in chemical reactions. Inositol is a highly stable compound.

POSITIVE FUNCTIONS

Myo-inositol is a structural component in living tissues. It has lipotropic action by preventing accumulation of cholesterol in one type of fatty liver disease and is involved with choline in maintaining normal lipid/metabolism. It is a growth-promoting substance for micro-organisms and prevents an alopecia in mice. It is a reserve carbohydrate in muscle as well as a major component of phosphoglycerides in animal tissues.

DEFICIENCY SYNDROME

Poor growth, increased gastric emptying time, oedema, dark colour and distended stomachs are symptoms observed in salmon, trout, carp and catfish held for long periods on inositol-deficient test rations. A 'spectacle-eye' condition described for rats has not been observed under the experimental conditions used in fish studies. The major deficiency sign is inefficiency in digestion and food utilization and concomitant poor growth leading to a population of fish with distended abdomens.

SOURCES AND PROTECTION

Myo-inositol occurs ubiquitously in large amounts wherever biological tissue is found. Wheat germ, dried peas and beans are rich sources. Brain, heart and glandular tissues are very good sources of biologically active inositol. Citrus fruit pulp and dried yeast also contain inositol. The compound is stable.

ANTIMETABOLITES AND INACTIVATION

Seven optically inactive and one optically active but biologically inactive stereo isomers occur. Since inositol is synthesized in the biologically active form by many microorganisms in the gut, only large quantities of chemically synthesized inactive isomers added to diets would interfere with the metabolism of inositol for growth and normal physiological function. Biologically inactive stereoisomers of myo-inositol do not compete for critical sites in metabolism. Methyl derivatives and mono-, di- and triphosphoric acid esters occur naturally. Salts of the hexaphosphate or phytin make the bound inositol practically unavailable to the animal.

CLINICAL ASSESSMENT

Assessment in fish has been based on lack of deficiency signs coupled with the most efficient food conversion. Salmon feeding actively in oceans show 1-1.5 mg of inositol/g of fresh liver tissue and young fingerlings raised in freshwater at 10-15°C had 600-700ωg/g of liver tissue. An alternate, better assessment may be based on a standard muscle section or whole carcass analysis for free or for bound inositol. Projection of inositol intake from normal fish diet ingredients should indicate an excess of this particular vitamin.

CHOLINE

Methylation as a basic metabolic process was postulated by Hoffmeister. Methyl transfer was shown in vivo by Thompson and the interrelationships between choline, methionine and homo-cystine were shown by du Vigneaud in 1939-42. Trout fed low choline rations developed haemorrhagic kidneys and salmon showed an aversion to food in choline-deficient diet.

Choline

Choline is a very strong organic base and forms many derivatives widely distributed in animal and vegetable tissue. One derivative, acetylcholine, is involved in the transmission of nerve impulses across synapses. Choline is very hygroscopic, very soluble in water and is stable to heat in acid, but decomposes in alkaline solutions.

Positive Functions

Choline acts as a methyl donor in trans-methylation reactions. It is a lipotropic and antihaemorrhagic factor preventing the development of fatty livers. It is involved in the synthesis of phospholipids and in fat transport. Acetylcholine transmits the excitory state across the ganglionic synapses and neuromuscular junctions. Choline is essential for growth and good food conversion in fish.

Deficiency Syndrome

Deficiency signs include poor growth and poor food conversion accompanied by impaired fat metabolism. Haemorrhagic kidneys and intestines have been reported in trout and increased gastric emptying time has been observed in salmon.

Sources and Protection

Rich sources of choline are wheat germ, beans, brain and heart tissue. Choline hydrochloride, the commercially available form, may inactivate -tocopherol and vitamin K when in direct contact with these vitamins and care should be exercised in selecting properly protected (gelatin coated) fat soluble vitamins in fish diet preparation.

Clinical Assessment

Choline status of fish can be estimated from the assay of choline content of the dietary ingredients with the absence of deficiency signs. Maximum liver storage may not be the best criteria to determine choline nutritional status but has been used to assess the tentative requirement listed for the two species of salmon.

p-AMINOBENZOIC ACID

Para-aminobenzoic acid is a white crystalline powder which is water soluble and heat and light stable in aqueous and mild alkaline solution. It is a growth promoting vitamin for micro-organisms which require the vitamin for folic acid synthesis. Large intake has been shown to counteract the antimetabolite effect of sulphonamides in bacterial culture. No positive function or deficiency signs have been observed in fish.

LIPOIC ACID

Lipoic acid is both fat soluble and water soluble. Its active form is the amide derivative known as lipoamide. Lipoic acid functions as a coenzyme in a -ketoacid decarboxylation. It was discovered independently in several laboratories during the period 1945-50 and shown to be an essential component of multienzyme, the pyruvate dehydrogenase complex. Pyruvate is converted to 'active acetaldehyde' which in turn is picked up by lipoamide. Subsequent oxidation of the aldehyde results in the reduction of lipoamide to a disulphydryl form. The multi-enzyme unit also includes thiamine pyrophosphate, coenzyme A and flavin adenine dinucleotide. Glandular tissues are good sources of lipoic acid. No requirements have been determined for fish.

FAT-SOLUBLE VITAMINS

Fat-soluble vitamins A, D, E and K differ from the water-soluble vitamins in their accumulative action. Little evidence has been recorded for hypervitaminosis with the water-soluble vitamins since these compounds are rapidly metabolized and excreted when intake exceeds liver or tissue storage capacity, but hypervitaminosis is a common occurrence in fish and other animals when large quantities of any one of the fat-soluble vitamins are ingested.

Toxicity symptoms involving vitamins A and D are indistinguishable from deficiency symptoms for the same vitamins.

On the other hand, symptoms of excess vitamins E and K intake are more discrete. Fish rations may often be enriched with fish oils to increase caloric density of the ration resulting in excessive intake of the fat-soluble vitamins.

VITAMIN A

A fat-soluble factor that promoted growth in rats was described by Hopkins and by Osborne and Mendel at the beginning of this century. McCollum and Simmonds cured xerophthalmia, a disease of the eye, with this material. The chemical structure of vitamin A and its relationship to -carotene was shown by Von Euler in 1928. Active vitamin A was synthesized in the mid-thirties.

The relationship of the vitamin A alcohols to naturally occurring -carotene is as follows:

Retinene, the aldehyde form of vitamin A, has been isolated from the retina of dark adapted eyes and is involved in vision in dim light. Retinoic acid, which is the oxidized form of vitamin A alcohol, has been shown to have some vitamin A activity. Vitamin A1 is found in saltwater fish, whereas vitamin A is more abundant in freshwater fish.

Interconversion of the two forms in living fish tissue has been demonstrated. Fish oils contain vitamin A as free alcohols or esters. Vitamin A alcohol occurs as a light coloured viscous oil which is heat labile and subject to air oxidation.

Beta-carotene occurs as an orange, crystalline compound which is more stable to heat and oxidation. Vitamin A is water insoluble but is soluble in fat and organic solvent.

POSITIVE FUNCTIONS

Vitamin A is essential in maintaining epithelial cells, preventing atrophy and keratinization of epithelial tissues. It is combined with a protein in visual purple and is important in night vision. Vitamin A also prevents xerophthalmia in rats and young children. Vitamin A promotes growth of new cells and aids in maintaining resistance to infection.

THE SYNDROME OF DEFICIENCY OR EXCESS

Hypovitaminosis A is characterized by poor growth, poor vision, keratinization of epithelial tissue, xerophthalmia, night blindness, haemorrhage in the anterior chamber of the eye, haemorrhage of the base of the fins and abnormal bone formation. Nerve degeneration has been reported in pigs, chickens, rats, rabbits and ducks, but only occasionally observed in fish after long periods of deficiency. Hypervitaminosis A has been described in fish and in other animals and involved enlargement of liver and spleen, abnormal growth, skin lesions, epithelial keratinization, hyperplasia. of head cartilage and abnormal bone formation resulting in ankylosis and fusion of vertebrae. Hypervitaminosis A is reflected in very high liver oil vitamin A content and elevated serum alkaline phosphatase. Removal of excess vitamin A from the diet promotes rapid recovery.

SOURCES AND PROTECTION

Cod liver oil is the best known source of vitamin A although black sea bass, swordfish, or ling cod oils contain much more of the vitamin. Whale liver oil contains kitol which has little or no biological activity until heated above 200°C when one molecule of biologically active vitamin A is generated per molecule of whale kitol.

This biologically inactive kitol may be deposited in the whale to avoid hypervitaminosis A during excessive vitamin A intake.

The possibility of hypervitaminosis A occurs when tuna, shark, or ling cod viscera are used in preparation of moist fish diets. Synthetic vitamin A preparations, such as vitamin A palmitate, are available and are often used to supplement rations low in fish meal, fish viscera, or carotenes. Some fish species seem able to utilize b-carotene as a vitamin A source, whereas others are unable to split the g-carotene molecule and vitamin A must be added to the diet.

CLINICAL ASSESSMENT

Vitamin A status can best be assessed by the absence of deficiency signs and by assay of liver oil for vitamin A content. Assay for the vitamin in blood or plasma has not been found to be useful.

VITAMIN D

Rickets was induced with test diets by Hopkins and

Mellanby cured the disease in dogs by adding cod liver oil to the ration.

Ultraviolet light exposure was shown to have an anti-rachitic effect and provitamin D activity had been ascribed to ergosterol. Crystalline vitamin D was isolated by Angus and activated 7-dehydrocholesterol was isolated by Windaus.

Vitamin D_2 or ergocalciferol, is one of several biologically active forms of vitamin D. Vitamin D_3 or activated 7-dehydrocholesterol has the chemical formula $C_{27}H_{44}O$ and contains a more simplified, unsaturated 8-carbon side chain. Vitamin D_3, also known as cholecalciferol, is formed in most animal tissue by the rupture of one of the ring bonds of 7-dehydrocholesterol by ultraviolet radiation.

Cholecalciferol is a white, crystalline compound soluble in fat and organic solvents and is stable to heat and oxidation in mild alkali or acid solution.

POSITIVE FUNCTIONS

Vitamin D is essential for maintaining calcium and inorganic phosphate homeostasis. It is involved in alkaline phosphatase activity, promotes intestinal absorption of calcium and influences the action of parathyroid hormone on bone. Fish may sequester calcium from water through the gill membrane; thus the major function of vitamin D for other animals may not be necessary to satisfy calcium requirements for fish.

SYNDROME OF DEFICIENCY OR EXCESS

Hypovitaminosis D for fish has not been described. Rickets and abnormal bone formation described in detail for animals has not been observed in fish fed low vitamin D diets. However little work has been done under carefully controlled conditions with young growing fish and only alteration in alkaline phosphatase activity has been reported on different vitamin D diet intake.

Hypervitaminosis D, however, has been reported. Brook trout fed large doses of vitamin D showed impaired growth, lethargy and dark colouration. High intake of vitamin D mobilizes phosphorus and calcium from the bone and tissues and may result in fragile bones, poor growth and poor appetite related to the nausea described in man afflicted with hypervitaminosis D. Arterial and kidney lesions due to excessive intake of vitamin D reported for rats and dogs have not been described histologically for fish nor has hypercalcaemia been described.

This area needs to be explored because of the potential for hypervitaminosis D in fish fed diets containing various fish viscera which might contain large amounts of the vitamin D. Tuna liver oil may contain, for example, 100 to 1 000 times as much active vitamin D as cod liver oil.

SOURCES AND PROTECTION

Vitamin D requirements of many animals can be met by ultraviolet irradiation of the skin where cholesterol derivatives are converted to calciferol. Since the vitamin is fat soluble and accumulates in lipid stores, fish liver oil is a rich source of the material.

Content varies tremendously in liver oil, however, with values of about 25 I.U./g present in shark liver oil and over 200 000 I.U./g in albacore tuna liver oil. Cod liver oil contains from 100-500 I.U./g and animal liver contains some vitamin D. One international unit (I.U.) is equal to 0.025 mg of crystalline vitamin D.

CLINICAL ASSESSMENT

Absorption maxima in the ultraviolet region can be used to detect provitamins D in the non-saponifiable fraction of oils. Concentrated preparations of vitamin D can be assayed by the Carr-Price antimony trichloride reaction when assay is necessary to determine biologically active materials in liver oil of fish on different treatments.

The chick assay may not apply to fish liver storage levels because the biologically active form for fish has not been determined. Therefore, clinical assessment for hypervitaminosis A must rely on crude methods for determination of vitamin D in liver oil samples by the Carr-Price reaction or by measuring absorption in the UV spectrum.

VITAMIN E

The existence of an antisterility vitamin was postulated by Evans and Bishop. This factor was named 'vitamin E' by Sure. The active tocopherol was isolated, characterized and synthesized by Karrer.

Vitamin E is composed of a class of compounds known as tocopherols. One of the most important tocopherols is -tocopherol.

Eight naturally occurring tocopherol derivatives have been isolated and all belong to the D series. Synthetic -tocopherol is a racemic DL-tocoperhol mixture. The derivatives of tocol or of tocotrienol are named, alpha, beta, gamma, delta, epsilon, eta, zeta1 and zeta2-tocopherol The pure tocopherols are fat-soluble oils which are capable of esterification to form crystalline compounds.

The tocopherols are stable to heat and acids in the absence of oxygen, but are rapidly oxidized in the presence of nascent oxygen, peroxides, or other oxidizing agents. The tocopherols are sensitive to ultraviolet light and are excellent antioxidants in the free form.

The esters are more stable and are commonly used as dietary supplements-anticipating hydrolysis in the gut and absorption of the free alcohol to act as an active intra- and intercellular antioxidant. Ethyl derivatives on the aromatic ring are also active. Oxidation products of a-tocopherol can be reduced with hydrosulphite to -tocopherylhydroquinone or, in the presence of ascorbic acid, to -tocopherol.

Table: The Tocopherols

Tocopherol (alpha)	5,7,8-trimethyltocol
Tocopherol (beta)	5,8-dimethyltocol
Tocopherol (gamma)	7,8-dimethyltocol
Tocopherol ($zeta_2$)	5,7-dimethyltocol
Tocopherol (eta)	7-methyltocol
Tocopherol (delta)	8-methyltocol
Tocopherol (epsilon)	5,8-dimethyltocotrienol
Tocopherol ($zeta_1$)	5,7,8-trimethyltocotrienol

POSITIVE FUNCTIONS

The tocopherols act as extra and intracellular antioxidants to maintain homeostasis of labile metabolites in the cell and tissue plasma. As physiological antioxidants, these usually protect oxidizable vitamins and labile unsaturated fatty acids. Vitamin E prevents encephalomalacia in chicks, erythrocyte haemolysis in several animals and steatitis in mink, pigs and farm animals.

The tocopherols also prevent exudative diathesis in chicks, white muscle disease in lambs and calves and dietary liver necrosis in rats. They act with selenium and with vitamin C for normal reproductive activity and are involved in prevention of nutritional muscular dystrophy in the chick, the yellowtail and carp. The tocopherols act as free radical traps to stop the chain reaction during peroxide formation and stabilize unsaturated carbon bonds of polyunsaturated fatty acids and other long-chain labile compounds.

Vitamin E is involved in the maintenance of normal blood capillary permeability and the integrity of heart muscle. It was first shown to be involved in prevention of sterility and foetal resorption in rats. It may likewise affect embryo membrane permeability and hatch-ability of fish eggs.

SYNDROME OF DEFICIENCY OR EXCESS

One of the first signs is erythrocyte fragility closely followed by anaemia, ascites, xerophthalmia, poor growth, poor food conversion, epicarditis and ceroid deposits in spleen and liver. Muscle dystrophy and xerophthalmia have been described in yellowtail and carp.

Non-specific forms of degenerative conditions have been described in several species of fish fed large quantities of polyunsaturated fatty acids with inadequate tocopherol in the ration. Hypervitaminosis E results in poor growth, toxic liver reaction and death.

SOURCES AND PROTECTION

Vegetable oils are rich sources of tocopherols. Synthetic - tocopherol in esterfied acetate or phosphate form is commonly used as a diet supplement. These esters are much more stable than the free form which is rapidly lost by air oxidation or in the presence of labile compounds like the polyunsaturated fish oils. Addition of antioxidants such as BHT (butylated hydroxyanisole) and BHT (butylated hydroxytoluene) protects fats and other labile compounds in the ration from oxidation, but these antioxidants have no vitamin E activity.

CLINICAL ASSESSMENT

The erythrocyte fragility test indicates the physiological state of fish. Absence of histologically detectable ceroid in liver and spleen from representative samples in the population is a good clue on the presence of adequate amounts of physiological antioxidants in the fish.

A barbituric acid test for oxidation of components has not been applied to fish tissues as a clinical tool for assessment of nutrition state except when liver oils become saturated with very labile polyunsaturated fatty acids such as when feeding squid or saury oils. These assays are good indicators of the state of oxidation of the finished ration, but the absence of deficiency signs and normal erythrocyte fragility are better clinical tests. Analysis for tocopherol is difficult and time-consuming and only applicable under critical research experiment situation.

VITAMIN K

The name vitamin K (for 'koagulation') was proposed by Dam. Dam isolated the vitamin from alfalfa and from fish meal in 1939. It was synthesized later that year.

The K vitamins K_2 contain 6,7, or 9 isoprene units in the side chain which varies from 30-45 carbon atoms. Many isosteres of vitamin K have been identified in animal tissues, plant tissues and micro-organisms. The structure of phthiocol and menadione (K_3) are shown below:

Although fairly stable compounds, they are destroyed upon oxidation and exposure to ultraviolet radition. Menadione is very reactive and is subject in aqueous media to chemical interaction.

POSITIVE FUNCTIONS

Vitamin K is involved in the hepatic synthesis of blood clotting proteins - prothrombin and proconvertin.

Substituted forms of vitamin K are strongly bacteriostatic and may serve as an alternate defence mechanism for bacterial infections.

Vitamin K is structurally related to flavo compounds which function as electron carriers in the electron transport system. The primary role of vitamin K is to maintain a fast normal blood clotting rate which is so important to fish living in a water environment.

SYNDROME OF DEFICIENCY OR EXCESS

Prothrombin time in salmon fed diets devoid of vitamin K was increased three to five times and, during prolonged deficiency states, anaemia and haemorrhagic areas appeared in the gills, eyes and vascular tissues. Increased blood clotting time has also been reported for other fish reared on diets with low vitamin K content.

Interrelationships with other vitamins have not been documented in fish where the primary deficiency signs are slow blood clotting and haemorrhage, severe anaemia, or death in wounded fish. Haemorrhagic areas often appear in fragile tissues such as the gills.

SOURCES AND PROTECTION

Vitamin K sources are green, leafy vegetables. Alfalfa leaves are among the best sources of vitamin K. Low levels are found in soybeans and animal liver. Synthetic menadione is also available. Although vitamin K in ground alfalfa is fairly stable, synthetic material should be protected from exposure to ultraviolet light and to excessive oxidizing or reducing conditions.

The use of rapidly cured dehydrated alfalfa is essential to minimize formation of the physiological antagonist, dicumarol. The diet should be kept dry, prepared with minimum exposure to air oxidation and fed as soon as practicable after manufacture to minimize vitamin K loss.

ANTIMETABOLITES AND INACTIVATION

Two vitamin K antagonists, dicumarol and warfarin are shown below:

Dicumarol was isolated from spoiled sweet clover hay and shown to prevent the normal function of vitamin K in maintenance of normal blood clotting times. Early work was with cattle and hogs but was soon extended to rats, other experimental animals and man.

Dicumarol is an anticoagulant and has been used to prevent thrombosis in animals and man. It is not an antimetabolite competing for vitamin K sites but plays another role in preventing normal blood clotting. Vitamin K counteracts the dicumarol effect quantitatively. Another anticoagulant, warfarin, is a common rat poison which has five to ten times the anticoagulant activity of dicumarol. The effects of warfarin can also be reversed by administration of vitamin K.

CLINICAL ASSESSMENT

Vitamin K status of fish is determined by measuring blood clotting time. Chemical methods for determining menadione content of active metabolite after storage are available but have not been applied to measure vitamin K activity in the liver or tissues of young, growing fish.

Index

C

D

E

F

G

H

I

R

S

T

V

W

Z